This book presents an up-to-date, unified treatment of research in bounded arithmetic and complexity of propositional logic with emphasis on independence proofs and lower bound proofs. The author discusses the deep connections between logic and complexity theory and lists a number of intriguing open problems.

An introduction to the basics of logic and complexity theory is followed by discussion of important results in propositional proof systems and systems of bounded arithmetic. Then more advanced topics are treated, including polynomial simulations and conservativity results, various witnessing theorems, the translation of bounded formulas (and their proofs) into propositional ones, the method of random partial restrictions and its applications, direct independence proofs, complete systems of partial relations, lower bounds to the size of constant-depth propositional proofs, the method of Boolean valuations, the issue of hard tautologies and optimal proof systems, combinatorics and complexity theory within bounded arithmetic, and relations to complexity issues of predicate calculus.

ENCYCLOPEDIA OF MATHEMATICS AND ITS APPLICATIONS

EDITED BY G.-C. ROTA

Volume 60

Bounded Arithmetic, Propositional Logic, and Complexity Theory

ENCYCLOPEDIA OF MATHEMATICS AND ITS APPLICATIONS

4 W. Miller, Jr. *Symmetry and separation of variables*
6 H. Minc *Permanents*
11 W. B. Jones and W. J. Thron *Continued fractions*
12 N. F. G. Martin and J. W. England *Mathematical theory of entropy*
18 H. O. Fattorini *The Cauchy problem*
19 G. G. Lorentz, K. Jetter, and S. D. Riemenschneider *Birkhoff interpolation*
21 W. T. Tutte *Graph theory*
22 J. R. Bastida *Field extensions and Galois theory*
23 J. R. Cannon *The one-dimensional heat equation*
25 A. Salomaa *Computation and automata*
26 N. White (ed.) *Theory of matroids*
27 N. H. Bingham, C. M. Goldie, and J. L. Teugels *Regular variation*
28 P. P. Petrushev and V. A. Popov *Rational approximation of real functions*
29 N. White (ed.) *Combinatorial geometries*
30 M. Pohst and H. Zassenhaus *Algorithmic algebraic number theory*
31 J. Aczel and J. Dhombres *Functional equations containing several variables*
32 M. Kuczma, B. Chozewski, and R. Ger *Iterative functional equations*
33 R. V. Ambartzumian *Factorization calculus and geometric probability*
34 G. Gripenberg, S.-O. Londen, and O. Staffans *Volterra integral and functional equations*
35 G. Gasper and M. Rahman *Basic hypergeometric series*
36 E. Torgersen *Comparison of statistical experiments*
37 A. Neumaier *Interval methods for systems of equations*
38 N. Korneichuk *Exact constants in approximation theory*
39 R. A. Brualdi and H. J. Ryser *Combinatorial matrix theory*
40 N. White (ed.) *Matroid applications*
41 S. Sakai *Operator algebras in dynamical systems*
42 W. Hodges *Model theory*
43 H. Stahl and V. Totik *General orthogonal polynomials*
44 R. Schneider *Convex bodies*
45 G. Da Prato and J. Zabczyk *Stochastic equations in infinite dimensions*
46 A. Bjorner, M. Las Vergnas, B. Sturmfels, N. White, and G. Ziegler *Oriented matroids*
47 G. A. Edgar and L. Sucheston *Stopping times and directed processes*
48 C. Sims *Computation with finitely presented groups*
49 T. Palmer *Banach algebras and the general theory of *-algebras*
50 F. Borceux *Handbook of Categorical Algebra I*
51 F. Borceux *Handbook of Categorical Algebra II*
52 F. Borceux *Handbook of Categorical Algebra III*
54 A. Katok and B. Hasselblatt *Introduction to the modern theory of dynamical systems*
58 R. Gardner *Geometric tomography*

ENCYCLOPEDIA OF MATHEMATICS AND ITS APPLICATIONS

Bounded Arithmetic, Propositional Logic, and Complexity Theory

JAN KRAJÍČEK

Academy of Sciences of the Czech Republic

CAMBRIDGE
UNIVERSITY PRESS

Published by the Press Syndicate of the University of Cambridge
The Pitt Building, Trumpington Street, Cambridge CB2 1RP
40 West 20th Street, New York, NY 10011-4211, USA
10 Stamford Road, Oakleigh, Melbourne 3166, Australia

First published 1995

Printed in the United States of America

Library of Congress Cataloging-in-Publication Data
Krajíček, Jan.
Bounded arithmetic, propositional logic, and complexity theory / Jan Krajíček.
p.   cm. – (Encyclopedia of mathematics and its applications; v. 60)
Includes bibliographical references (p. 000–000) and indexes.
ISBN 0-521-45205-8
1. Constructive mathematics.   2. Proposition (Logic).
3. Computational complexity.   I. Title.   II. Series.
QA9.56.K73   1995
511.3 – dc20   94-47054
                                                            CIP

A catalog record for this book is available from the British Library.

ISBN 0-521-45205-8 hardback

To Karel Tesař

CONTENTS

Preface   *page* **xi**

Acknowledgments   **xiv**

1 Introduction   **1**

2 Preliminaries   **3**
   2.1 Logic   3
   2.2 Complexity theory   5

3 Basic complexity theory   **8**
   3.1 The P versus NP problem   8
   3.2 Bounded arithmetic formulas   17
   3.3 Bibliographical and other remarks   22

4 Basic propositional logic   **23**
   4.1 Propositional proof systems   23
   4.2 Resolution   25
   4.3 Sequent calculus   31
   4.4 Frege systems   42
   4.5 The extension and the substitution rules   53
   4.6 Quantified propositional logic   57
   4.7 Bibliographical and other remarks   60

5 Basic bounded arithmetic   **62**
   5.1 Theory $I\Delta_0$ 63
   5.2 Theories S_2 and T_2 68
   5.3 Theory PV   75

5.4   Coding of sequences   79
5.5   Second order systems   83
5.6   Bibliographical and other remarks   92

6   Definability of computations   **93**
6.1   Polynomial time with oracles   94
6.2   Bounded number of queries   97
6.3   Interactive computations   97
6.4   Bibliographical and other remarks   101

7   Witnessing theorems   **102**
7.1   Cut-elimination for bounded arithmetic   102
7.2 Σ_i^b-definability in S_2^i and oracle polynomial time   105
7.3 Σ_{i+2}^b- and Σ_{i+1}^b-definability in S_2^i and bounded queries   113
7.4 Σ_{i+2}^b-definability in T_2^i and counterexamples   120
7.5 Σ_1^b-definability in T_2^1 and polynomial local search   121
7.6   Model-theoretic constructions   126
7.7   Bibliographical and other remarks   131

8   Definability and witnessing in second order theories   **132**
8.1   Second order computations   132
8.2   Definable functionals   134
8.3   Bibliographical and other remarks   138

9   Translations of arithmetic formulas   **139**
9.1   Bounded formulas with a predicate   139
9.2   Translation into quantified propositional formulas   144
9.3   Reflection principles and polynomial simulations   158
9.4   Model-theoretic constructions   172
9.5   Witnessing and test trees   180
9.6   Bibliographical and other remarks   183

10 Finite axiomatizability problem   **185**
10.1 Finite axiomatizability of S_2^i and T_2^i 185
10.2 T_2^i versus S_2^{i+1} 186
10.3 S_2^i versus T_2^i 193
10.4 Relativized cases   194
10.5 Consistency notions   202
10.6 Bibliographical and other remarks   208

11 Direct independence proofs   **210**
11.1 Herbrandization of induction axioms   210
11.2 Weak pigeonhole principle   213

11.3 An independence criterion   220
11.4 Lifting independence results   225
11.5 Bibliographical and other remarks   231

12 Bounds for constant-depth Frege systems   **232**
12.1 Upper bounds   232
12.2 Depth d versus depth $d + 1$ 236
12.3 Complete systems   243
12.4 k-evaluations   252
12.5 Lower bounds for the pigeonhole principle and for counting
      principles   258
12.6 Systems with counting gates   266
12.7 Forcing in nonstandard models   272
12.8 Bibliographical and other remarks   277

13 Bounds for Frege and extended Frege systems   **279**
13.1 Counting in Frege systems   279
13.2 An approach to lower bounds   286
13.3 Boolean valuations   289
13.4 Bibliographical and other remarks   297

14 Hard tautologies and optimal proof systems   **299**
14.1 Finitistic consistency statements and optimal proof systems 299
14.2 Hard tautologies   304
14.3 Bibliographical and other remarks   307

15 Strength of bounded arithmetic   **308**
15.1 Counting   308
15.2 A circuit lower bound   312
15.3 Polynomial hierarchy in models of bounded arithmetic   316
15.4 Bibliographical and other remarks   324

References   **327**

Subject index   **335**

Name index   **339**

Symbol index   **341**

PREFACE

The central problem of complexity theory is the relation of deterministic and nondeterministic computations: whether P equals NP, and generally whether the polynomial time hierarchy PH collapses. The famous *P versus NP problem* is often regarded as one of the most important and beautiful open problems in contemporary mathematics, even by nonspecialists (see, for example, Smale [1992]).

The central problem of bounded arithmetic is whether it is a finitely axiomatizable theory. That amounts to deciding whether there is a model of the theory in which the polynomial time hierarchy does not collapse.

The central problem of propositional logic is whether there is a proof system in which every tautology has a proof of size polynomial in the size of the tautology. In this generality the question is equivalent to asking whether the class NP is closed under complementation. Particular cases of the problem, to establish lower bounds for usual calculi, are analogous to constructing models of associated systems of bounded arithmetic in which NP $\neq$ coNP.

Notions, problems, and results about complexity (of predicates, functions, proofs, ...) are deep-rooted in mathematical logic, and (good) theorems about them are among the most profound results in the field. Bounded arithmetic and propositional logic are closely interrelated and have several explicit and implicit connections to the computational complexity theory around the P versus NP problem. Central computational notions (Turing machine, Boolean circuit) are crucial in the metamathematics of the logical systems, and models of these systems are natural structures for concepts of computational complexity.

Moreover, the only approach in sight universal enough to have a chance of producing lower bounds to the size of general Boolean circuits needed for the first problem is the method of approximations, which is a version of the ultraproduct construction (and of forcing); forcing bears a relation to the second and the third problems, and a general framework for the last problem is in terms of Boolean valuations (Sections 3.1, 9.4, 12.7, and 13.3).

Much of the contemporary research in computational complexity theory concentrates on proving weaker versions of P $\neq$ NP, for example, on proving lower bounds to the size of restricted models of circuits, and some deep results (although telling little about the P versus NP problem) have been obtained.

It is, however, possible to approach the same problem differently and to try to prove statement P $\neq$ NP first for other structures than natural numbers N, in particular for nonstandard models of systems of bounded arithmetic.

Such an approach is, in fact, common in mathematics, where for example a number-theoretic conjecture about the field of rational numbers is first tested for function fields that share many properties with the rationals. Similarly, we can try to prove that P $\neq$ NP holds in a model of a system of bounded arithmetic. Nonstandard models of systems of bounded arithmetic are not ridiculously pathological structures, and a part of the difficulty in constructing them stems exactly from the fact that it is hard to distinguish these structures, by the studied properties, from natural numbers.

Methods (all essentially combinatorial) used for known circuit lower bounds are demonstrably inadequate for the general problem. It is to be expected that a nontrivial combinatorial or algebraic argument will be required for the solution of the P versus NP problem. However, I believe that the close relations of this problem to bounded arithmetic and propositional logic indicate that such a solution should also require a nontrivial insight into logic. For example, recent strong lower bounds for constant-depth proof systems needed a reinterpretation of logical validity (Section 12.4).

The relations among bounded arithmetic, propositional logic, and complexity theory are not ad hoc but are reflected in numerous more specific relations, ranging from intertranslatability of arithmetic and propositional proofs and computations of machines, to characterizations of provably total functions in various subsystems of bounded arithmetic in terms of familiar computational models, correspondence in definability of predicates by restricted means and their decidability in a particular computational model, to proof methods based on analogous combinatorial backgrounds in all three areas, and finally to formalizability of basic concepts and methods of complexity theory within bounded arithmetic. It is the main aim of this book to explain these relations.

The last several years have seen important developments in areas of complexity theory, as well as in bounded arithmetic and complexity of propositional logic, and other deep relations between these areas have been established. Although there are several monographs on computational complexity theory and very good survey articles covering the main fields of research, many recent results in bounded arithmetic and propositional logic are scattered in research articles, and some important facts, such as relations between various theorems and methods, are only a part of unpublished folklore or appeared in a longer but now less significant, and hence less read, work.

To my knowledge there are three published monographs treating, at least partially, bounded arithmetic and its relation to complexity theory: Wilkie (1985), Buss (1986), and the last part of Hájek and Pudlák (1993) (Chapter 5, pp. 267–408). Although these are very interesting books, the first two contain none of the developments of the last several years (obviously) and none of the three treats propositional logic.

This book is not intended to be a textbook of either logic or complexity theory. It merely wants to present the main aspects of contemporary research in bounded arithmetic and complexity of propositional logic in a coherent way and to illustrate topics pointed out at the beginning of this Preface. It is aimed at research mathematicians, computer scientists, and graduate students. No previous knowledge of the topics is required, but it is expected that the reader is willing to learn what is needed along the way. My hope is that the book will stimulate more people to contribute to this fascinating area.

Prague   Jan Krajíček
July 28, 1994

ACKNOWLEDGMENTS

I thank my colleagues Petr Hájek, Pavel Pudlák, Jiří Sgall, Antonín Sochor, and Vítězslav Švejdar from our logic seminar at the Mathematical Institute of the Academy of Sciences at Prague for creating an extremely stimulating and encouraging research environment. In particular, I have learned a lot from extensive collaboration with Pavel Pudlák.

I am also indebted to Gaisi Takeuti (Urbana), Sam Buss (San Diego), and Peter Clote (Boston), with whom I had the privilege of collaborating on various research projects. Gaisi Takeuti never failed to provide an inspiration during my two years at the Department of Mathematics of the University of Illinois at Urbana.

I also wish to thank Steve Cook from the Department of Computer Science of the University of Toronto, where I wrote a large part of the first version of this book during my stay in spring semester 1993, for many inspiring discussions concerning topics related to this book.

Finally I thank the following people for comments on parts of the manuscript: M. Baaz (Vienna), S. R. Buss (San Diego), M. Chiari (Parma), S. A. Cook (Toronto), F. Pitt (Toronto), P. Pudlák (Prague), A. A. Razborov (Moscow), G. Takeuti (Urbana), and D. Zambella (Amsterdam).

1

Introduction

Ten years ago I had the wonderful opportunity to attend a series of lectures given by Jeff Paris in Prague on his and Alec Wilkie's work on bounded arithmetic and its relations to complexity theory. Their work produced fundamental information about the strength and properties of these weak systems, and they developed a variety of basic methods and extracted inspiring problems.

At that time Pavel Pudlák studied sequential theories and proved interesting results about the finitistic consistency statements and interpretability (Pudlák 1985,1986,1987). A couple of years later Sam Buss's Ph.D. thesis (Buss 1986) came out with an elegant proof–theoretic characterization of the polynomial time computations. Then I learned about Cook (1975), predating the above developments and containing fundamental ideas about the relation of weak systems of arithmetic, propositional logic, and feasible computations. These ideas were developed already in the late 70s by some of his students but unfortunately remained, to a large extent, unavailable to a general audience. New connections and opportunities opened up with Miki Ajtai's entrance with powerful combinatorics applied earlier in Boolean complexity (Ajtai 1988).

The work of these people attracted other researchers and allowed, quite recently, further fundamental results.

It appears to me that with a growing interest in the field a text surveying some basic knowledge could be helpful. The following is an outline of the book.

Chapter 2 lists notions and results from logic and complexity theory the reader is expected to have heard about. Chapter 3 overviews basic Boolean complexity and basic facts about predicates definable by bounded arithmetic formulas. Sketches of a few proofs are offered there, but mostly I refer the reader to other survey texts. All later chapters contain all necessary proofs.

Chapters 4 and 5 present basic information about the main propositional proof systems and complexity of proofs issues, and about the main first order systems of bounded arithmetic and their strength and mutual relations.

Chapters 6 and 7 survey the characterizations of definable functions in the systems of arithmetic known as *witnessing theorems*. Chapter 8 treats the second order systems of bounded arithmetic using *RSUV isomorphism* and transfers some results of the previous three chapters to these systems.

Chapter 9 defines and studies propositional translations of arithmetic formulas and propositional simulations of arithmetic proofs with applications to polynomial simulation results.

Chapter 10 is devoted to the fundamental problem of whether bounded arithmetic S_2 is finitely axiomatizable, the central problem in the area. It surveys all relevant results known (to me) to date.

Chapter 11 studies direct combinatorial arguments allowing separation of the lowest relativized subsystems of bounded arithmetic.

Chapters 12 and 13 concern the central question of propositional logic, whether there is a proof system admitting polynomial size proofs of all tautologies, for *Frege* and extended *Frege* systems and for constant-depth systems. The main results are several exponential lower bounds for the constant-depth systems and a certain conceptual framework for the unrestricted system.

Chapter 14 presents finitistic consistency statements and studies the issue of hard tautologies and optimal proof systems.

The final chapter, Chapter 15, develops some combinatorics and Boolean complexity theory within bounded arithmetic and studies several model-theoretic constructions relevant to all the basic questions studied earlier in the book.

I have made an attempt to present the chosen material as completely and as up to date as possible but I did not try to compile a handbook of the whole field (hence the Bibliography also does not attempt to list the whole literature in the field). Open problems are occasionally mentioned in the text (see the Index), but I refer the reader to Clote and Krajíček (1993) for a comprehensive annotated list of open problems in the area.

In the main text I give explicit credit only for main ideas and results. The chapters end with a section of bibliographical and other remarks where complete bibliographical information is given, and where I briefly comment on related but not covered topics.

Finally I want to comment on material that is *not* covered in the book. This includes, in particular, intuitionistic versions of bounded arithmetic systems, functional interpretations of these theories (and the issue of feasible functionals in general), equational theories and machine-independent characterizations of various computational classes, and modifications of the basic systems relating them to a variety of subclasses of the polynomial time. Some of this material is omitted as I do not feel familiar with it; some is omitted because I think that it – although technically difficult and innovative – builds on basic ideas already apparent from earlier results presented in the book.

2

Preliminaries

In this chapter we briefly review the basic notions and facts from logic and complexity theory whose knowledge is assumed throughout the book. We shall always sketch important arguments, both from logic and from complexity theory, and so a determined reader can start with only a rough familiarity with the notions surveyed in the next two sections and pick the necessary material along the way.

For those readers who prefer to consult relevant textbooks we recommend the following books: The best introduction to logic are parts of Shoenfield (1967); for elements of structural complexity theory I recommend Balcalzár, Diáz, and Gabbarró (1988, 1990); for NP-completeness Garey and Johnson (1979); and for a Boolean complexity theory survey of lower bounds Boppana and Sipser (1990) or the comprehensive monograph Wegener (1987). A more advanced (but self-contained) text on logic of first order arithmetic theories is Hájek and Pudlák (1993).

2.1. Logic

We shall deal with first order and second order theories of arithmetic. The second order theories are, in fact, just two-sorted first order theories: One sort are numbers; the other are finite sets. This phrase means that the underlying logic is always the first order predicate calculus; in particular, no set-theoretic assumptions are a part of the underlying logic.

From basic theorems we shall use Gödel *completeness* and *incompleteness* theorems, Tarski's *undefinability of truth*, and, in arithmetic, constructions of *partial truth definitions*.

A prominent theory is *Peano arithmetic* (PA), in the *language of arithmetic* $L_{PA} = \{0, 1, +, \cdot, <, =\}$ axiomatized by *Robinson's arithmetic Q*

1. $a + 1 \neq 0$

2. $a + 1 = b + 1 \to a = b$
3. $a + 0 = a$
4. $a + (b + 1) = (a + b) + 1$
5. $a \cdot 0 = 0$
6. $a \cdot (b + 1) = (a \cdot b) + a$
7. $a \neq 0 \to \exists x, x + 1 = a$

see Tarski, Mostowski, and Robinson (1953), and by the *induction scheme* IND

$$\left(\phi(0, \overline{a}) \wedge \forall x (\phi(x, \overline{a}) \to \phi(x + 1, \overline{a})) \right) \to \forall x \phi(x, \overline{a})$$

for every formula $\phi(x, \overline{a})$ in the language L_{PA}.

We shall use the letters $x, y, z, \ldots$ mostly for bounded variables; the letters $a, b, c, \ldots$ will be reserved for free variables (also called parameters). Free variables in axioms are assumed to be universally quantified; for example, the first axiom given is equivalent to the formula $\forall x, x + 1 \neq 0$.

There are other schemes that can equivalently replace the induction scheme, for example, *the least number principle* LNP scheme

$$\phi(b, \overline{a}) \to \exists x \forall y \left(\phi(x, \overline{a}) \wedge (y < x \to \neg \phi(y, \overline{a})) \right).$$

The *standard model N* of PA is the set of natural numbers with the symbols of L_{PA} interpreted with the usual meaning. A crucial fact about PA is that there are *nonstandard* models (models not isomorphic with N) of PA and indeed of the theory of N, Th(N). Natural numbers N are isomorphic to a unique initial substructure of any nonstandard model M and we shall usually simply assume that $N \subset M$.

A *cut* in a nonstandard model M is any nonempty $I \subseteq M$ satisfying

1. $a < b \wedge b \in I \to a \in I$, all $a, b \in M$
2. $a \in I \to a + 1 \in I$, all $a \in M$.

For example, N is a cut in every nonstandard model. Cuts in nonstandard models of PA closed under both addition and multiplication have special prominence as they are particular models of *bounded arithmetic* $I\Delta_0$: They satisfy induction for all *bounded arithmetic formulas* Δ_0, which are formulas in the language L_{PA} with all quantifiers bounded (Section 3.2 is devoted to bounded formulas).

Nonstandard models of PA and even of its proper subtheories are difficult to construct; it is a theorem of Tennenbaum (1959) that there are no countable recursive nonstandard models of PA (and, indeed, of a weak subtheory IE_1 with the induction just for bounded existential formulas, cf. Paris (1984). In particular, these results show that every nonstandard countable model of IE_1 has a nonstandard cut that is a model of whole PA; hence, in a sense, the model theory of bounded arithmetic is as complex as that of PA. Consult Hájek and Pudlák (1993), Kaye (1991), or Smorynski (1984) for the model theory of PA.

From proof theory we shall use theorems of Gentzen and Herbrand in various versions. The reader is advised to refer to Takeuti (1975) for Gentzen's sequent calculus.

We close this section with some remarks on notation. Logical connectives we shall use are the standard $\neg, \vee, \wedge, \rightarrow$, and $\equiv$ with the usual meaning – negation, disjunction, conjunction, implication, and equivalence – and the constants 1, 0 for *truth* and *falsity*.

The symbols $\subset$ and $\subseteq$ are used in the sense of *proper inclusion* and *inclusion*.

The symbols $f(n) = O(g(n))$, $f(n) = \Omega(g(n))$ and $f(n) = \Theta(g(n))$ denote that eventually $f(n) \leq cg(n)$, $f(n) \geq cg(n)$, and $c_1 g(n) \leq f(n) \leq c_2 g(n)$ where c, c_1, and c_2 are positive constants, and $f(n) = o(g(n))$ means that $f(n)/g(n) \rightarrow 0$.

2.2. Complexity theory

I assume that the reader is acquainted with such notions as *Turing machine, oracle Turing machine*, and *time* and *space* complexity measures. We adopt the multi-tape version of Turing machines with a read-only input tape and with a finite but arbitrarily large alphabet.

The basic relations between classes of languages Time(f) and Space(f) recognized by a deterministic Turing machine in time (respectively space) bounded by $f(n)$, n the length of the input, and their nondeterministic versions NTime(f) and NSpace(f) are

1. Time($f(n)$) $\subseteq$ NTime($f(n)$) $\subseteq$ Space($f(n)$)
2. Space($f(n)$) $\subseteq \bigcup_c$ Time($c^{f(n)}$)
3. (Hartmanis and Stearns 1965) Time($f(n)$) $=$ Time($c \cdot f(n)$) and Space($f(n)$) $=$ Space($c \cdot f(n)$) whenever $n = o(f(n))$ and $n \leq f(n)$
4. (Hartmanis and Stearns 1965, Hartmanis, Lewis, and Stearns 1965)

$$\text{Space}(f) \subset \text{Space}(g)$$

and

$$\text{Time}(f) \subset \text{Time}(g \log(g))$$

whenever $f = o(g(n))$.

5. (Savitch 1970)

$$\text{NSpace}(f) \subseteq \text{Space}(f^2)$$

whenever $f(n)$ is itself computable in space $f(n)$

6. (Szelepcsényi 1987, Immerman 1988)

$$\text{NSpace}(f) = \text{coNSpace}(f)$$

for $f(n) \geq \log(n)$ and f itself computable in nondeterministic space $f(n)$.

7. (Hopcroft, Paul, and Valiant 1975) For $n \leq f(n)$

$$\text{Time}(f(n)) \subseteq \text{Space}\left(\frac{f(n)}{\log f(n)}\right)$$

Particular bounds to time or space define the usual complexity classes

$$\text{LinTime} = \bigcup_c \text{Time}(cn)$$

$$P = \bigcup_c \text{Time}(n^c)$$

$$NP = \bigcup_c \text{NTime}(n^c)$$

$$L = \text{Space}(\log(n))$$

$$\text{PSpace} = \bigcup_c \text{Space}(n^c)$$

$$\text{LinSpace} = \text{Space}(n)$$

$$E = \bigcup_c \text{Time}(c^n)$$

$$\text{EXP} = \bigcup_c \text{Time}(2^{n^c})$$

Oracle computations allow one to define hierarchies of languages, the most important of which are the *linear time hierarchy* LinH of Wrathall (1978)

$$\Sigma_0^{\text{lin}} = \text{LinTime and } \Sigma_{i+1}^{\text{lin}} = \text{NLinTime}^{\Sigma_i^{\text{lin}}}$$

and the *polynomial time hierarchy* PH of Stockmeyer (1977)

$$\Sigma_0^p = P \quad \text{and} \quad \Sigma_{i+1}^p = \text{NP}^{\Sigma_i^p}$$

The class of complements of languages from class X is denoted co$X$, and special classes of this form coΣ_i^{lin} and coΣ_i^p are denoted Π_i^{lin} and Π_i^p, respectively.

The class $\Box_{i+1}^p$ is the class of functions computable by a polynomial-time machine with access to an oracle from the class Σ_i^p.

Some important facts about these classes include the following: $\Sigma_i^{\text{lin}} \subset \Sigma_i^p$ (and generally more resource in the "same" computational class properly increases the class; see Žák 1983 for a general diagonalization technique), and LinH contains L and is, in fact, equal to the class of *rudimentary predicates* as defined by Smullyan (1961) (cf. Wrathall 1978). It is also known that LinH also equals the class of predicates definable by Δ_0-formulas; we shall prove that in Section 3.2. Also note that either LinH $\neq$ PH or LinH does not collapse (i.e., LinH $\neq \Sigma_i^{\text{lin}}$ for all i).

The notion of NP-completeness, Cook's theorem, and the *P versus NP problem* are central to complexity theory, as well as to the connections with logic, and in Section 3.1 we shall review more basics, in particular some facts from circuit complexity.

Many interesting problems and notions arise in connection with *counting functions*. For $R(x, y)$ a binary predicate with the property that for every x there are only finitely many y's satisfying $R(x, y)$ defines the function

$$\#R(x) := \text{ the number of } y\text{'s such that } R(x, y)$$

Class #P consists of all functions $\#R(x)$ with the polynomial time computable relation $R(x, y)$ and satisfying the preceding finiteness property in a stronger form (cf. Valiant 1979):

$$R(x, y) \rightarrow |y| \leq |x|^{O(1)}$$

An important result of Toda (1989) is that every language in PH is polynomial–time reducible to a function in #P.

Nonuniform versions of the preceding classes are defined with the help of *advice functions*. *Polynomially bounded advice* is a function $f : N \rightarrow \{0, 1\}^*$ such that:

$$|f(n)| = n^{O(1)}$$

The class P/poly, a nonuniform version of P, is the class of all sets A such that there are a set $B \in P$ and a polynomially bounded advice function f for which it holds

$$x \in A \text{ iff } (x, f(|x|)) \in B$$

The classes NP/poly, L/poly, and so on, are defined analogously (see the paragraph after Theorem 3.1.4).

3

Basic complexity theory

We shall survey the basic notions and results of Boolean complexity (Section 3.1) and bounded formulas (Section 3.2) in this chapter. Most of the results in the first section are stated without a proof; some proofs appear later (Chapter 15, in particular) formalized in bounded arithmetic. In the second section, most proofs are at least sketched.

3.1. The P versus NP problem

The central problem in complexity theory, and a major problem of contemporary logic and mathematics, is whether the class P equals the class NP, the famous *P versus NP problem* (Cook 1971). By Cook's theorem the problem is equivalent to asking whether there is a polynomial time deterministic algorithm recognizing the set of satisfiable propositional formulas, or equivalently, such an algorithm recognizing the set of propositional tautologies.

One approach to this problem is via investigating the circuit-complexity of Boolean functions. Some interesting, although only preliminary, results were obtained in Boolean complexity.

Definition 3.1.1. *A* Boolean function *with n inputs and m outputs is a function*

$$f : \{0, 1\}^n \mapsto \{0, 1\}^m.$$

Examples of Boolean functions are obtained from any language $Z \subseteq \{0, 1\}^*$: For any n, define the Boolean function $Z_n : \{0, 1\}^n \mapsto \{0, 1\}$ to be the characteristic function of $Z \cap \{0, 1\}^n$.

On the other hand, a sequence of Boolean functions

$$f_n : \{0, 1\}^n \mapsto \{0, 1\}, \ n = 0, 1, \ldots$$

defines a language

$$\bigcup_n \{w \in \{0, 1\}^n \mid f_n(w) = 1\}$$

Definition 3.1.2.

(a) *A Boolean* connective *is a Boolean function with one output. A* basis *is a finite set of connectives.*

(b) *A Boolean circuit with input variables $x_1, \ldots, x_n$; output variables $y_1, \ldots, y_m$; and basis of connectives $\Omega = \{g_1, \ldots, g_k\}$ is a labeled acyclic directed graph whose out-degree 0 nodes are labeled by y_j's, in-degree 0 nodes are labeled by x_i's or by constants from Ω, and whose in-degree $\ell \geq 1$ nodes are labeled by functions from Ω of arity ℓ.*

(c) *A Boolean* formula *is a Boolean circuit in which every node has the out-degree at most 1.*

We shall mostly consider the *de Morgan* basis $\Omega = \{0, 1, \neg, \vee, \wedge\}$.

A Boolean circuit with input variables $x_1, \ldots, x_n$ naturally computes a Boolean function with domain $\{0, 1\}^n$: given input $\bar{\epsilon} \in \{0, 1\}^n$ evaluate consecutively the nodes of the circuit by 0, 1 where a node gets the value computed by the connective labeling the node from the values at the incoming nodes. The requirement that a circuit is acyclic guarantees that this can be done consistently (and uniquely).

Definition 3.1.3.

(a) *The* size *of a circuit is the number of its nodes.*

(b) *The* depth *of a circuit is the maximum length of a directed path in the circuit.*

(c) *For a Boolean function f, $C_\Omega(f)$ denotes the minimal size of a circuit with basis Ω computing f, $Depth_\Omega(f)$ denotes the minimal depth of a circuit with basis Ω computing f, and $L_\Omega(f)$ denotes the minimal size of a formula with basis Ω computing f.*

When Ω is the de Morgan *basis then the index Ω is usually omitted.*

The following theorem is the stimulus for investigating the circuit complexity of Boolean functions.

Theorem 3.1.4. *Let $Z \subseteq \{0, 1\}^*$ be a polynomial-time recognizable language $Z \in P$. Then there exist a polynomial $p(x)$ and a sequence $\{C_n\}_n$ of circuits in* de Morgan *basis with one output such that for all n*

1. *Z_n is computed by C_n*
2. *the size of C_n is at most $p(n)$.*

In other words, $C(Z_n) \leq p(n)$, for all n.

Note that the languages Z with polynomially bounded $C(Z_n)$ are exactly those from the class P/poly: circuits C_n can act as advice for inputs of length n (as the *evaluation* of a circuit can be performed by a polynomial time algorithm), and, for each n, an algorithm computing whether $(x, f(|x|)) \in B$ (see the end of Section 2.2) can be turned into a polynomial size circuit.

Corollary 3.1.5. *Assume that for some $Z \in$ NP the function $C(Z_n)$ is not bounded by any polynomial in n.*
   Then $P \neq NP$.

Hence the following problem is a fundamental one.

Fundamental problem. *Is there a language $Z \in$ NP with superpolynomial circuit-size complexity?*

The next theorem shows that even the unexpected negative answer has important corollaries.

Theorem 3.1.6 (Karp and Lipton 1982). *Assume that every NP language can be computed by a family of polynomial size circuits; that is, NP $\subseteq$ P/poly.*
   Then the polynomial time hierarchy PH collapses to its second level

$$PH = \Sigma_2^p = \Pi_2^p.$$

Theorem 3.1.7 (Shannon 1949, Muller 1956). *For every n there are Boolean functions with n inputs and one output having the circuit complexity $\Omega(2^{n-\log n})$.*

Proof. There are 2^{2^n} Boolean functions with n unknowns and one output, whereas there are at most

$$(n + m)^{O(m)}$$

circuits of size $\leq m$, which is less than 2^{2^n} for $m \leq (2^n/c \cdot n)$ and c a sufficiently large constant.   Q.E.D.

Next we give an account of the *method of approximations* of Razborov (1989), following to some extent an exposition of Karchmer (1993) but stressing the ultra-product interpretation of the construction. We include this material here because it is the only framework that can be, at least in principle, applied to unrestricted circuits.

   Let $f : \{0, 1\}^n \to \{0, 1\}$ be a Boolean function of n inputs and one output, and $U := f^{-1}(0)$.

   Let C be a circuit in n inputs and of size m. We shall denote the nodes of C by $z_1, \ldots, z_m$ where the first n nodes are labeled by $x_1, \ldots, x_n$ and the last one is the output node y. We identify node z_i with the Boolean function of n inputs computed by the subcircuit ending in z_i and we shall occasionally write x_i instead of z_i if $i \leq n$ and y if $i = m$.

   The idea is to take the set of all computations of C on inputs $u \in U$ and produce by "ultraproduct" a new computation on some $w \notin U$. As all original computations are rejecting (assuming that C computes f), the new computation will also be rejecting. Hence $w \notin U$ will mean that C cannot compute f correctly on all inputs.

Let B be the Boolean algebra of subsets of U. For any $g : \{0, 1\}^n \to \{0, 1\}$ put

$$||g|| := \{u \in U | g(u) = 1\}$$

If $F \subseteq B$ were an ultrafilter then we would define a new computation by labeling node z_i by 1 if $||z_i|| \in F$, and by 0 otherwise. This would produce a correct rejecting computation, but the particular input defined by the choice of F would be from U as all ultrafilters on a finite set are principal. Hence we have to relax the notion of ultrafilter if anything nontrivial should be achieved.

A subset $F \subseteq B$ is *closed upward* if $a \in F$ and $a \subseteq b$ implies $b \in F$. F *preserves* the pair (a, b) iff

$$a \in F \wedge b \in F \to a \cap b \in F$$

W_F is the set of all $w \in \{0, 1\}^n$ satisfying

$$w_i = 1 \to ||x_i|| \in F \text{ and } w_i = 0 \to ||\neg x_i|| \in F$$

If exactly one of each $||x_i||$ and $||\neg x_i||$ is in F, then W_F consists of one 0–1 vector denoted w_F.

Let $\rho(f)$ be the minimal t such that there exist t pairs $(a_i, b_i)_{i \le t}$ of elements of B having the property that if $F \subseteq B$ is closed upward, $\emptyset \notin F$ and F preserves all pairs $(a_i, b_i)_{i \le t}$ then $W_F \subseteq U$.

Theorem 3.1.8 (Razborov 1989).

$$\frac{1}{2}\rho(f) \le C(f) = O(\rho^3(f)) + O(n^3)$$

Proof. We first prove the easier of the two inequalities: $\frac{1}{2}\rho(f) \le C(f)$. Let C be an optimal circuit for f and assume all $\neg$ are at the bottom level (we may assume that by *de Morgan* rules possibly increasing the size of the circuit twice). Let $(c_i, d_i)_{i \le t}$ be all pairs of nodes of C such that c_i and d_i fan into a common node labeled by $\wedge$. That is, if (g_i, h_i) is a pair of Boolean functions computed at nodes c_i, d_i then the conjunction $g_i \wedge h_i$ is computed at some node of C too (and $(g_i, h_i)_{i \le t}$ exhaust all such pairs).

Set $a_i := ||g_i||$ and $b_i := ||h_i||$. Assume that $F \subseteq B$ is closed upward and that it preserves all pairs $(a_i, b_i)_{i \le t}$.

Claim 1. *Let g be a function computed by a subcircuit of C and let $w \in W_F$. If $g(w) = 1$ then $||g|| \in F$.*

The claim is readily established by induction on the size of the subcircuit computing g: It holds for x_i and $\neg x_i$ by the definition of W_F, it holds for subcircuits with top gate $\vee$ as F is closed upward, and it holds for circuits with top gate $\wedge$ as F preserves all $(a_i, b_i)_{i \leq t}$: that is, all possible conjunctions in C.

Now assume that for some $w \in W_F$, $w \notin U$: that is, $f(w) = 1$. By the claim then $||f|| \in F$. But $||f|| = \emptyset$: that is, $\emptyset \in F$. This proves the first inequality.

For the second inequality assume that we have $(a_i, b_i)_{i \leq t}$ such that for any $F \subseteq B$ closed upward such that $\emptyset \notin F$ and that preserves all $(a_i, b_i)_{i \leq t}$, we have $W_F \subseteq U$.

Claim 2. *For any* $w \in \{0, 1\}^n$, $f(w) = 1$ *iff* $\emptyset \in F_w$, *where* F_w *is the minimal subset of* B *that is closed upward, preserves all pairs* $(a_i, b_i)_{i \leq t}$, *and satisfies* $||x_i|| \in F_w$ *(resp.* $||\neg x_i|| \in F_w$) *for* $w_i = 1$ *(resp.* $w_i = 0$).

To see the claim assume first $f(w) = 1$, from which $W_{F_w} \not\subseteq U$ follows, as by the definition of F_w, $w \in W_{F_w}$. Hence $\emptyset \in F_w$.

Now assume $f(w) = 0$: that is, $w \in U$. Take $F = \{X \subseteq U | w \in X\}$. Then $F_w \subseteq F$, F is closed upward and preserves all $(a_i, b_i)_{i \leq t}$, but $\emptyset \notin F$. Hence also $\emptyset \notin F_w$.

We are ready to construct a circuit computing the function f based on an idea that for $w \in \{0, 1\}^n$ the circuit will try to prove that $\emptyset \in F_w$.

Let

$$A = \{a_1, b_1, a_1 \cap b_1, \ldots, a_t, b_t, a_t \cap b_t\} \cup \{||x_1||, ||\neg x_1||, \ldots, ||x_n||, ||\neg x_n||\} \cup \{\emptyset\}$$

The size of A is at most $3t + 2n + 1$.

For any $a \in A$ and $k \leq 3t + 2n + 1$ consider a function v_a^k inductively introduced by

$$v_a^0 = \begin{cases} 1 & \text{if } a = ||x_i|| \wedge w_i = 1, \quad \text{or} \quad \text{if } a = ||\neg x_i|| \wedge w_i = 0 \\ 0 & \text{otherwise} \end{cases}$$

and

$$v_a^{k+1} := \bigvee_{b \subseteq a, b \in A} v_b^k \vee \bigvee_{j \in J_a} (v_{a_j}^k \wedge v_{b_j}^k) \vee \bigvee_{i \leq n} \left(v_{||x_i||}^k \wedge v_{||\neg x_i||}^k \right)$$

where $J_a = \{j \leq t | a_j \cap b_j = a\}$.

Obviously, if for some r

$$\forall a \in A; \qquad v_a^r = v_a^{r+1}$$

then for all $s > r$ and all $a \in A$: $v_a^r = v_a^s$ and $A \cap F_w = \{a | v_a^r = 1\}$. Hence $\emptyset \in F_w$ iff $v_\emptyset^k = 1$ for $k = 3t + 2n + 1$, and the definition of $v_\emptyset^k$ constitutes a definition of a circuit of size $O((3t + 2n + 1)^3) = O(t^3) + O(n^3)$ computing whether $\emptyset \in F_w$, that is, computing $f(w)$.   Q.E.D.

In principle thus, one can establish a lower bound t to $C(f)$ by showing that for each set of $2t$ pairs $(a_i, b_i)_{i \leq t}$ of subsets of U there is a nontrivial F closed upward and preserving all pairs $(a_i, b_i)_{i \leq t}$ such that $W_F \not\subseteq U$.

The method has been successfully applied in the monotone case (i.e., for monotone functions and circuits in the basis $\{0, 1, \vee, \wedge\}$); (cf. Razborov 1985). It seems that that was possible as the condition posed on a vector to be in W_F is in the monotone case relaxed to one implication, $w_i = 1 \rightarrow ||x_i|| \in F$, only.

The following function has $\binom{n}{2}$ variables x_{ij}. Denote by $G(x_{ij})$ the undirected graph with vertices $V = \{1, \ldots, n\}$ and edges

$$E = \{\{i, j\} \mid x_{ij} = 1\}$$

A *clique* in a graph is a complete subgraph.

The clique function

$$\text{CLIQUE}_{n,k}(x_{ij}) = \begin{cases} 1 & \text{if } G(x_{ij}) \text{ has a clique of size at least } k \\ 0 & \text{otherwise} \end{cases}$$

Let $C^+(f)$ denote the minimal size of a circuit in the basis $\{0, 1, \vee, \wedge\}$ computing a monotone function f. $C^+(f)$ is sometimes also denoted $C^m(f)$.

Theorem 3.1.9 (Razborov 1985, Alon and Boppana 1987). *For $k \leq n^{1/4}$:*

$$C^+(CLIQUE_{n,k}) = n^{\Omega(\sqrt{k})}$$

Now we turn our attention to another restricted model of circuits: constant-depth circuits. To obtain a nontrivial model of computation we have to allow $\bigvee$ and $\bigwedge$ of unbounded arity. Note that any Boolean function can then be computed by a depth 2 circuit: Take its disjunctive or conjunctive normal form.

It is easy to see that any depth d size m circuit can be rewritten as depth d size $\leq m^d$ formula; hence there is no essential difference in studying constant-depth circuits versus formulas.

There are three simple Boolean functions of particular importance.

The parity function

$$\oplus(x_1, \ldots, x_n) = \begin{cases} 1 & \text{if } \sum_i x_i \text{ is odd} \\ 0 & \text{otherwise} \end{cases}$$

The majority function

$$\text{MAJ}(x_1, \ldots, x_n) = \begin{cases} 1 & \text{if } (n/2) \leq \sum_i x_i \\ 0 & \text{otherwise} \end{cases}$$

The Sipser function of depth d

$$S_d(x_{i_1 \ldots i_d}) = \bigvee_{i_1 \leq n} \bigwedge_{i_2 \leq n} \cdots x_{i_1 \ldots i_d}$$

where i_j range over $\{1, \ldots, n\}$ and the d connectives $\bigvee, \bigwedge$ alternate.

Note that the disjunctive normal form of all these functions has exponential size (in $n^{\Omega(1)}$).

Theorem 3.1.10 (Yao 1985, Hastad 1989). *For any $d \geq 2$, any depth d circuit computing the parity function*

$$\oplus(x_1, \ldots, x_n)$$

must have size at least

$$2^{\Omega(n^{(1/(d-1))})}.$$

The proof of the optimal lower bound utilizes *the method of random restrictions* and the so-called Hastad's switching lemma (there are two, in fact); the idea of the structure of the lower bound proof goes back to Ajtai (1983); Furst, Saxe, and Sipser (1984); and Yao (1985).

A *restriction* of a circuit is a partial evaluation of its inputs

$$\rho : \{x_1, \ldots, x_n\} \mapsto \{0, 1, *\}$$

where the value $\rho(x) = *$ is a convenient abbreviation of "$\rho(x)$ is undefined." The restricted circuit then computes a function of the inputs that received $*$ by ρ.

The idea of the lower bound proofs by the method of restrictions is that if the circuit is small then there will be a restriction leaving some input variables unevaluated, such that the restricted circuit will compute a *constant* function. If a circuit computes the parity function, each restriction computes either the parity or its negation and specifically cannot be a constant function. Hence there cannot be small constant-depth circuits computing the parity function.

We shall now only state the *Hastad's switching lemma* and leave a proof of its variant to Chapter 15 (Lemma 15.2.2), where we shall formalize its proof within bounded arithmetic.

Lemma 3.1.11 (Hastad 1989). *Let $0 < p < 1$. Construct a restriction ρ by the following random process: for any x_i define $\rho(x_i)$ independently*

$$\rho(x_i) = \begin{cases} * & \text{with probability } p \\ 1 & \text{with probability } (1-p)/2 \\ 0 & \text{with probability } (1-p)/2 \end{cases}$$

Assume that a circuit C is a depth 2 circuit that is a disjunction of conjunctions of literals, with each conjunction of arity at most t.

Then with a probability of at least

$$1 - (5pt)^s$$

the function computed by the circuit restricted by a random restriction can also be

*computed by a depth 2 circuit that is a conjunction of disjunctions of literals, with
each disjunction of arity at most s.*

In applications t, s are of the form n^δ and p is of the form $n^{\epsilon-1}$. In that case the
bound from the lemma is very close to $1 : 1 - 2^{n^{-\xi}}$. Hence it is highly probable
that a disjunction of small conjunctions can be "switched" into a conjunction of
small disjunctions.

That this really simplifies the function is clarified by a simple lemma. First we
need a definition.

Definition 3.1.12.
   (a) *A* branching program *is a directed acyclic graph with one in-degree 0 node
   (the source); with all nodes of out-degree either 2 (the inner nodes) or 0
   (the leaves); the inner nodes labeled by variables with the two outgoing
   edges labeled by 0, 1, respectively; and the leaves labeled by elements of
   a set Y.*
   (b) *A* decision tree *is a branching program that is a tree with the edges directed
   from the root toward the leaves.*
   *Any evaluation α of variables determines a path $P(\alpha)$ through the program or
   the tree: The path uses the edge labeled 1 from a node labeled by x_i if and only if
   $\alpha(x_i) = 1$.*
   *We say that a branching program (a decision tree) computes a function
   $f(x_1, \ldots, x_n)$ if for every evaluation α*

$$f(\alpha(x_1), \ldots, \alpha(x_n)) = y \quad \text{iff path } P(\alpha) \text{ ends with a leaf labeled by } y$$

The size *of a branching program (a decision tree) is the number of nodes.*
   The height *of a decision tree is the maximum length of a path through it.*

Most often the set Y is just the set $\{0, 1\}$, in which case the branching program
(decision tree) computes a Boolean function.

For the first part of the next lemma note that any conjunction in a disjunction (of
conjunctions) computing f has to have a literal in common with every disjunction
in a conjunction (of disjunctions) computing f.

Lemma 3.1.13. *Assume that a function $f(x_1, \ldots, x_n)$ is computed by a depth 2
circuit that is a disjunction of conjunctions of arity at most s, and at the same time
also by a depth 2 circuit that is a conjunction of disjunctions of arity at most s.*
   Then f can be computed by a decision tree *of height at most s^2.*
   *On the other hand, if a function is computed by a decision tree of height t then
it can be computed by a depth 2 circuit that is a conjunction of disjunctions of arity
at most t and also by a depth 2 circuit that is a disjunction of conjunctions of arity
at most t.*

Modular counting functions are defined similarly to the parity function.

Modular counting function

$$\mathrm{MOD}_p(x_1, \ldots, x_n) = \begin{cases} 1 & \text{if } \sum_i x_i \text{ is not divisible by } p \\ 0 & \text{otherwise} \end{cases}$$

Theorem 3.1.14 (Razborov 1987, Smolensky 1987). *Let p, q be different primes and let $d \geq 2$ be arbitrary. Then any depth d circuit in the basis $\{0, 1, \neg, \vee, \wedge, \mathrm{MOD}_q\}$ computing the function $\mathrm{MOD}_p(x_1, \ldots, x_n)$ must have size at least*

$$2^{\Omega(n^{(1/(2d))})}.$$

In particular, any such circuit computing the majority function

$$\mathrm{MAJ}(x_1, \ldots, x_n)$$

must have size at least

$$2^{\Omega(n^{(1/(2d+1))})}$$

The bound to the majority function follows from the first part as it is easy to see that all MOD_p functions have depth 2 circuits of size $O(n)$ in the basis $\{0, 1, \neg, \vee, \wedge, \mathrm{MAJ}\}$.

We shall now turn our attention from the size of circuits to the depth of circuits. We consider circuits in the *de Morgan* basis again with the binary $\vee$ and $\wedge$. There are two simple but important facts. The first is about the relation of the circuit depth to the size of formulas. The idea for the proof of this theorem is used in several proofs in later chapters.

Theorem 3.1.15 (Spira 1971). *For any Boolean function f*

$$\mathrm{Depth}(f) = O(\log L(f)) \quad and \quad L(f) \leq 2^{1+\mathrm{Depth}(f)}$$

In particular, the notions "polynomial size formulas" and "$O(\log n)$-depth circuits" are equivalent concepts.

The second fact concerns the important notion of *communication complexity*. Consider a game played by two players A and B; player A receives $\bar{a} \in \{0, 1\}^n$ and player B receives $\bar{b} \in \{0, 1\}^n$ such that $f(\bar{a}) \neq f(\bar{b})$. Their task is to find i such that $a_i \neq b_i$. They send each other bits of information and the game ends when the players agree on an answer.

The *communication complexity* of function $f(\bar{x})$ is the minimal number of bits they need to exchange in the worst case before the game ends. We shall denote it $\mathrm{CC}(f)$. The following is an important characterization of the circuit-depth measure. The theorem is proved by a straightforward induction.

Theorem 3.1.16 (Karchmer and Wigderson 1988). *For any Boolean function f*

$$\text{Depth}(f) = \text{CC}(f)$$

As little is known about the size of formulas as about the circuit-size.

Theorem 3.1.17 (Andreev 1987, Hastad 1993). *There is a polynomial-time language Z such that*

$$L(Z_n) \geq n^{3-o(1)}$$

The language Z from the theorem is a rather artificial one. Earlier Chrapchenko (1971) showed

$$L(\oplus(x_1, \ldots, x_n)) \geq n^2$$

3.2. Bounded arithmetic formulas

We shall consider several languages of arithmetic as underlying languages for various systems of bounded arithmetic, but there are two basic ones: the language of *Peano arithmetic* L_{PA} defined in Section 2.1, and the language of the theory S_2, denoted simply L, which extends the language L_{PA} by three new function symbols

$$\left\lfloor \frac{x}{2} \right\rfloor \qquad |x| \qquad x\#y$$

The intended values of $|x|$ and $x\#y$ are $\lceil \log_2(x+1) \rceil$ for $x > 0$ and $|0| = 0$, and $2^{|x|\cdot|y|}$, respectively. Note that $|x|$ is the length of the binary representation of x, if $x > 0$.

We shall consider the class of bounded formulas in the language L_{PA} first. They were first defined by Smullyan (1961), who called sets defined by such formulas *constructive arithmetic sets*.

Definition 3.2.1.
1. $E_0 = U_0$ is the class of quantifier free formulas.
2. Class E_{i+1} is the class of formulas logically equivalent (i.e., in the predicate calculus) to a formula of the form

$$\exists x_1 < t_1(\overline{a}) \ldots \exists x_k < t_k(\overline{a})\phi(\overline{a}, \overline{x})$$

with the formula $\phi \in U_i$ and $t_i(\overline{a})$'s terms of the language L_{PA}
3. U_{i+1} is the class of formulas logically equivalent to a formula of the form

$$\forall x_1 < t_1(\overline{a}) \ldots \forall x_k < t_k(\overline{a})\phi(\overline{a}, \overline{x})$$

with the formula $\phi \in E_i$.

4. Class Δ_0 *of* bounded arithmetic formulas *is the union of classes* E_i *and* U_i

$$\Delta_0 = \bigcup_i E_i = \bigcup_i U_i$$

Note that both E_i and U_i are contained in both E_{i+1} and U_{i+1}.

For M a structure for language L_{PA} symbols $E_i(M^\ell)$, $U_i(M^\ell)$, and $\Delta_0(M^\ell)$, respectively, denote the classes of subsets of M^ℓ definable by the E_i, U_i, and Δ_0 formulas, respectively (we shall usually omit the superscript ℓ when it is obvious from the context). Already the class $E_1(M)$ can be quite nontrivial from the complexity-theoretic point of view, as according to Adleman and Manders (1977) the class $E_1(N)$ contains an NP-complete set

$$\{(a, b, c) \mid \exists x < c \exists y < c, ax^2 + by = c\}$$

There are several important characterizations of the class $\Delta_0(N)$. We start with the notion of *rudimentary sets* introduced by Smullyan (1961).

The intended structure for the language of rudimentary sets is the set of words over $\{0, 1\}$ or, via dyadic coding, the set of natural numbers.

The language of rudimentary sets consists of

   1. Λ: *the empty word,*
   2. $\frown$: *the concatenation,*
   3. 0, 1: *constants,*

and two special kind of quantifiers

   4. $\exists x \subseteq_p y$ and $\forall x \subseteq_p y$: *the part-of quantifiers,*
   5. $\exists |x| \leq |y|$ and $\forall |x| \leq |y|$: *the length-bounded quantifiers.*

The meaning of $x \subseteq_p y$ is that the word x is a part of the word y

$$\exists z_1, z_2; z_1 \frown x \frown z_2 = y$$

and the meaning of $|x| \leq |y|$ is obvious: the length of x is at most the length of y.

Definition 3.2.2 (Smullyan 1961).

   1. The class of rudimentary sets *RUD is the class of subsets of* N^ℓ *definable in the language of rudimentary sets with all quantifiers either* part-of *or* length-bounded.

   2. The class of strictly rudimentary sets *SRUD is the class of subsets of* N^ℓ *definable in the language of rudimentary sets with all quantifiers of the* part-of *type.*

   3. The class of positive rudimentary sets *RUD$^+$ is the class of subsets of* N^ℓ *definable in the language of rudimentary sets with all quantifiers are either* part-of *or* length-bounded, *and in which all quantifiers* $\exists |x| \leq |y|$ *appear positively and all quantifiers* $\forall |x| \leq |y|$ *appear negatively.*

4. *The class of strongly rudimentary sets strRUD is the class of sets that are positive rudimentary and whose complements are also positive rudimentary.*

Note that terms are allowed to appear in the quantifiers.

5. *A function $f : N^\ell \mapsto N$ is* rudimentary *if its graph is a rudimentary set and the function is majorized by a polynomial.*

Theorem 3.2.3 (Bennett 1962).

$$RUD = \Delta_0(N)$$

Proof (sketch). Clearly there are only two claims to be established:

Claim 1. *The graphs of addition and the multiplication are in RUD.*

Claim 2. *The graph of the operation of concatenation is in $\Delta_0(N)$.*

The idea of the proof of Claim 1 is in *Bennett*'s lemma saying that any function defined by bounded recursion on notation is rudimentary (see Lemma 3.2.4). We shall see a bit stronger argument of the same type in Theorem 3.2.8.

For Claim 2 note that

$$x \frown y = z \quad \text{iff} \quad \exists w < z, \ y < w \wedge x \cdot w + y = z \wedge \text{ “ } w \text{ is a power of two”}$$

where the last condition is expressed by

$$\forall 1 < u, \ v < w \ \exists t \le u, \ u \cdot v = w \rightarrow 2 \cdot t = u$$

Q.E.D.

Lemma 3.2.4 (Bennett 1962). *Assume that a function f is defined from two rudimentary functions g and h by bounded recursion on the notation*

1. $f(\overline{0}, \overline{y}) = h(\overline{y})$
2. $f(x_1 \frown \epsilon_1, \ldots, x_n \frown \epsilon_n, \overline{y}) = g(\overline{x}, \overline{\epsilon}, \overline{y}, f(\overline{x}, \overline{y}))$ *for all $\overline{\epsilon} \in \{0, 1\}^n$*

and satisfies the condition

3. $|f(\overline{x}, \overline{y})| \le O\left(\sqrt{\sum_i |x_i| + \sum_j |y_j|}\right)$.

Then the function f is rudimentary too.

Theorem 3.2.5 (Wrathall 1978).

$$\operatorname{Lin}H = RUD$$

Proof (sketch). Using the natural coding of computations of machines by 0–1 strings one verifies that $\Sigma_0^{\lin} \subseteq RUD$, from which $\operatorname{Lin}H \subseteq RUD$ follows immediately.

The opposite inclusion is obvious. Q.E.D.

The possibility of coding in $\Delta_0(N)$ merits further discussion. We shall now mention two results and return to this topic again in Section 5.4.

Theorem 3.2.6 (Bennett 1962). *The graph of exponentiation*

$$\{(x, y, z) \mid x^y = z\}$$

is rudimentary.

Theorem 3.2.7 (Wrathall 1978). *All context-free languages are rudimentary and hence in $\Delta_0(N)$.*

The last theorem finds a root in an important theorem of Nepomnjascij (1970), generalizing Lemma 3.2.4.

The term TimeSpace($f(n)$, $g(n)$) denotes the class of languages recognized by a Turing machine working simultaneously in time $f(n)$ and space $g(n)$.

Theorem 3.2.8 (Nepomnjascij 1970). *Let $c > 0$ and $1 > \epsilon > 0$ be two constants. Then*

$$\text{TimeSpace}(n^c, n^\epsilon) \subseteq \Delta_0(N)$$

Proof (sketch). We shall give an idea of the proof. By induction on k prove that

$$\text{TimeSpace}(n^{k \cdot (1-\epsilon)}, n^\epsilon) \subseteq \Delta_0(N)$$

If $k = 1$ then the sequence consisting of the instantaneous descriptions of a TimeSpace($n^{k \cdot (1-\epsilon)}, n^\epsilon$) computation has size $O(n)$, and hence its code is bounded by a polynomial in input x, $|x| = n$.

Assume we have

$$\text{TimeSpace}(n^{k \cdot (1-\epsilon)}, n^\epsilon) \subseteq \Delta_0(N)$$

and let

$$L \in \text{TimeSpace}(n^{(k+1) \cdot (1-\epsilon)}, n^\epsilon)$$

Then $x \in L$ if and only if there exists a sequence $w = (w_0, \ldots, w_r)$ such that $w_0 = x$, each w_{i+1} is an instantaneous description obtained from the instantaneous description w_i by a TimeSpace($n^{k \cdot (1-\epsilon)}, n^\epsilon$) computation, w_r is a halting accepting position, and $r \leq n^{1-\epsilon}$.

The length of any such w is again $O(n)$ and the conditions defining it are Δ_0-definable by the induction assumption.   Q.E.D.

Corollary 3.2.9.

$$L \subseteq \Delta_0(N)$$

The main problem about $\Delta_0(N)$ is whether the hierarchy collapses, which is the same as whether LinH collapses: that is, whether

$$\Delta_0(N) = E_i(N)$$

for some i.

The only partial result is the following *weak hierarchy theorem* of Wilkie and Woods.

Theorem 3.2.10 (Wilkie 1980, Woods 1986). *Denote by $V_k(N)$ the class of subsets of N definable by a Δ_0-formula $\phi(x)$ with at most k quantifiers bounded by $\leq x$.*

Then for all k

$$V_k(N) \subset V_{k+1}(N)$$

The rest of this section is devoted to bounded formulas in the language L.

Definition 3.2.11 (Buss 1986).

1. *The class $\Sigma_0^b = \Pi_0^b$ of sharply bounded formulas consists of formulas in which all quantifiers have the form*

$$\exists x < |t| \quad or \quad \forall x < |t|$$

   That is, the quantifiers are bounded by the length of a term.

2. *For $0 \leq i$ the classes Σ_{i+1}^b and Π_{i+1}^b are the smallest classes satisfying*

   (a) $\Sigma_i^b \cup \Pi_i^b \subseteq \Sigma_{i+1}^b \cap \Pi_{i+1}^b$

   (b) *both Σ_{i+1}^b and Π_{i+1}^b are closed under sharply bounded quantification, disjunction $\vee$, and conjunction $\wedge$*

   (c) Σ_{i+1}^b *is closed under bounded existential quantification*

   (d) Π_{i+1}^b *is closed under bounded universal quantification*

   (e) *the negation of a Σ_{i+1}^b-formula is Π_{i+1}^b, and the negation of a Π_{i+1}^b-formula is Σ_{i+1}^b.*

3. *The class Σ_∞^b of bounded L-formulas is the union $\bigcup_i \Sigma_i^b = \bigcup_i \Pi_i^b$.*

4. *A Σ_i^b-formula is Δ_i^b (respectively Δ_i^b in a theory T) iff it is equivalent to a Π_i^b-formula in predicate logic (respectively in T).*

In words: The complexity of bounded formulas in language L is defined by counting the number of alternations of bounded quantifiers, ignoring the sharply bounded ones, analogously to the definition of levels of the *arithmetical hierarchy* where one counts the number of alternations of quantifiers, ignoring the bounded ones.

Theorem 3.2.12.   *The subsets of N defined by Σ_∞^b-formulas are exactly the sets from the polynomial-time hierarchy PH.*

   In fact, for $i \geq 1$ the Σ_i^b-formulas exactly define the Σ_i^p-predicates.

Proof (sketch).   The only difference from Lemma 3.2.4 and Theorem 3.2.5 is that now we need to code computations of length $n^{O(1)}$, $n = |x|$. If $|y| \leq n^{O(1)}$ then $y \leq x\#\dots\#x$, which is a term of L; hence such y's can appear in bounded quantifiers.   Q.E.D.

We should note that Bennett (1962) also considered a class of the *extended rudimentary* sets, which are defined similarly to the rudimentary sets except that the language is augmented by a function of the growth rate of the function #. It is then a straightforward extension of Theorem 3.2.5 that the extended rudimentary sets are exactly those from the polynomial time hierarchy PH.

3.3. Bibliographical and other remarks

For the history of results and ideas from Section 3.1 the reader should consult Boppana and Sipser (1990), Wegener (1987), and Sipser (1992). Important topics omitted are NP-*completeness*, for which Garey and Johnson (1979) is a good source, and the completeness results for other classes, in particular, the completeness of *directed st-connectivity* for class NL and the completeness of *undirected st-connectivity* for class L/poly (Aleliunas et al. 1979). Karchmer and Wigderson (1988) and Raz and Wigderson (1990) study the depth of monotone circuits for connectivity and matching.

   Other interesting facts, but not used later in the book, concern branching programs: Barrington (1989) characterized Boolean functions with polynomial size formulas as those computed by width 5, polynomial-size branching programs, and a relation of space bounded Turing computations to size of branching programs: L/poly = BP (cf. Wegener 1987).

   Very important but unfortunately unpublished is Bennett's Ph.D. thesis (Bennett 1962), containing either explicitly or implicitly most later definability results such as Cobham (1965) and Nepomnjascij (1970). Paris and Wilkie (1981b) study rudimentary sets explicitly.

4

Basic propositional logic

This chapter will present *basic propositional calculus*. By that I mean properties of propositional calculus established by direct combinatorial arguments as distinguished from *high level* arguments involving concepts (or motivations) from other parts of logic (bounded arithmetic) and complexity theory.

Examples of the former are various simulation results or the lower bound for *resolution* from Haken (1985). Examples of the latter are the simulation of the *Frege system with substitution* by the *extended Frege system* (Lemma 4.5.5 and Corollary 9.3.19), or the construction of the provably hardest tautologies from the finitistic consistency statements (Section 14.2).

We shall define basic propositional proof systems: *resolution R, extended resolution ER, Frege system F, extended Frege system EF, Frege system with the substitution rule SF, quantified propositional calculus G*, and Gentzen's *sequent calculus LK*. We begin with the general concept of a *propositional proof system*.

4.1. Propositional proof systems

A property of the usual textbook calculus is that it can be checked in deterministic polynomial time whether a string of symbols is a proof in the system or not. This is generalized into the following basic definition of Cook and Reckhow (1979).

Definition 4.1.1. *Let TAUT be the set of propositional tautologies in the language with propositional connectives: constants 0 (FALSE) and 1 (TRUE), ¬ (negation), ∨ (disjunction), and ∧ (conjunction), and atoms $p_1, p_2, \ldots$.*

A propositional proof system is a polynomial time function P whose range is the set TAUT.

For a tautology $\tau \in TAUT$, any string w such that $P(w) = \tau$ is called a P-proof of τ.

The size *of a string w is the total number of occurrences of symbols in it; it is denoted* $|w|$.

Note that standard calculi, like those mentioned earlier, are all covered by this definition as one can define a function P

$$P(w) := \begin{cases} \tau & w \text{ is a proof of } \tau \text{ in the calculus} \\ 1 & w \text{ is not a proof in the calculus} \end{cases}$$

This function is polynomial-time, the P-proofs of τ are precisely the original proofs (except when $\tau = 1$), and the range of P is the whole set TAUT as all these calculi are complete (resolution and extended resolution after a suitable encoding of all formulas by formulas in the disjunctive normal form).

The following problem is one of the most fundamental open problems of logic.

Fundamental problem. *Is there a propositional proof system P in which every tautology has a polynomial size proof? That is, are there a proof system P and a polynomial* $p(x)$ *such that any tautology* τ *has a P-proof of size at most* $p(|\tau|)$?

Theorem 4.1.2 (Cook and Reckhow 1979). *There exists a propositional proof system in which every tautology has a polynomial size proof if and only if the class NP is closed under complementation: NP =* coNP.

Proof. The "only if" part follows as $\exists w(|w| \leq p(|\tau|); P(w) = \tau$ defines a non-deterministic acceptor for the coNP-complete set TAUT whenever all tautologies have size $\leq p(|\tau|)$ P-proofs.

For the "if part" let M be a nondeterministic polynomial time acceptor for TAUT and define

$$P(w) := \begin{cases} \tau & w \text{ is an accepting computation of } M \text{ on } \tau \\ 1 & \text{otherwise} \end{cases}$$

                                                                    Q.E.D.

Not surprisingly, then, there are no lower bounds known for the size of proofs in a general propositional proof system. The current research activity concentrates rather on proving lower bounds for particular natural propositional calculi or on comparing the efficiency of various systems. For the latter task the following definition is basic.

Definition 4.1.3. *Let P and Q be two propositional proof systems.*

(a) *Let* $g : N \to N$ *be a function. We say that a system P has a* speed-up g *over a system Q iff:*

$$P(w) = \tau \to \exists w_1, |w_1| \leq g(|w|) \wedge Q(w_1) = \tau$$

holds for all w and τ.

We say that P has a polynomial speed-up over Q iff it has an n^k speed-up
(k a constant), and we say that it has a superpolynomial speed-up (resp.
an exponential) iff it has no n^k speed-up for all k (resp. if it has no 2^{n^ϵ}
speed-up, some $\epsilon > 0$).

We denote by $P \leq Q$ the fact that P has a polynomial speed-up over Q.

(b) *A polynomial simulation of P by Q (or briefly a p-simulation) is a poly-*
nomial time function $f(w, \tau)$ such that for all w and τ

$$P(w) = \tau \rightarrow Q(f(w, \tau)) = \tau$$

We write $P \leq_p Q$ if there is a polynomial simulation of P by Q.

An immediate corollary of the definition is that both $\leq$ and $\leq_p$ are quasi-orderings of propositional proof systems, and that $\leq_p$ is a finer quasi-ordering than $\leq$. Note that if tautologies have polynomial size P-proofs then they have polynomial size proofs in every system $\leq$-greater than P. A system P admitting polynomial size proofs for all tautologies would be, in particular, a maximal element in the quasi-ordering $\leq$. It is unknown, however, whether there are any maximal elements in this quasi-order (cf. Krajíček and Pudlák 1989a).

4.2. Resolution

The *resolution system* was introduced by Blake (1937) and developed in Davis and Putnam (1960) and Robinson (1965). The system operates with atoms p_1, $p_2, \ldots$ and their negations $\neg p_1, \neg p_2, \ldots$, but has no other logical connectives. The basic object is a *clause*, a finite (possibly empty) set of *literals*, that is, atoms or negated atoms. We think of a clause as of a disjunction of the literals in it. A truth assignment

$$\alpha : \{p_1, p_2, \ldots\} \rightarrow \{0, 1\}$$

satisfies a clause C if and only if it satisfies at least one literal in C. This will be denoted by $\alpha \models C$. It follows that no assignment satisfies the empty clause.

The *resolution rule* allows us to derive new clause $C_1 \cup C_2$ from two clauses $C_1 \cup \{p\}$ and $C_2 \cup \{\neg p\}$

$$\frac{C_1 \cup \{p\} \quad C_2 \cup \{\neg p\}}{C_1 \cup C_2}$$

There are no restrictions on occurrences of literals p, $\neg p$ in C_1 or C_2. An obvious property of the resolution rule is that if a truth assignment satisfies both upper clauses of the rule then it also satisfies the lower clause. This is *the soundness* of the resolution rule.

A *resolution refutation* of a set $C = \{C_1, \ldots, C_k\}$ of clauses is a *sequence* of clauses $D_1, \ldots, D_t$ such that each D_i is either an element of C or derived by the resolution rule from some earlier D_u, D_v, $u, v < i$, and such that the last clause D_t is the empty clause.

Although the system does not allow one to work with propositional formulas directly, there is an indirect way called *limited extension* similar to reducing the general satisfiability problem to the satisfiability of sets of clauses. Let θ be a formula in atoms $p_1, \ldots, p_n$. Introduce for each subformula ϕ of θ, including θ itself, a new atom q_ϕ and a set $\mathrm{Ext}(\theta)$ of all clauses of the form

  1. $\{q_\phi, \neg p_i\}$, $\{\neg q_\phi, p_i\}$, if ϕ is atom p_i
  2. $\{q_\phi, q_\psi\}$, $\{\neg q_\phi, \neg q_\psi\}$ if $\phi = \neg\psi$
  3. $\{\neg q_\phi, q_{\psi_1}, q_{\psi_2}\}$, $\{q_\phi, \neg q_{\psi_1}\}$, $\{q_\phi, \neg q_{\psi_2}\}$ if $\phi = \psi_1 \vee \psi_2$
  4. $\{\neg q_\phi, q_{\psi_1}\}$, $\{\neg q_\phi, q_{\psi_2}\}$, $\{q_\phi, \neg q_{\psi_1}, \neg q_{\psi_2}\}$ if $\phi = \psi_1 \wedge \psi_2$

Clearly the set $\mathrm{Ext}(\theta)$ is satisfiable if and only if the formula θ is satisfiable.

Theorem 4.2.1. *A set of clauses is unsatisfiable: that is, there is no truth assignment satisfying simultaneously all clauses in the set, if and only if there is a resolution refutation of the set.*

Proof. Any truth assignment satisfying all clauses in set C would have to satisfy, by the soundness of the resolution rule, all clauses in a resolution refutation of C. In particular, the empty clause would also be satisfied, and that is impossible. This proves the "if part" of the theorem.

Assume now that C is unsatisfiable and let $p_1, \ldots, p_n$ be all atoms such that only the literals $p_1, \neg p_1, \ldots, p_n, \neg p_n$ appear in C. We shall prove by induction on n that for any such C there is a resolution refutation of C.

Decompose C into four disjoint sets $C_{00} \cup C_{01} \cup C_{10} \cup C_{11}$, of those clauses which contain no p_n and no $\neg p_n$, no p_n, but do contain $\neg p_n$, do contain p_n but not $\neg p_n$ and contain both $p_n, \neg p_n$, respectively.

Let $C_{01} \times C_{10}$ be the set of clauses obtained by the resolution rule applied to all pairs of clauses $C_1 \cup \{\neg p_n\}$ from C_{01} and to $C_2 \cup \{p_n\}$ from C_{10}; these new clauses do not contain either p_n or $\neg p_n$. Form a new set of clauses

$$C' = C_{00} \cup (C_{01} \times C_{10})$$

The set C' is unsatisfiable too. This is because any assignment $\alpha' : \{p_1, \ldots, p_n\} \to \{0, 1\}$ satisfies either all clauses C_1 such that $C_1 \cup \{\neg p_n\} \in C_{01}$, or all clauses C_2 such that $C_2 \cup \{p_n\} \in C_{10}$.

Apply the same procedure to atoms $p_{n-1}, p_{n-2}, \ldots$ to produce sets C'', C''', $\ldots$ which are all unsatisfiable and derivable by the resolution rule from the original set C. The nth such set will contain just the empty clause: that is, we get a resolution refutation of set C.   Q.E.D.

As a corollary of the completeness proof we get a simple upper bound to the number of clauses appearing in a resolution refutation of an unsatisfiable set of k clauses.

Corollary 4.2.2. *Let C be an unsatisfiable set of k clauses formed from literals built from n atoms. Then for every $0 \leq t < n$ there is a resolution refutation of C with at most*

$$4^{n-t} + \sum_{s=0}^{t} \frac{k^{2^s}}{4^s}$$

clauses.

Proof. By the completeness of R any unsatisfiable set of clauses formed from ℓ atoms or their negations has a resolution refutation with $\leq 4^{\ell}$ clauses as 4^{ℓ} is the number of all such clauses.

In the proof of Theorem 4.2.1 we start with k clauses in n atoms; after the first step we create an unsatisfiable set of at most $(k^2/4)$ new clauses in $n - 1$ atoms, and after t steps we have $k + (k^2/4) + \ldots (k^{2^t}/4^t)$ clauses with the last set being an unsatisfiable set in $n - t$ atoms, that is, with a refutation with at most 4^{n-t} clauses.
Q.E.D.

We shall see later that this upper bound cannot be much improved.

The following *search problem* is associated with an unsatisfiable set $C = \{C_1, \ldots, C_k\}$: given a truth assignment α find $C_i \in C$ false under α. This search problem can be solved by a *branching program* (cf. Definition 3.1.12), with leaves labeled by the clauses from C. A branching program solves the search problem if and only if for every truth assignment α, the leaf of the path $P(\alpha)$ determined by α is labeled by a clause false under α.

Theorem 4.2.3. *Let t be the minimal number of clauses in a* regular *resolution refutation of C, where "regular" means that on every path through the refutation every atom is resolved at most once.*

Let s be the minimal number of nodes in a read-once *branching program solving the search problem associated with C, where "read-once" means that on every path through the branching program every atom occurs at most once as a label of a node.*

Then

$$t = s$$

Proof. Let π be a regular resolution refutation of C. Construct a branching program as follows: The last clause in π (the empty one) becomes the root, the initial clauses from C become the leaves, and all other clauses in π become out-degree 2 nodes with the two edges directed from a clause to the two clauses that form the hypotheses of the inference giving the clause. If the resolved atom in that inference is p_i then the edge to the clause containing p_i is labeled by 0, and vice versa.

Now it is straightforward to verify that all clauses on a path determined by a truth assignment α are false under α: That is, the branching program solves the search problem and it is read-once as π was regular.

For the opposite direction let σ be a read-once branching program of the minimal size. With every node v in σ associate a clause C_v having the property that every assignment determining a path going through v falsifies C_v.

If v is a leaf then C_v is the clause from $\mathcal{C}$ labeling v in σ. Assume that the node v is labeled by atom p_i and the edge (v, v_1) is labeled by 1, and (v, v_0) by 0.

We claim that C_{v_1} does not contain p_i and C_{v_0} does not contain $\neg p_i$. This is because σ is read-once and so no path reaching v (and at least one path does reach v as s is minimal possible size) determines the value of p_i. Hence we could prolong such a path by giving value 1 to p_i if $p_i \in C_{v_1}$ or value 0 if $\neg p_i \in C_{v_0}$. This new path would satisfy C_{v_0} or C_{v_1} respectively, contradicting the previous assumption.

It follows that either one of the clauses C_{v_1}, C_{v_0} contains none of $p_i, \neg p_i$, or that C_{v_0} contains p_i but not $\neg p_i$ and C_{v_1} contains $\neg p_i$ but not p_i. In the former case define C_v to be the clause containing none of $p_i, \neg p_i$, and in the latter case define C_v to be the resolution of clauses C_{v_1} and C_{v_0} with respect to (w.r.t.) atom p_i.

It is easy to verify (using an argument similar to the earlier one) that no path through v satisfies C_v.

The root of σ has to be assigned the empty clause as all paths go through it. Hence the constructed object is a regular resolution refutation.   Q.E.D.

Note that the first part of the proof showed that $s \leq t$ even without the restriction to regular and read-once.

In the rest of the section we prove a strong lower bound to the number of clauses in resolution refutations of particular sets of clauses having clear combinatorial meaning: *the weak pigeonhole principle.*

Let $m > n$ be two natural numbers. Consider set $\neg \mathrm{PHP}_n^m$ of clauses with atoms $p_{ij}, i = 0, \ldots, m - 1$ and $j = 0, \ldots, n - 1$

$$\{\neg p_{ik}, \neg p_{jk}\}, \quad \text{all } i \neq j \text{ and } k$$

$$\{p_{i0}, \ldots, p_{i(n-1)}\}, \quad \text{all } i$$

Interpreting conditions $p_{ij} = 1$ as defining a relation $\subseteq \{0, \ldots, m - 1\} \times \{0, \ldots, n - 1\}$ the clauses say that the relation is a graph of an injective map from $\{0, \ldots, m - 1\}$ to $\{0, \ldots, n - 1\}$. Hence the set is unsatisfiable and has (by Theorem 4.2.1) a resolution refutation. The following theorem was first proved by Haken (1985) for $m = n + 1$.

Theorem 4.2.4 (Haken 1985, Buss and Turán 1988). *In any resolution refutation of set $\neg PHP_n^m$ at least*

$$2^{\Omega(\frac{n^2}{m})}$$

different clauses must occur.

Proof. A 1–1 truth assignment to atoms p_{ij} is a truth assignment α for which the relation

$$\{(i, j) \in m \times n \mid \alpha(p_{ij}) = 1\}$$

is a partial 1–1 function from m to n. We shall denote by $\mathrm{dom}(\alpha)$ and $\mathrm{rng}(\alpha)$ its domain and range.

A maximal 1–1 assignment is a 1–1 assignment α that assigns n ones.

Let π be a resolution refutation of $\neg\, \mathrm{PHP}_n^m$ and assume n is divisible by 4.

Claim 1. *For every maximal 1–1 truth assignment α there is a clause C_α in π satisfying*

1. *α does not satisfy C_α*
2. *for every $i \notin \mathrm{dom}(\alpha)$ there are at most $(n/2)$ j's such that $p_{ij} \in C_\alpha$*
3. *there is exactly one $i \notin \mathrm{dom}(\alpha)$ such that there are exactly $(n/2)$ j's such that $p_{ij} \in C_\alpha$*

To see the claim consider the unique path P_α in π consisting of clauses $C_0, C_1, \dots, C_t$ such that

(i) C_0 is the end clause of π, that is, the empty clause
(ii) C_{i+1} is one of the two clauses that are the hypothesis of the resolution inference giving C_i
(iii) α does not satisfy any of C_i
(iv) $C_t \in \neg\, \mathrm{PHP}_n^m$

As α is maximal 1–1, clause C_t must have the form

$$C_t = \{p_{i0}, \dots, p_{i(n-1)}\}$$

for some $i \notin \mathrm{dom}(\alpha)$. Let C be the first such clause in $C_0, \dots, C_t$ such that for some $i \notin \mathrm{dom}(\alpha)$, there are at least $(n/2)$ j's such that $p_{ij} \in C$. This C satisfies all conditions 1–3.

For α a maximal 1–1 assignment let C_α be the first clause in π satisfying 1–3.

For β a 1–1 assignment, not necessarily maximal, let C^β be the first clause C_α for some maximal α extending β.

Claim 2. *Assume that β is a 1–1 assignment which assigns $(n/4)$ ones. Then there are at least $(n/4) + 1$ i's for which* either *there is j such that $\neg p_{ij} \in C^\beta$ or there are at least $(n/2)$ j's such that $p_{ij} \in C^\beta$.*

To prove the claim let $C^\beta = C_\alpha$ for a maximal 1–1 assignment α. Define:
$$\mathrm{COL}^- := \{i \mid \exists j, \neg p_{ij} \in C^\beta\}$$
$$\mathrm{COL}^+ := \{i \mid i \in \mathrm{dom}(\alpha) \wedge (\forall j, \neg p_{ij} \notin C^\beta) \wedge \exists^{\geq(n/2)} j, p_{ij} \in C^\beta\}$$
$$i_0 := \text{the } i \notin \mathrm{dom}(\alpha) \text{ such that for exactly } (n/2) \text{ } j\text{'s}, p_{ij} \in C^\beta$$
$$A := \{p_{ij} \mid \alpha(p_{ij}) = 1 \wedge \beta(p_{ij}) = 0\}$$

The sets $\mathrm{COL}^-, \mathrm{COL}^+$ and $\{i_0\}$ are mutually disjoint (as $\mathrm{COL}^- \cap (n \setminus \mathrm{dom}(\alpha)) = \emptyset$ holds because C^β is not satisfied by α). Point i_0 satisfies the condition of the claim, and as, by the definition, for all $i \notin \mathrm{dom}(\alpha)$, $i \neq i_0$ there are less than $(n/2)$ j's for which $p_{ij} \in C^\beta$, it remains to show

$$|\mathrm{COL}^-| + |\mathrm{COL}^+| \geq \frac{n}{4}$$

Assume otherwise; then there must be $p_{ij} \in A$ for which $\{p_{i_0 j}, \neg p_{i_0 j}\} \cap C^\beta = \emptyset$ and $i \notin \mathrm{COL}^- \cup \mathrm{COL}^+$. This is true because the former condition excludes only $(n/2)$ elements of A and the latter excludes less then $(n/4)$ elements of A: That is, both conditions exclude less than $(3n/4) = |A|$ elements of A. Now take such an atom p_{ij} and define a new maximal 1–1 assignment α' from α by changing the value of p_{ij} from 1 to 0 and of $p_{i_0 j}$ from 0 to 1. Clearly $\beta \subset \alpha'$ and α' does not satisfy C^β either, but for all $u \notin \mathrm{dom}(\alpha')$ there are less than $(n/2)$ v's for which $p_{uv} \in C^\beta$ (as we replaced i_0 by i). It follows that the clause $C_{\alpha'}$ precedes C^β in the refutation, contradicting the definition of C^β.

We are ready to prove the bound. Let $h(n, m)$ be the number of 1–1 assignments α with $|\mathrm{dom}(\alpha)| = (n/4)$ and let $g(n, m)$ be the maximal number of such 1–1 assignments β sharing the clause C^β. Clearly then the ratio

$$\frac{h(n, m)}{g(n, m)}$$

is a lower bound to the number of different clauses in the refutation of $\neg \mathrm{PHP}_n^m$.

To simplify the expression put $k := (n/4)$. Then clearly

$$h(n, m) = \binom{m}{k}\binom{n}{k}k!$$

but $h(n, m)$ can also be expressed differently. For fixed clause $C = C^\beta$ let X, $|X| = k + 1$, be a set of some $k + 1$ i's satisfying the property of Claim 2. If t denotes the number of atoms p_{ij}, $\beta(p_{ij}) = 1$, such that $i \in X$ then we also have

$$h(n, m) = \sum_{t=0}^{k} \binom{k + 1}{t}\binom{m - k - 1}{k - t} \frac{n!}{(n - k)!}$$

For each $i \in X$ there are at most $(n/2)$ j's for which it is possible that $\beta(p_{ij}) = 1$, for a 1–1 assignment β such that $C = C^\beta$, $|\mathrm{dom}(\beta)| = k$. Thus

$$g(n, m) \leq \sum_{t=0}^{k} \binom{k + 1}{t}\binom{m - k - 1}{k - t} \left(\frac{n}{2}\right)^t \frac{(n - t)!}{(n - k)!}$$

Therefore we obtain a lower bound to the ratio

$$\frac{h(n, m)}{g(n, m)} \geq \frac{\sum_{t=0}^{k} \binom{k+1}{t}\binom{m-k-1}{k-t}}{\sum_{t=0}^{k} \binom{k+1}{t}\binom{m-k-1}{k-t}(n/2)^t((n - t)!/n!)} \geq \frac{\sum_{t=0}^{k} \binom{k+1}{t}\binom{m-k-1}{k-t}}{\sum_{t=0}^{k} \binom{k+1}{t}\binom{m-k-1}{k-t}(2/3)^t}$$

because

$$\left(\frac{n}{2}\right)^t \frac{(n-t)!}{n!} \leq \left(\frac{2}{3}\right)^t$$

holds for $t \leq k = (n/4)$. The last ratio is $2^{\Omega(n^2/m)}$.   Q.E.D.

Note that when $m \leq n^{1-\epsilon}$ the bound is exponential and for $m = o(n^2/\log(n))$ it is still superpolynomial.

Open problem.   *Are there resolution refutations of $\neg PHP_n^m$, $m = n^2$, with polynomially many clauses?*

4.3. Sequent calculus

Gentzen's sequent calculus is the most elegant proof system, for both propositional and predicate logic, allowing sharp proof-theoretic analysis. In this section we shall treat the basics of the propositional part of Gentzen's system LK, which we shall also denote by LK as there is no danger of confusion. The full predicate system LK and its modifications for bounded arithmetic are considered in the next chapter. For the details of Gentzen's system we refer the reader to Takeuti (1975).

We shall confine ourselves to the same language as in Section 4.1: 0, 1, $\neg$, $\vee$, $\wedge$, and atoms. The formulas are built by using the connectives from atoms and constants. First we define the notion of the depth of a formula (a proof).

Definition 4.3.1.
  1. *The* logical depth *of formula ϕ, denoted $\ell dp(\phi)$, is the maximum nesting of connectives in ϕ. Precisely:*
     (a) $\ell dp(0) = \ell dp(1) = \ell dp(p) = 0$, *any atom p*
     (b) $\ell dp(\neg\psi) = 1 + \ell dp(\psi)$, *any ψ*
     (c) $\ell dp(\eta \circ \psi) = 1 + \max(\ell dp(\eta), \ell dp(\psi))$, *for any η, ψ and $\circ = \vee, \wedge$*
  2. *The* depth *of formula ϕ, denoted $dp(\phi)$, is the maximum number of alternations of connectives in ϕ. Precisely:*
     (a) $dp(\psi) = 0$ *iff $\ell dp(\psi) = 0$*
     (b) $dp(\neg\psi) = dp(\psi)$ *if the outmost connective of ψ is $\neg$, and*
        $dp(\neg\psi) = 1 + dp(\psi)$ *otherwise*
        (i) $dp(\eta \vee \psi) = \max(dp(\eta), dp(\psi))$ *if the outmost connective in both η, ψ is $\vee$,*
        (ii) $dp(\eta \vee \psi) = 1 + \max(dp(\eta), dp(\psi))$ *if none of η, ψ has $\vee$ as the outmost connective,*
        (iii) $dp(\eta \vee \psi) = \max(1 + dp(\eta), dp(\psi))$ *if $\vee$ is the outmost connective of ψ but not of η,*

(iv) $dp(\eta \vee \psi) = \max(dp(\eta), 1 + dp(\psi))$ *if* $\vee$ *is the outmost connective of* η *but not of* ψ.

3. *The* depth *of proof* π *(resp. the* logical depth*) is the maximal depth of a formula in* π *(resp. the maximal logical depth). It is denoted* $dp(\pi)$ *or* $\ell dp(\pi)$, *respectively.*

The basic object of the sequent calculus is a *sequent*, an ordered pair of two finite sequences (possibly empty) of formulas written as

$$\phi_1, \ldots, \phi_u \longrightarrow \psi_1, \ldots, \psi_v.$$

The symbol $\longrightarrow$ separates the *antecedent* $\phi_1, \ldots, \phi_u$ from the *succedent* $\psi_1, \ldots,$ ψ_v. We shall use letters Γ, Δ, Π, ... to denote finite sequences of formulas, also called *cedents*.

The truth definition is extended from formulas to sequents by the following: An assignment α satisfies a sequent $\Gamma \longrightarrow \Delta$ if and only if α satisfies a formula from the succedent Δ or the negation of a formula from the antecedent Γ. In particular, the empty sequent $\emptyset \longrightarrow \emptyset$, also written simply $\longrightarrow$, is unsatisfiable.

Definition 4.3.2. *An LK-proof is a* sequence *of sequents in which every sequent either is an* initial sequent, *that is, a sequent having one of the forms*

$$A \longrightarrow A, \quad 0 \longrightarrow, \quad \longrightarrow 1$$

with A an atom, or is derived from previous sequents by one of the following rules:

1. weakening rules

$$\textbf{left } \frac{\Gamma \longrightarrow \Delta}{A, \Gamma \longrightarrow \Delta} \quad and \quad \textbf{right } \frac{\Gamma \longrightarrow \Delta}{\Gamma \longrightarrow \Delta, A}$$

2. exchange rules

$$\textbf{left } \frac{\Gamma_1, A, B, \Gamma_2 \longrightarrow \Delta}{\Gamma_1, B, A, \Gamma_2 \longrightarrow \Delta} \quad and \quad \textbf{right } \frac{\Gamma \longrightarrow \Delta_1, A, B, \Delta_2}{\Gamma \longrightarrow \Delta_1, B, A, \Delta_2}$$

3. contraction rules

$$\textbf{left } \frac{\Gamma_1, A, A, \Gamma_2 \longrightarrow \Delta}{\Gamma_1, A, \Gamma_2 \longrightarrow \Delta} \quad and \quad \textbf{right } \frac{\Gamma \longrightarrow \Delta_1, A, A, \Delta_2}{\Gamma \longrightarrow \Delta_1, A, \Delta_2}$$

4. $\neg$: introduction rules

$$\textbf{left } \frac{\Gamma \longrightarrow \Delta, A}{\neg A, \Gamma \longrightarrow \Delta} \quad and \quad \textbf{right } \frac{A, \Gamma \longrightarrow \Delta}{\Gamma \longrightarrow \Delta, \neg A}$$

5. $\wedge$: introduction rules

$$\textbf{left } \frac{A, \Gamma \longrightarrow \Delta}{A \wedge B, \Gamma \longrightarrow \Delta} \quad and \quad \frac{A, \Gamma \longrightarrow \Delta}{B \wedge A, \Gamma \longrightarrow \Delta}$$

$$and \quad \textbf{right } \frac{\Gamma \longrightarrow \Delta, A \quad \Gamma \longrightarrow \Delta, B}{\Gamma \longrightarrow \Delta, A \wedge B}$$

6. $\lor$: introduction rules

$$\textbf{left} \quad \frac{A, \Gamma \longrightarrow \Delta \quad B, \Gamma \longrightarrow \Delta}{A \lor B, \Gamma \longrightarrow \Delta} \quad and$$

$$\textbf{right} \quad \frac{\Gamma \longrightarrow \Delta, A}{\Gamma \longrightarrow \Delta, A \lor B} \quad and \quad \frac{\Gamma \longrightarrow \Delta, A}{\Gamma \longrightarrow \Delta, B \lor A}$$

7. cut-rule

$$\frac{\Gamma \longrightarrow \Delta, A \quad A, \Gamma \longrightarrow \Delta}{\Gamma \longrightarrow \Delta}$$

The new formula introduced in a rule is called the principal formula *of the rule; formulas from which it is inferred are called the* minor formulas *of the rule. All other formulas are called* side formulas.

For a formula in Δ or Γ in the lower sequent of a rule, the same occurrence in the upper sequent(s) is called the immediate ancestor *of the formula. The immediate ancestor(s) of a principal formula of a rule are the minor formulas of the rule.*

An ancestor *of a formula is any formula obtained by repeating the immediate ancestor step.*

Theorem 4.3.3. *The system LK is complete. That is, whenever a sequent $\Gamma \longrightarrow \Delta$ is satisfied by every truth assignment then there is an LK-proof of $\Gamma \longrightarrow \Delta$.*

Proof. We shall define a tree with nodes labeled by sequents by the following process:

1. The root is labeled by $\Gamma \longrightarrow \Delta$.
2. If a node is labeled by

$$\Pi \longrightarrow \Sigma, \neg\phi$$

where $\neg\phi$ is a formula with the maximal logical depth in the sequent, then the node has only one successor labeled by

$$\phi, \Pi \longrightarrow \Sigma$$

3. If a node is labeled by

$$\Pi \longrightarrow \Sigma, \phi \land \psi$$

where $\phi \land \psi$ has the maximal logical depth, then the node has exactly two successors labeled by

$$\Pi \longrightarrow \Sigma, \phi \quad and \quad \Pi \longrightarrow \Sigma, \psi$$

respectively.

4. If a node is labeled by

$$\Pi \longrightarrow \Sigma, \phi \vee \psi$$

where $\phi \vee \psi$ has the maximal logical depth, then it has one successor labeled by

$$\Pi \longrightarrow \Sigma, \phi \vee \psi, \phi \vee \psi$$

which has one successor labeled by

$$\Pi \longrightarrow \Sigma, \phi, \phi \vee \psi$$

and this node also has one successor labeled by

$$\Pi \longrightarrow \Sigma, \phi, \psi$$

5. The cases where a formula of the maximal logical depth is only in the antecedent are defined dually, exchanging the roles of $\vee$ and $\wedge$.

Claim. *All sequents appearing as labels are valid.*

The claim follows from the assumption that the sequent $\Gamma \longrightarrow \Delta$ labeling the root is valid.

It follows that the leaves of the tree must be labeled by a sequent consisting of atoms in which an atom appears in both the antecedent and the succedent, or containing 1 (resp. 0) in the succedent (resp. in the antecedent).

Such sequents can be obtained from the initial sequents by the weakening rules. The rest of the tree then defines an LK-derivation of $\Gamma \longrightarrow \Delta$.   Q.E.D.

Definition 4.3.4.
   (a) An LK-proof is called treelike *if every sequent in the proof is used as a hypothesis of an inference at most once.*
   (b) An LK-proof is called cut-free *iff the cut rule is not used in the proof.*
   (c) The height *of an LK-proof is the maximal number h such that there is a sequence $S_0, \ldots, S_h$ of sequents of the proof in which S_i is a hypothesis of the inference yielding S_{i+1}, all $i < h$.*

Note that if the LK-proof is treelike its height is just the usual height of the proof tree.

Corollary 4.3.5. *Assume that $\Gamma \longrightarrow \Delta$ is a valid sequent of size n. Then there is a cut-free, treelike LK-proof of $\Gamma \longrightarrow \Delta$ of size at most $O(n \cdot 2^n)$.*

Proof. Let $f(n)$ be the least upper bound to the size of treelike LK-derivations of valid sequents of size n. Then the proof of the previous theorem shows that f

satisfies inequality

$$f(n) = O(n) + 2f(n - 1)$$

which yields

$$f(n) = O(n2^n)$$

Q.E.D.

Corollary 4.3.6. *Every valid sequent*

$$\phi_1, \ldots, \phi_u \longrightarrow \psi_1, \ldots, \psi_v$$

has a (cut-free) LK-proof in which every formula is a subformula of one of the formulas $\phi_1, \ldots, \phi_u, \psi_1, \ldots, \psi_v$.

Proof. By Theorem 4.3.3 every valid sequent has a cut-free proof. Observe that in every rule except the cut-rule all formulas appearing in the upper sequents of the rule also appear as subformulas in the lower one. Q.E.D.

Important applications of cut-free proofs are the Craig interpolation theorem and the Beth definability theorem.

Theorem 4.3.7 (Craig 1957). *Let the sequent*

$$\Gamma(\overline{p}, \overline{q}) \longrightarrow \Delta(\overline{p}, \overline{r})$$

have a cut-free, treelike LK-proof of height h.

Then there is a formula $I(\overline{p})$, an interpolant *of the sequent, satisfying the following conditions:*

1. *$I(\overline{p})$ contains no atoms $\overline{q}$ or $\overline{r}$*
2. *both sequents*

$$\Gamma \longrightarrow I \quad and \quad I \longrightarrow \Delta$$

   are valid
3. *$\ell dp(I) \leq h + 1$*

Moreover, both sequents from condition 2 have a cut-free, treelike proof of height $O(h)$.

Proof. Let $\Pi \longrightarrow \Sigma$ be a sequent in the proof. Assume that $\Pi = \Pi_1 \cup \Pi_2$ and $\Sigma = \Sigma_1 \cup \Sigma_2$ where Π_1, Σ_1 are the ancestors of Γ and Π_2, Σ_2 are the ancestors of Δ. We shall show by induction on h', the height of the subproof yielding $\Pi \longrightarrow \Sigma$, that there is a formula I' with no atoms $\overline{q}, \overline{r}$ such that both sequents

$$\Pi_1 \longrightarrow \Sigma_1, I \quad and \quad I, \Pi_2 \longrightarrow \Sigma_2$$

are valid, and such that $\ell dp(I') \leq h' + 1$.

For $h' = 0$ the sequent $\Pi \longrightarrow \Sigma$ is an initial sequent. If it is $0 \longrightarrow$ put

$$I' := \begin{cases} 0 & \text{if } 0 \in \Pi_1 \\ 1 & \text{if } 0 \in \Pi_2 \end{cases}$$

and dually in the case of the sequent $\longrightarrow 1$

$$I' := \begin{cases} 1 & \text{if } 1 \in \Sigma_2 \\ 0 & \text{if } 1 \in \Sigma_1 \end{cases}$$

If $\Pi \longrightarrow \Sigma$ is the initial sequent $p_i \longrightarrow p_i$ define

$$I' := \begin{cases} p_i & \text{if } \Pi_2 = \Sigma_1 = \emptyset \\ 0 & \text{if } \Pi_2 = \Sigma_2 = \emptyset \\ 1 & \text{if } \Pi_1 = \Sigma_1 = \emptyset \\ \neg p_i & \text{if } \Pi_1 = \Sigma_2 = \emptyset \end{cases}$$

If $\Pi \longrightarrow \Sigma$ is the initial sequent $q_i \longrightarrow q_i$ then both occurrences of q_i must be ancestors of Γ and we put $I' := 0$. Dually, if $\Pi \longrightarrow \Sigma$ is the initial sequent $r_i \longrightarrow r_i$ then we put $I' := 1$.

The only case of an initial sequent when the logical depth of I' is not 0 is when $I' = \neg p_i$, that is, always $\ell dp(I') \le 1 = h' + 1$.

For $h' > 0$ we distinguish several cases according to which inference rule was used to derive $\Pi \longrightarrow \Sigma$ and according to which of the sets $\Pi_1, \Pi_2, \Sigma_1, \Sigma_2$ contain the principal formula of the inference.

Let us consider just the case of the $\wedge : right$ inference of $\Pi \longrightarrow \Sigma$ from $\Pi \longrightarrow \Sigma', \alpha$ and $\Pi \longrightarrow \Sigma', \beta$ where $\alpha \wedge \beta$ is the principal formula and I_1 and I_2 are two interpolants associated with the upper sequents. Define:

$$I' := \begin{cases} I_1 \wedge I_2 & \text{if } \alpha \wedge \beta \in \Sigma_2 \\ I_1 \vee I_2 & \text{if } \alpha \wedge \beta \in \Sigma_1 \end{cases}$$

Hence $\ell dp(I) \le 1 + \max(\ell dp(I_1), \ell dp(I_2)) \le 1 + h'$.

The cases of the other rules are treated analogously. We also leave it to the reader to verify that all sequents

$$\Pi_1 \longrightarrow \Sigma_1, I \quad \text{and} \quad I, \Pi_2 \longrightarrow \Sigma_2$$

have cut-free, treelike proofs of the height $O(h')$.   Q.E.D.

Theorem 4.3.8 (Beth 1959). *Let the sequent*

$$\Gamma(\overline{p}, \overline{q}), \Gamma(\overline{p}, \overline{r}), q_1 \longrightarrow r_1$$

with $\overline{q} = (q_1, \ldots, q_m), \overline{r} = (r_1, \ldots, r_m)$ have a cut-free, treelike LK-proof of height h.

Then there is formula $E(\overline{p})$, an explicit definition *of q_1, satisfying the following conditions:*

1. $E(\overline{p})$ *contains no $\overline{q}$ or $\overline{r}$*
2. *both sequents*

$$\Gamma(\overline{p},\overline{q}), q_1 \longrightarrow E(\overline{p}) \quad and \quad \Gamma(\overline{p},\overline{q}), E(\overline{p}) \longrightarrow q_1$$

   are valid
3. $\ell dp(E) \leq h + 1.$

Moreover, both sequents in 2 have treelike, cut-free LK-proofs of height $O(h)$.

Proof. Let E be the interpolant constructed in the previous proof for the sequent $\Pi_1, \Pi_2 \longrightarrow \Sigma_1, \Sigma_2$ with $\Pi_1 = \Gamma(\overline{p},\overline{q}), q_1$, $\Pi_2 = \Gamma(\overline{p},\overline{r})$, $\Sigma_1 = \emptyset$ and $\Sigma_2 = r_1$. Then

$$\Gamma(\overline{p},\overline{q}), q_1 \longrightarrow E$$

and

$$\Gamma(\overline{p},\overline{r}), E \longrightarrow r_1$$

from which immediately follows by renaming $\overline{r}$ into $\overline{q}$

$$\Gamma(\overline{p},\overline{q}), E \longrightarrow q_1$$

Q.E.D.

It is an open problem whether the assumption that the proof is cut-free can be dropped from the hypothesis of the last two theorems; see Krajíček (1995b) and Krajíček and Pudlák (1995).

The last but one theorem of this section is a lower bound to the worst case increase of the size of proofs after cut-elimination, showing that the upper bounds from Corollary 4.3.5 are good even if we would strengthen the hypothesis of 4.3.5 to the assumption that the sequent has a short LK-derivation.

Theorem 4.3.9. *There is a sequent S of size $|S| = m$ such that every cut-free, treelike LK-proof of S has at least $2^{m^{\Omega(1)}}$ sequents.*

Moreover, the sequent S has an LK-proof of size $m^{O(1)}$.

Proof. Let S be the sequent whose succedent is empty and whose antecedent consists of $n + 1 + \binom{n+1}{2} \cdot n$ disjuncts $\bigvee C$ (arbitrarily bracketed), one for each clause C of $\neg \mathrm{PHP}_n^{n+1}$ from Section 4.2.

Let π be a cut-free, treelike LK-proof of S. By Lemma 4.3.6 every sequent Z in π has the form

$$D_1, \ldots, D_u \longrightarrow q_1, \ldots, q_v$$

where D_r are disjunctions of literals p_{ij} or $\neg p_{ij}$ and q_s are atoms among p_{ij}. By induction on the number k of inferences in the subproof of π yielding the sequent we show that there is a resolution derivation of the clause

$$\{q_1, \ldots, q_v\}$$

from the clauses $D_1, \ldots, D_u$ (we identify a disjunction of literals with the clause consisting of those literals) with at most k resolution inferences.

If Z is an initial sequent in π or derived by a structural rule, then there is nothing to prove. The only other inferences in π can be a $\bigvee : \textit{left}$ inference or a $\neg : \textit{left}$ inference. In the former case,

$$\frac{\Gamma, E \longrightarrow \Delta \qquad \Gamma, F \longrightarrow \Delta}{\Gamma, E \vee F \longrightarrow \Delta}$$

where $\Gamma, E \vee F \longrightarrow \Delta$ is the sequent Z, assume that σ_0 and σ_1 are two resolution derivations of Δ from Γ, E and from Γ, F, respectively. Construct a new derivation σ of Δ from $\Gamma, E \vee F$ by first replacing E in σ_0 by $E \vee F$, getting a derivation of Δ, F, and then joining this with σ_1, where F is replaced by Δ, F.

In the latter case

$$\frac{\Gamma \longrightarrow \Delta, q}{\Gamma, \neg q \longrightarrow \Delta}$$

we have by the induction assumption a resolution derivation of Δ, q from Γ. Prolong this derivation by a resolution inference applied to Δ, q, and $\{\neg q\}$.

Clearly the number of resolution inferences is bounded by the number inferences in π; we use the assumption that π is treelike to maintain this bound during the simulation of a $\bigvee : \textit{left}$ inference.

By Theorem 4.2.4 every resolution refutation of $\neg\, \mathrm{PHP}_n^{n+1}$ must contain at least $2^{\Omega(n)}$ clauses. Hence π must contain $2^{\Omega(n)}$ inferences, which is $2^{m^{\Omega(1)}}$ for $m = |S| = O(n^3)$.

By Corollary 13.1.8 and Lemma 4.4.15 the sequent S has a size $m^{O(1)}$ LK-proof.
$$\text{Q.E.D.}$$

We shall see in Section 12.5 that Theorem 12.5.3 has a much stronger lower bound for the sequent S.

The next proposition will show that every treelike proof can be balanced with only a polynomial increase in size. The lemma itself looks rather technical so let us first motivate it. Assume that we have a proof π of sequent S from sequents $S_1, \ldots, S_k$ (additional axioms: initial sequents). Let α be a truth assignment not satisfying S. It determines a path $P(\alpha)$ through π: a sequence $Z_0, \ldots, Z_t$ of sequents such that $Z_0 = S$, each Z_i is false under α, Z_{i+1} is a hypothesis of the inference yielding Z_i, and Z_t is initial. Obviously $Z_t \in \{S_1, \ldots, S_k\}$ and hence π acts as a branching program solving the search problem to find false S_i (cf. Theorem 4.2.3). Z_{i+1} is uniquely determined by Z_i alone in all rules except in $\bigwedge : \textit{right}, \bigvee : \textit{left}$ and in the cut-rule. It is thus of interest to estimate the number of these inferences on any path through π as such an estimate gives information about the complexity of the search problem.

We introduce two new derived rules:

1. *special* $\bigwedge$ *: right*

$$\frac{\Gamma \longrightarrow \Delta, A}{\Gamma \longrightarrow \Delta, \gamma_1 \wedge \ldots \wedge \gamma_a \wedge \neg\delta_1 \wedge \ldots \wedge \neg\delta_b \wedge A}$$

2. *special* $\bigvee$ *: left*

$$\frac{A, \Gamma \longrightarrow \Delta}{\neg\gamma_1 \vee \ldots \vee \neg\gamma_a \vee \delta_1 \vee \ldots \vee \delta_b \vee A, \Gamma \longrightarrow \Delta}$$

where $\Gamma = (\gamma_1, \ldots, \gamma_a)$ and $\Delta = (\delta_1, \ldots, \delta_b)$. Their advantage, in connection with the preceding discussion, is that they have only one hypothesis. We call a proof *special* if it uses *special* $\bigwedge$ *: right* and *special* $\bigvee$ *: left* but does not use $\wedge$ *: right* or $\vee$ *: left*. Note that the only inference in special proofs with more than one hypothesis is the cut.

Lemma 4.3.10. *Let π be a treelike LK-proof of sequent S. Assume that $d = dp(\pi)$ is the maximal depth of a formula in π and that π has size m.*

Then there is a treelike special LK-proof π' of S from a set $\{T_1, \ldots, T_k\}$ of some tautological sequents such that

1. *$dp(\pi') \leq 1 + d$*
2. *the size of any cut formula in π' is $O(m^3)$*
3. *the number of cuts on any path through π' is at most $O(\log m)$*

Proof. Assume π satisfies the hypothesis of the lemma. Construct a proof σ_π of S by replacing in π any $\wedge$ *: right* with a principal formula $A \wedge B$ (resp. $\vee$ *: left* with a principal formula $A \vee B$) by two cuts with the sequent $A, B \longrightarrow A \wedge B$ (resp. with $A \vee B \longrightarrow A, B$); let these new sequents form the set $T = \{T_1, \ldots, T_k\}$ of tautological sequents used as axioms in σ_π. Clearly $|\sigma_\pi| = O(m)$.

Denote by $H(\sigma)$ the number of cuts in a proof σ and by $h(\sigma)$ the maximum number of cuts on a path through σ.

The following claim borrows from the idea of the proof of Theorem 3.1.15 from Spira (1971).

Claim 1. *Let σ be an LK-proof and assume*

$$(3/2)^{t-1} < H(\sigma) \leq (3/2)^t$$

Then there is a subproof σ_0 of σ ending with cut-inference

$$\frac{\Gamma \longrightarrow \Delta, A \quad A, \Gamma \longrightarrow \Delta}{\Gamma \longrightarrow \Delta}$$

and with two immediate subproofs σ_1 *and* σ_2 *(with end-sequents* $\Gamma \longrightarrow \Delta, A$ *and* $A, \Gamma \longrightarrow \Delta$, *resp.) such that for* $i = 1$ *or* $i = 2$ *it holds*

$$\frac{1}{2}\left(\frac{3}{2}\right)^{t-1} \leq H(\sigma_i) \leq \left(\frac{3}{2}\right)^{t-1}$$

and

$$H(\sigma) - H(\sigma_i) \leq (3/2)^{t-1}$$

The lemma follows from *Claim 2* for $\sigma = \sigma_\pi$ as $dp(\sigma_\pi) = dp(\pi)$ and $|\sigma_\pi| = O(|\pi|)$.

Claim 2. *Let* σ *be a treelike special LK-proof from axioms* T. *Let* d *be the maximum depth of a cut-formula in* σ *and* $n(\sigma)$ *the maximum size of a sequent in* σ. *Let* $t := \lceil \log_{3/2}(H(\sigma)) \rceil$.

Then there is a treelike special proof σ' *with the same end-sequent as* σ *such that*

1. *$dp(\sigma') \leq 1 + d$*
2. *the size of any cut-formula in* σ' *is* $\leq 2^t \cdot n(\sigma)$
3. *$h(\sigma') \leq 1 + t = O(\log(|\sigma|))$*

We prove *Claim 2* by induction on t: Assume σ satisfies these inequalities. By *Claim 1* there is a subproof σ_0 ending with a cut-inference

$$\frac{\Gamma \longrightarrow \Delta, A \quad A, \Gamma \longrightarrow \Delta}{\Gamma \longrightarrow \Delta}$$

with two subproofs σ_1, σ_2. Assume without loss of generality (w.l.o.g.) $i = 1$ in *Claim 1*. Denote by $\eta(\sigma)$ the maximum size of a cut-formula in σ.

By the induction hypotheses applied to σ_1 we get a special proof σ_1' of

$$\Gamma \longrightarrow \Delta, A$$

such that $\eta(\sigma_1') = 2^{t-1} \cdot n(\sigma_1')$, $dp(\sigma_1') \leq 1 + d$ and $h(\sigma_1') \leq 1 + t - 1 \leq t$. Prolonging σ_1' by several $\neg$: *right* and $\vee$: *right* and contractions we get a special proof σ_1'' of

$$\longrightarrow \neg\gamma_1 \vee \ldots \neg\gamma_a \vee \delta_1 \vee \ldots \vee \delta_b \vee A$$

where γ_i's and δ_j's are as in the definition of the special rules.

After the proof σ_2 insert one *special* $\bigvee$: *left* to derive from $A, \Gamma \longrightarrow \Delta$

$$\neg\gamma_1 \vee \ldots \neg\gamma_a \vee \delta_1 \vee \ldots \vee \delta_b \vee A, \Gamma \longrightarrow \Delta$$

and then continue as in σ carrying the extra side formula $\neg\gamma_1 \vee \ldots \neg\gamma_a \vee \delta_1 \vee \ldots \vee \delta_b \vee A$ in the antecedents. This yields a special proof σ_2', $n(\sigma_2') \leq 2n(\sigma_2)$, and $H(\sigma_2') = H(\sigma) - H(\sigma_1)$.

Apply the induction hypothesis to σ_2' to get a special proof σ_2'' of

$$\neg\gamma_1 \vee \ldots \neg\gamma_a \vee \delta_1 \vee \ldots \vee \delta_b \vee A, \Pi \longrightarrow \Sigma$$

where $\Pi \longrightarrow \Sigma$ is the end-sequent of σ, $dp(\sigma_2'') \leq 1 + d$, $h(\sigma_2'') \leq t$, and $\eta(\sigma_2'') \leq 2^{t-1} \cdot n(\sigma_2') \leq 2^t \cdot n(\sigma_2)$.

Joining the proofs σ_1'' and σ_2'' by one cut-inference gets the required special proof σ', estimating the size of the last cut-formula by $2n(\sigma_2) < 2^t \cdot m$.   Q.E.D.

There are other forms of this lemma, depending on a particular formulation of the proof system and on a particular purpose (see 12.2.4).

In later chapters we shall frequently talk about *constant-depth* LK-proofs: the proofs in which the depth of formulas is bounded by an independent constant. For such systems it is more elegant to allow the connectives $\bigwedge$ and $\bigvee$ to have an unbounded arity and change Rules 5 and 6 of Definition 4.3.2 to allow introduction of these new conjunctions and disjunctions from several hypotheses in one step. We will conclude this section by the definition of these constant-depth systems and by a simple statement.

Definition 4.3.11. *Unbounded arity connectives $\bigwedge$ and $\bigvee$ are connectives such that $\bigwedge(A_1, \ldots, A_n)$ and $\bigvee(A_1, \ldots, A_n)$ (written briefly as $\bigwedge_{i \leq n} A_i$ or $\bigvee_{i \leq n} A_i$) are formulas whenever A_i's are formulas.*

Rules for these connectives are extensions of Rules 5 and 6 of Definition 4.3.2:

5.' $\bigwedge$: introduction rules

$$\textbf{left} \quad \frac{A, \Gamma \longrightarrow \Delta}{\bigwedge_{i \leq n} A_i, \Gamma \longrightarrow \Delta}$$

$$\textbf{right} \quad \frac{\Gamma \longrightarrow \Delta, A_1 \quad \ldots \quad \Gamma \longrightarrow \Delta, A_n}{\Gamma \longrightarrow \Delta, \bigwedge_{i \leq n} A_i}$$

6.' $\bigvee$: introduction rules

$$\textbf{left} \quad \frac{A_1, \Gamma \longrightarrow \Delta \quad \ldots \quad A_n, \Gamma \longrightarrow \Delta}{\bigvee_{i \leq n} A_i, \Gamma \longrightarrow \Delta}$$

$$\textbf{right} \quad \frac{\Gamma \longrightarrow \Delta, A}{\Gamma \longrightarrow \Delta, \bigvee_{i \leq n} A_i}$$

where A is among $A_1, \ldots, A_n$ in $\bigwedge$: left and $\bigvee$: right.

The depth d *LK-system is the system LK with Rules 5' and 6' and allowing in proofs only formulas of depth at most d.*

The next lemma follows from the proof of Theorem 4.3.3.

Lemma 4.3.12. *Let $\Gamma \longrightarrow \Delta$ be a valid sequent consisting of formulas of depth at most d.*

Then $\Gamma \longrightarrow \Delta$ is provable in the depth d LK-system.

4.4. Frege systems

A *Frege system* is a propositional calculus based on finitely many axiom schemes and inference rules that is implicationally complete and sound. We shall restrict ourselves to the language $0, 1, \neg, \vee,$ and $\wedge$ as before, but we consider the general case at the end of the section. Most of the propositions in this section hold identically for all Frege systems.

Definition 4.4.1. *A Frege rule is a $k + 1$-tuple of formulas $A_0, \ldots, A_k$ in atoms $p_1, \ldots, p_n$ written as*

$$\frac{A_1, \ldots, A_k}{A_0}$$

such that any assignment $\alpha : \{p_1, \ldots, p_n\} \rightarrow \{0, 1\}$ satisfying all formulas $A_1, \ldots, A_k$ also satisfies formula A_0 (the soundness of the rule).

A Frege rule in which $k = 0$ is called a Frege axiom scheme.

Formula ϕ_0 is inferred by the rule from formulas $\phi_1, \ldots, \phi_k$ if there exists a substitution $p_i := \psi_i$ for which $\phi_j = A_j(p_i/\psi_i)$.

A well-known example of a Frege rule is modus ponens

$$\frac{A \quad A \rightarrow B}{B}$$

where $A \rightarrow B$ is an abbreviation for $\neg A \vee B$.

Definition 4.4.2. *Let F be a finite collection of Frege rules.*

1. *A* Frege proof *(or, briefly, an F-proof) of formula η_0 from formulas $\{\eta_1, \ldots, \eta_u\}$ is a finite sequence $\theta_1, \ldots, \theta_k$ of formulas such that every θ_i is either an element of $\{\eta_1, \ldots, \eta_u\}$ or inferred from some earlier θ_j's ($j < i$) by a rule from F, and such that the last formula θ_k is η_0.*

2. *F is implicationally complete if and only if for any set $\{\eta_1, \ldots, \eta_u\}$ such that every assignment satisfying all formulas in this set also satisfies η_0, the formula η_0 is F-provable from $\{\eta_1, \ldots, \eta_u\}$.*

3. *F is a* Frege proof system *if and only if it is implicationally complete.*

A Frege proof system that is often used consists of modus ponens and finitely many axiom schemes.

We shall also define several complexity measures, besides *the size*, which make sense for the *sequent calculus, Frege systems*, and systems treated in a future section but not for a general propositional proof system. We shall introduce these measures now as we want to understand some later results also in the perspective of these complexity measures.

Definition 4.4.3.

1. *The* number of steps *in a proof $\theta_1, \ldots, \theta_k$ is k; it is the number of inferences. The minimal number of steps in a proof of tautology τ is denoted $k_P(\tau)$ (P denotes the propositional proof system considered).*

2. *The* number of distinct formulas in a proof π *is the minimal number ℓ such that there are formulas $\phi_1, \ldots, \phi_\ell$, such that any formula occurring as a subformula in π is identical to one of $\phi_1, \ldots, \phi_\ell$. The minimal number of formulas in a proof of tautology τ is denoted $\ell_P(\tau)$.*

The next several lemmas clarify to some extent the relations among the size, the depth, and these measures. Recall that $|w|$ denotes the *size* of string w; in particular $|\pi|$ denotes the size of proof π. The following lemma is a particular corollary of a more general result valid for predicate calculus.

Lemma 4.4.4. *For every Frege system F there is a constant $c > 0$ such that every tautology τ has an F-proof π such that*

$$\ell dp(\pi) \leq c \cdot k_F(\tau) + \ell dp(\tau)$$

In fact, for any F-proof $\phi_1, \ldots, \phi_k$ there is another F-proof $\theta_1, \ldots, \theta_k$ such that

1. *$\ell dp(\theta_i) \leq c \cdot k$, all $i \leq k$*
2. *there is a substitution δ such that:*

$$\delta(\theta_i) = \phi_i$$

for all $i \leq k$.

Proof. Think of formulas as of labeled binary trees: The inner nodes are labeled by connectives and the leaves by atoms or constants. If formula β is a substitution instance of formula α then the tree representing α can be uniquely embedded into the tree representing β, mapping the root of α onto the root of β and the sons of a node onto the sons of the image of the node. Moreover, the labels of the inner nodes are preserved by such a map.

We say that the subformula γ of β is in *the image of α in β* if the root of γ is in the image of α in the embedding.

Let

$$\frac{A_1, \ldots, A_k}{A_0}$$

be a Frege rule and

$$\frac{\psi_1, \ldots, \psi_k}{\psi_0}$$

its instance. The Z-part of the instance of the rule consists of all subformulas of ψ_i that are in the image of A_i, $i = 0, \ldots, k$.

The Z-part of a proof $\phi_1, \ldots, \phi_k$ is the union of the Z-*parts* of all inferences used in the proof. Note that the Z-part of a proof with k steps has cardinality at most $c \cdot k$, where constant c is the maximal number of subformulas occurring in a Frege rule of the system.

Call those subformulas in the proof *special* whose identical copy is in the Z-part. Those occurrences of special formulas α form the *border*, which has two properties:

(i) if α is a subformula of β then β is special.

(ii) at least one of immediate subformulas of α is not special.

Define from $\phi_1, \ldots, \phi_k$ a new sequence of formulas $\theta_1, \ldots, \theta_k$ by replacing every border occurrence of a formula α by *new* atom p_α.

Claim. *Sequence $\theta_1, \ldots, \theta_k$ is a Frege proof and $\ell dp(\theta_i) \leq c \cdot k$.*

The sequence is a Frege proof as each θ_i can be derived from $\theta_{i_1}, \ldots, \theta_{i_j}$ by the same rule as ϕ_i was derived from $\phi_{i_1}, \ldots, \phi_{i_j}$. The logical depth of θ_i is at most $c \cdot k$ as any proper chain of subformulas formed from special formulas may contain at most the cardinality of Z members.

Finally, note that the substitution

$$\delta \; : \; p_\alpha \mapsto \alpha$$

gives the wanted substitution: $\delta(\theta_i) = \phi_i$.   Q.E.D.

Lemma 4.4.5. *Let τ be a tautology and assume that π is a Frege proof of τ of the minimal possible size. Then*

$$\ell dp(\pi) \leq O(\sqrt{|\pi| \cdot \ell dp(\tau)})$$

Proof. Let $\pi = \phi_1, \ldots, \phi_k$ be a proof of τ of the *minimal* size. Take a step ϕ_j of π and write

$$\ell dp(\phi_j) = t \cdot \sqrt{|\pi|}, \quad \text{some } t > 0$$

Let the set S of subformulas occurring in π consist of those in the Z-part of π and of those occurring in the last formula. Note that any proper chain of subformulas (each one included in the previous one) from S can have cardinality at most $\max(c, \ell dp(\tau))$, with c the constant associated to the system in Lemma 4.4.4.

Let $\theta_1, \ldots, \theta_k$ be the sequence constructed in Lemma 4.4.4 and let δ_0 be a *minimal* substitution such that $\delta_0(\theta_k) = \phi_k$, but not necessarily: $\delta_0(\theta_i) = \phi_i$ for $i \neq k$. But by the hypothesis π is of the minimal size, so, in fact, $\delta_0(\theta_i) = \phi_i$ must hold for all i, as otherwise $\delta_0(\theta_1), \ldots, \delta_0(\theta_k)$ would be a proof of τ of size smaller than $|\pi|$.

From the construction in the previous proof it follows that for every subformula α of ϕ_j there exists an identical subformula in the set S. Put

$$d := 1 + \max(c, \ell dp(\tau))$$

Hence every proper chain of subformulas from S has cardinality less than d.

Assume $\alpha_s, \ldots, \alpha_0$ is a maximal sequence of subformulas of ϕ_i such that
(i) α_i is a subformula of α_{i+1}
(ii) the depth of the root of α_i in the tree ϕ_i is $(s - i) \cdot d$
Thus we have

$$s = \frac{t\sqrt{|\pi|}}{d} - \frac{u}{d}, \quad \text{some } 0 \le u < d$$

Let $\alpha'_s, \ldots, \alpha'_0$ be subformulas from S such that α'_i is identical to α_i. Let P_i denote the path in the tree ϕ_i starting with α_i on which $\alpha_i, \ldots, \alpha_0$ lies and let P'_i be the corresponding path originating from α'_i. The cardinality of P'_i is

$$|P'_i| = d \cdot i + u + 1$$

Now for $i \ne j$: $\alpha'_i \notin P'_i$ by the property that any proper chain of subformulas from S has cardinality less than d. This implies that all paths P'_i are disjoint, so

$$|\pi| \ge \sum_{j=0}^{s} |P'_i| \ge \sum_{j=0}^{s} (d \cdot j + u + 1)$$

Substituting into this inequality the value of s (and using $0 \le u < d$) we obtain

$$|\pi| \ge \frac{t^2 \cdot |\pi|}{2d}, \quad \text{i.e., } t \le \sqrt{2d}$$

Hence:

$$\ell dp(\phi_v) \le \sqrt{2|\pi|d} = O(\sqrt{|\pi| \cdot \ell dp(\tau)})$$

Q.E.D.

This bound is, in fact, also optimal (see Krajíček 1989a).

The following lemma shows that the measures $k_F(\tau)$ and $\ell_F(\tau)$ are essentially the same.

Lemma 4.4.6. *For every Frege system F there is a constant $c > 0$ such that for every tautology τ*

$$k_F(\tau) \le \ell_F(\tau) \le c \cdot k_F(\tau) + |\tau|$$

Proof. Let $\phi_1, \ldots, \phi_k$ be a proof of τ, and let the constant c and the formulas $\theta_1, \ldots, \theta_k$ be those from Lemma 4.4.4. Let δ_0 be a minimal substitution such that $\delta_0(\theta_k) = \phi_k = \tau$. Then sequence $\delta_0(\theta_1), \ldots, \delta_0(\theta_k)$ is a proof of τ and it contains at most $c \cdot k + |\tau|$ distinct formulas, as any subformula in it must be identical to one from the Z-part of the original proof or to a subformula of τ.

This proves the second inequality. The first one is trivial. Q.E.D.

The following three lemmas are simple structural properties of Frege proofs.

Definition 4.4.7. *A Frege proof $\theta_1, \ldots, \theta_k$ is treelike if and only if every step θ_i is a hypothesis of at most one inference in the proof.*

Lemma 4.4.8. *Assume that τ has a Frege proof with k steps, of depth d and with size m.*

Then τ has a treelike F-proof π such that:
1. $dp(\pi) \le d + c$
2. $|\pi| \le c \cdot \log(k) \cdot m$
3. π has at most $c \cdot k \cdot \log(k)$ steps

for some constant $c > 0$ depending only on F.

Proof. Let $\theta_1, \ldots, \theta_k$ be an F-proof of τ satisfying the assumptions. Define $\phi_i := \theta_1 \wedge \ldots \wedge \theta_i$, brackets balancing the conjunction into a binary tree of depth at most $O(\log(i))$. We show that ϕ_{i+1} has a treelike proof from ϕ_i with a $O(\log(i))$ number of steps, size $O(i \cdot |\phi_i|)$, and depth $d + O(1)$. This follows from the next claim.

Claim. *For $j \le i$, any θ_j can be proved from ϕ_i by a treelike proof with $O(\log(i))$ steps, size $O(\log(i) \cdot |\phi_i|)$, and depth $dp(\phi_i) + O(1)$.*

Then the whole proof has $\sum_i O(\log(i)) = O(k \cdot \log(k))$ steps and size $\sum_i O(\log(i) \cdot |\phi_i|) \le O(\log(k) \cdot m)$.   Q.E.D.

Lemma 4.4.9. *Let $\tau(p_1, \ldots, p_n)$ be a tautology. Then there are constants k and c such that for every $\phi_1, \ldots, \phi_n$ the tautology*

$$\tau(\phi_1, \ldots, \phi_n)$$

has a treelike proof with k steps, depth at most $c + \max_i(dp(\phi_i))$, and size at most $c \cdot (\sum_i |\phi_i|)$.

Proof. Let π be any fixed treelike proof of τ. Then a proof obtained from π by substituting ϕ_i for p_i satisfies all the requirements.   Q.E.D.

The following lemma provides a bound for the deduction lemma.

Lemma 4.4.10. *Let ϕ be F-provable from ψ by a proof with k steps, size m, and depth d. Then the implication $\psi \to \phi$ has an F-proof with $O(k)$ steps, size $O(k \cdot m)$, and depth $d + O(1)$.*

Proof. Let $\theta_1, \ldots, \theta_k$ be an F-proof of ϕ from $\psi (= \theta_1)$. Construct successively the proofs of implications $\psi \to \theta_i$.

If θ_i was inferred from $\theta_{j_1}, \ldots, \theta_{j_r}$ in the original proof then the implication

$$(\psi \to \theta_{j_1}) \to ((\psi \to \theta_{j_2}) \to (\ldots((\psi \to \theta_{j_r}) \to (\psi \to \theta_i))\ldots)$$

is a substitution instance of tautology

$$((q \to A_{j_1}) \to (\ldots \to (q \to A_i))\ldots)$$

where

$$\frac{A_{j_1}, \dots, A_{j_r}}{A_i}$$

is an F-rule. Hence by Lemma 4.4.9 the implication has a treelike proof with $O(1)$ steps, size $O(|\psi| + |\theta_i| + \sum_{s \le r} |\theta_{j_s}|) = O(m)$, and depth $d + O(1)$.

From this nested implication and from the implications

$$\psi \to \theta_{j_s}, \quad s \le r$$

proved earlier the desired implication

$$\psi \to \theta_i$$

follows in $O(1)$ steps, size $O(m)$, and depth $d + O(1)$.

Hence the total number of steps in the proof of $\psi \to \phi$ will be $O(k)$, the size will be $O(k \cdot m)$, and the depth $d + O(1)$.   Q.E.D.

The knowledge of optimal bounds to the complexity of proofs of a tautology measured by any of the measures is poor, except for the depth, which follows from 4.3.6.

Lemma 4.4.11. *Let τ be a tautology with logical depth d. Then τ has an F-proof of logical depth at most $d + O(1)$.*

For the number of steps and the size only the following trivial consequence of Lemma 4.4.4 is known.

Lemma 4.4.12. *Let τ be a tautology which is not a substitution instance of any shorter tautology. Let m be the sum of the sizes of all subformulas of τ (including τ).*

Then any F-proof of τ must have at least $\Omega(\ell dp(\tau))$ steps and size at least $\Omega(m)$.

In particular, any F-proof of $\tau_n := \neg \dots \neg 1$, negation $\neg$ $2n$-times, must have $\Omega(n)$ steps and size $\Omega(n^2)$.

Proof. Let $\pi = \phi_1, \dots, \phi_k$ be an F-proof of τ and let constant c and proof $\theta_1, \dots, \theta_k$ be those constructed in Lemma 4.4.4. As τ is not a substitution instance of any shorter tautology we must have: $\theta_k = \phi_k = \tau$ and $\theta_1, \dots, \theta_k$ is a proof of τ. Hence also $\ell dp(\tau) \cdot c^{-1} \le k$.

It follows from the construction of $\theta_1, \dots, \theta_k$ that any subformula has an identical occurrence in the Z-part of $\phi_1, \dots, \phi_k$ and, in particular, every subformula of τ is *special* in every F-proof of τ. A subformula from a Z-part has at most c subformulas also in the Z-part; this condition entails that the size of any F-proof of τ must be at least $\Omega(m)$.   Q.E.D.

I have postponed till the end of this section the most important result of Reckhow (1976). To appreciate it we now allow the language of Frege systems to be arbitrary *complete* finite set of propositional connectives, where complete means that any Boolean function can be expressed by a formula in this language.

Recall Definition 4.1.3. In this definition we considered only the case when P and Q had the same language. For the following theorem we extend the definition to P, Q, which do not necessarily have the same language, by replacing the implication in part (b) of Definition 4.1.3 by the implication

$$P(w) = g(\tau) \rightarrow Q(f(w, \tau)) = \tau$$

where g is a polynomial-time function translating formulas in the language of Q into the language of P. It is assumed that if τ is a tautology then $g(\tau)$ is also a tautology.

Theorem 4.4.13 (Reckhow 1976). *Any Frege system polynomially simulates any other Frege system. Moreover, if the language of F_1 contains the language of F_2 then g can be the identity function and F_1 has at most a polynomial speed-up over F_2, and the simulation f can be chosen so that both the number of steps and the size increase at most proportionally and the depth increases by a constant.*

Proof (sketch).   Instead of giving the full proof we shall outline the easy part and point out the difficulty with the general case. The way how to overcome the difficulty is then described separately in Lemma 4.4.14.

Let F be a *Frege* system in the language $\{0, 1, \neg, \vee, \wedge\}$, with finitely many axiom schemes and the modus ponens as the only rule of inference.

Let F$'$ be any other *Frege* system. We shall show that F and F$'$ mutually polynomially simulate each other.

This is rather easy to see when the language of F$'$ is also $\{0, 1, \neg, \vee, \wedge\}$. Any rule of F$'$ has the form

$$\frac{A_1(\overline{p}), \ldots, A_k(\overline{p})}{A_0(\overline{p})}$$

and is sound, so the formula

$$A_1(\overline{p}) \rightarrow (A_2(\overline{p}) \rightarrow (\ldots \rightarrow (A_k(\overline{p}) \rightarrow A_0(\overline{p}))\ldots)$$

is a tautology and hence any of its instances of size m has size $O(m)$ F-proof (cf. Lemma 4.4.9). Thus k applications of modus ponens simulate the inference in F$'$ using the rule, and this part of the simulation has size $O(m)$.

On the other hand, all instances of axiom schemes of F have short F$'$-proofs, by Lemma 4.4.9 again. System F$'$ is implicationally complete so there is a fixed size F$'$-proof of q from $\{p, p \rightarrow q\}$. Thus any instance of modus ponens can be simulated by a linear size F$'$-proof too.

To see the difficulty that may arise when F$'$ has a different language consider the case when $\equiv$ (the *equivalence* connective) is in the language of F$'$. In the language of F the equivalence $p \equiv q$ is defined as $(p \wedge q) \vee (\neg p \wedge \neg q)$, and, in fact, there is no equivalent definition in which a variable would occur only once.

Consider the formula

$$p_1 \equiv (p_2 \equiv (p_3 \equiv \ldots (p_{n-1} \equiv p_n) \ldots)$$

which has size $O(m)$. If we translate it into the language of F by using the preceding definition we obtain a formula of size $\Omega(2^m)$; generally if the nesting of $\equiv$'s is k, then the translation will have size $\Omega(2^k)$.

This shows that we cannot just translate the connectives in one language into the other one and literally translate proofs. However, it also suggests that a statement analogous to the Spira theorem 3.1.15 might help: First transform the size m F'-proof into a size $m^{O(1)}$ F'-proof in which all formulas have the logical depth $O(\log m)$, and then apply the straightforward translation to construct size $m^{O(1)}$ F-proof.

We extract this idea as a particular lemma explicitly in 4.4.14, and leave the rest of the details to the reader. Q.E.D.

In the next lemma we again use the language $0, 1, \vee, \wedge, \neg$.

Lemma 4.4.14. *Assume that τ has an F-proof of size m. Then τ has a proof of size $m^{O(1)}$ and of logical depth at most $O(\log(m)) + \ell dp(\tau)$.*

Proof. We first state a claim allowing inductive proof of Spira's theorem. The claim is verified by induction on the logical complexity of α.

Claim 1. *Let $\alpha(\overline{p})$ be a formula of size n. Then there are a formula $\beta(\overline{p}, q)$ with one occurrence of atom q and a formula $\gamma(\overline{p})$ such that*

1. *$\alpha = \beta(q/\gamma)$*
2. *$(1/3)n < |\beta|, |\gamma| \leq (2/3)n$*

We assume that among several such decompositions $\beta(\gamma)$ we pick one in some canonical way, and we shall call it the *canonical decomposition*. We also assume that $\neg$ is pushed by *de Morgan* rules into literals and that $\neg \psi$ is an abbreviation for the formula arising from ψ by interchanging $\vee$ and $\wedge$, 0 and 1, and atoms and their negations.

Then we define the *balanced form of α*, denoted α^*, by induction on the size of α:

1. $\alpha^* := \alpha$, if $|\alpha| \leq f$
2. $\alpha^* := (\beta^*(q/1) \wedge \gamma^*) \vee (\beta^*(q/0) \wedge \neg \gamma^*)$, where $|\alpha| > f$ and $\beta(\gamma)$ is the canonical decomposition of α.

The constant f is used for convenience and we shall specify it later.

The following claim is readily verified by induction on the logical depth of formula β^*.

Claim 2. *The equivalence*

$$((\beta^*(q/1) \wedge \gamma^*) \vee (\beta^*(q/0) \wedge \neg \gamma^*)) \equiv \beta^*(\gamma^*)$$

is valid and has an F-proof of size $O(|\beta^*(\gamma^*)|^2)$ *and of logical depth* $\ell dp(\beta^*(\gamma^*)) +$ $O(1)$.

The next claim is crucial.

Claim 3. *Let* $\alpha = \beta(q_1/\gamma_1, \ldots, q_k/\gamma_k)$ *where each* q_i *has exactly one occurrence in* β. *Then the equivalence*

$$\alpha^* \equiv \beta^*(q_1/\gamma_1^*, \ldots, q_k/\gamma_k^*)$$

has an F-proof of size $O(|\alpha|^{O(1)})$ *and logical depth* $\ell dp(\alpha^*) + O(1)$.

In proving the claim we shall proceed by induction on $r := \lceil \log_{3/2}(|\alpha|) \rceil$ and we shall distinguish two cases:

1. The canonical decomposition of $\alpha = \rho(t/\sigma)$ has the form

$$\rho = \phi(q_1/\gamma_1, \ldots, q_\ell/\gamma_\ell, t)$$

and

$$\sigma = \psi(q_{\ell+1}/\gamma_{\ell+1}, \ldots, q_k/\gamma_k)$$

where

$$\beta = \phi(q_1, \ldots, q_\ell, t/\psi(q_{\ell+1}, \ldots, q_k))$$

2. The canonical decomposition of $\alpha = \rho(t/\sigma)$ has the form

$$\rho = \beta(q_1/\gamma_1, \ldots, q_{k-1}/\gamma_{k-1}, q_k/\phi(t))$$

where

$$\gamma_k = \phi(t/\sigma)$$

Let us consider the first case, leaving the analogous treatment of the second one to the reader. For notational simplicity we shall consider a completely general case with $k = 2, \ell = 1$, denoting γ_1 by γ and γ_2 by δ.

By the definition α^* is

$$((\phi(\gamma, t/1))^* \wedge (\psi(\delta))^*) \vee ((\phi(\gamma, t/0))^* \wedge \neg(\psi(\delta))^*)$$

As $\lceil \log_{3/2}(|\phi(\gamma, t)|) \rceil, \lceil \log_{3/2}(|\psi(\delta)|) \rceil \leq r - 1$, by the induction hypothesis this is (shortly provably) equivalent to

$$(\phi^*(\gamma^*, 1) \wedge \psi^*(\delta^*)) \vee (\phi^*(\gamma^*, 0) \wedge \neg \psi^*(\delta^*))$$

which is by Claim 2 (shortly provably) equivalent to

$$\phi^*(\gamma^*, \psi^*(\delta^*))$$

Now we would like to show the equivalence

$$\beta^* = \phi^*(q_1, \psi^*(q_2))$$

which would yield the wanted equivalence of α^* with

$$\beta^*(\gamma^*, \delta^*)$$

We cannot apply the induction hypothesis yet as it might hold that $\lceil \log_{3/2}(|\phi(q_1, \psi)|) \rceil = r$, but we note that $\lceil \log_{3/2}(|\phi|) \rceil$, $\lceil \log_{3/2}(|\psi|) \rceil \leq r - 1$. Instead we shall simply take the canonical decomposition $\theta(\omega)$ of $\phi(q_1, \psi)$ and use this to prove the equivalence.

There are three cases to distinguish (the substitution is always for only one occurrence of an atom)

(a) for some θ_0

$$\theta = \phi(\theta_0) \quad \text{and} \quad \psi = \theta_0(\omega)$$

(b) for some ω_0

$$\phi = \theta(\omega_0) \quad \text{and} \quad \omega = \omega_0(\psi)$$

(c) neither (a) nor (b) holds: That is, the occurrence of ω in $\phi(\psi)$ is disjoint with ψ.

We shall consider the first case only; the other two are identical.

By the definition $(\phi(\psi))^*$ is the formula

$$\left((\phi(\theta_0(1)))^* \wedge \omega^* \right) \vee \left((\phi(\theta_0(0)))^* \wedge \neg\omega^* \right)$$

which is, by the induction hypothesis applied to $\phi(\theta_0)$, (shortly provably) equivalent to

$$\left((\phi^*(\theta_0^*(1))) \wedge \omega^* \right) \vee \left((\phi^*(\theta_0^*(0))) \wedge \neg\omega^* \right)$$

and by Claim 2 to

$$\phi^* \left(\theta_0^*(\omega^*) \right)$$

As $\lceil \log_{3/2}(|\theta_0(\omega)|) \rceil = \lceil \log_{3/2}(|\psi|) \rceil \leq r - 1$, by the induction hypothesis this is (shortly provably) equivalent to

$$\phi^* \left((\theta_0(\omega))^* \right)$$

which is just

$$\phi^*(\psi^*)$$

Let $f(m)$ denote the smallest size of an F-proof of $(\nu(\kappa))^* \equiv \nu^*(\kappa^*)$ for $m = |\nu(\kappa)|$. As the induction hypothesis is in the preceding argument applied to formulas of length $\leq (2/3)m$ we have (using the bound from Claim 2 too)

$$f(m) = O\left(f\left(\frac{2}{3}m\right) + m^2\right)$$

which yields $f(m) = m^{O(1)}$.

This proves Claim 3.

Choose constant f to be a number bigger than the size of any formula A_i in a Frege rule of system F (cf. Definition 4.4.1). Let $\pi = \theta_1, \ldots, \theta_t$ be an F-proof of size n of formula $\tau (= \theta_t)$. We construct F-proofs of θ_i^*, $i = 1, \ldots, t$ by induction on i.

Take $i = i_0$ and assume θ_{i_0} was inferred from $\theta_{i_1}, \ldots, \theta_{i_k}$, $i_1, \ldots, i_k < i_0$ by the rule

$$\frac{A_1, \ldots, A_k}{A_0}$$

such that $\theta_{i_j} = A_j(\psi_1, \ldots, \psi_\ell)$. We already have F-proofs of $\theta_{i_j}^*$ $(j = 1, \ldots, k)$ and hence by Claim 3 also of

$$A_j^*(\psi_1^*, \ldots, \psi_\ell^*)$$

As $A_j^* = A_j$ by the choice of f, applying the same inference rule yields

$$A_0^*(\psi_1^*, \ldots, \psi_\ell^*)$$

from which we deduce (by Claim 3 again) formula $\theta_{i_0}^*$.

By Claim 3 the prolongation of the constructed proof needed to get $\theta_{i_0}^*$ is $\sum_{j \leq k} |\theta_{i_j}|^{O(1)} = m^{O(1)}$, and hence we eventually get size $m^{O(1)}$ F-proof of formula τ^*, which is equivalent (Claim 3 again) – with a short proof – to τ.

The logical depth of the proof of τ^* is $\leq \max_i \ell dp(\theta_i^*) + O(1) = O(\log m)$, and of τ from τ^* is $\leq \ell dp(\tau) + O(1)$. Hence the logical depth of the new proof is $O(\log m) + \ell dp(\tau)$, as required.   Q.E.D.

We conclude the section by stating a straightforward relation between the systems LK and F.

Lemma 4.4.15. *The system LK and F mutually polynomially simulate each other. In particular, if a formula ϕ has an F-proof of size m then the sequent*

$$\longrightarrow \phi$$

has an LK-proof of size $O(m^2)$, and vice versa: If a sequent

$$\psi_1, \ldots, \psi_u \longrightarrow \phi_1, \ldots, \phi_v$$

has an LK-proof of size m then the formula

$$\left(\bigvee_{i \leq u} \neg \psi_i\right) \vee \left(\bigvee_{j \leq v} \phi_j\right)$$

(disjunctions bracketed arbitrarily) has an F-proof of size $O(m^2)$.

Moreover, the depth of proofs in these simulations increases at most by a constant.

The proof is omitted.

4.5. The extension and the substitution rules

In this section we consider two extensions of system F. As before we confine the discussion to the language 0, 1, $\vee$, $\wedge$, $\neg$.

Definition 4.5.1. *The substitution rule allows simultaneous substitution of formulas for atoms in one inference step*

$$\frac{\theta(p_1, \ldots, p_n)}{\theta(\phi_1, \ldots, \phi_n)}$$

A Frege system F augmented by the substitution rule will be denoted SF.

Another useful rule is the *extension rule*, whose version we saw already in Section 4.2. Its introduction is a little less straightforward, and rather than define a system, we define its proofs.

Definition 4.5.2. *Let F be a Frege system. An* extended Frege proof *is a sequence of formulas $\theta_1, \ldots, \theta_k$ such that every θ_i either is obtained from some previous θ_j's by a rule from F or has the form*

$$q \equiv \psi$$

where:

1. *the atom q appears in neither ψ nor θ_j for some $j < i$*
2. *the atom q does not appear in θ_k*
3. *$\alpha \equiv \beta$ abbreviates $(\alpha \wedge \beta) \vee (\neg\alpha \wedge \neg\beta)$*

A formula of this form is called an extension axiom *and the atom q is called an* extension atom.

An extended Frege system EF *is the proof system whose proofs are the extended Frege proofs.*

The first two lemmas are trivial.

Lemma 4.5.3. *For every τ*

$$k_{EF}(\tau) \leq k_F(\tau) = O\left(k_{EF}(\tau)\right)$$

Proof. The first inequality is obvious. For the second one, replace an EF-proof with the extension axioms

$$q_1 \equiv \psi_1(\overline{p}), \qquad q_2 \equiv \psi_2(\overline{p}, q_1), \quad \ldots, \qquad q_r \equiv \psi_r(\overline{p}, q_1, \ldots, q_{r-1})$$

first successively q_r by ψ_r, q_{r-1} by $\psi_{r-1}, \ldots$, and so on. This is an F-proof of τ from axioms of the form $\psi \equiv \psi$ that remained from the extension axioms; each has an F-proof with constant number of steps by Lemma 4.4.9.   Q.E.D.

Lemma 4.5.4. *A Frege system with the substitution rule SF polynomially simulates any Frege system with the extension rule EF.*

Proof. Let $q_1 \equiv \psi_1, \ldots, q_r \equiv \psi_r$ be all extension axioms introduced in an EF-proof $\theta_1, \ldots, \theta_k$ of τ of size m. We may assume that these r formulas are the first r steps of the proof. Note that none of $q_1, \ldots, q_r$ occurs in τ but that q_j may occur in $\psi_{j+1}, \ldots, \psi_r$.

By the deduction Lemma 4.4.10 there is an F-proof of size $O(m^2)$ of the implication of size $O(m)$

$$q_r \equiv \psi_r \to (q_{r-1} \equiv \psi_{r-1} \to (\ldots (q_1 \equiv \psi_1)) \ldots) \to \tau .$$

Apply the substitution rule to this formula by substituting ψ_r for q_r

$$\psi_r \equiv \psi_r \to (q_{r-1} \equiv \psi_{r-1} \to (\ldots (q_1 \equiv \psi_1)) \ldots) \to \tau$$

and then separately derive $\psi_r \equiv \psi_r$ (by a proof of size $O(|\psi_r|)$: Lemma 4.4.9), and by the modus ponens thus infer

$$q_{r-1} \equiv \psi_{r-1} \to (\ldots (q_1 \equiv \psi_1) \ldots) \to \tau$$

By repeating this r-times we derive τ by a proof of the total size $O(m^2)$. Q.E.D.

The next lemma is the converse of Lemma 4.5.4 and is considerably more difficult.

Lemma 4.5.5. *Any extended Frege system EF polynomially simulates any Frege system with the substitution rule SF.*

Proof. We shall assume for simplicity that the only rules in an SF system are the modus ponens and the substitution rule (plus any number of axiom schemes).

Let $\phi_1(\overline{p}), \ldots, \phi_k(\overline{p})$ be an SF-proof where $(p_1 \ldots, p_m) = \overline{p}$ are all atoms appearing in the proof, and let $(\overline{q}_1 \ldots, \overline{q}_k)$ be tuples of mutually distinct atoms such that $\overline{q}_k := \overline{p}$.

A formula ψ_i denotes $\phi_i(\overline{q}_i)$, for $i \leq k$. For $j \leq k$ define a tuple $\overline{\beta}_j$ of m formulas as follows:

1. if $\phi_j(\overline{p})$ is an axiom or has been inferred by modus ponens then set $\overline{\beta}_j :=$ $\overline{q}_j$
2. if $\phi_j(\overline{p})$ has been inferred by the substitution rule (say by substitution α) from formula $\phi_i(\overline{p})$, $\phi_j(\overline{p}) = \phi_i(\alpha(\overline{p}))$, then set $\overline{\beta}_j = \alpha(\overline{q}_j)$

Denote by

$$\Psi_{ij} = \psi_i \wedge \ldots \wedge \psi_j, \qquad 0 < i - 1 \leq j < k$$

with $\Psi_{i,i-1}$ denoting the constant 1.

The simulation of the original proof in system EF proceeds as follows. First introduce variables $\overline{q}_{k-1}, \ldots, \overline{q}_1$ by the extension rule

$$q_{i,\ell} := (\Psi_{i+1,i} \wedge \neg\psi_{i+1} \wedge \beta_{i+1,\ell}) \vee \ldots \vee (\Psi_{i+1,k-1} \wedge \neg\psi_k \wedge \beta_{k,\ell})$$

Obviously

$$\Psi_{i+1,j-1} \wedge \neg\psi_j \rightarrow q_{i,\ell} \equiv \beta_{j,\ell}, \quad i < j$$

is valid and can be derived by a polynomial size proof. Hence by 4.4.9 we have polynomial size proofs of

$$\Psi_{i+1,j-1} \wedge \neg\psi_j \rightarrow \phi_i(\overline{q}_i) \equiv \phi_i(\overline{\beta}_j), \quad i < j$$

which can also be written as

$$\Psi_{i+1,j-1} \wedge \neg\psi_j \rightarrow \psi_i \equiv \psi_i(\overline{\beta}_j), \quad i < j$$

Now we prove consecutively $\psi_1, \psi_2, \ldots, \psi_k$ and since $\psi_k = \phi_k$ we obtain a polynomial simulation.

Assume that the formulas $\psi_1, \ldots, \psi_{j-1}$ have been proved. Consider two possibilities: Either ϕ_j is an axiom, in which case so is ψ_j, or ϕ_j follows from ϕ_u, ϕ_v, $u, v < j$, by modus ponens, or ϕ_j has been inferred by the substitution rule.

In the case of modus ponens, first derive $\Psi_{u+1,j-1}$ and $\Psi_{v+1,j-1}$ and thus get from the previous implication (with $i = u$ and $i = v$)

$$\neg\psi_j \rightarrow \phi_u(\overline{\beta}_j) \wedge \phi_v(\overline{\beta}_j)$$

Applying modus ponens to $\phi_u(\overline{\beta}_j)$ and $\phi_v(\overline{\beta}_j)$ we get $\neg\psi_j \rightarrow \phi_j(\overline{\beta}_j)$, which is $\neg\psi_j \rightarrow \psi_j$, and hence ψ_j follows.

In the case that ϕ_j was inferred by the substitution α from ϕ_i we obtain $\neg\psi_j \rightarrow \phi_i(\overline{\beta}_j)$ in the same way as in the preceding formula. But by the definition then:

$$\phi_i(\overline{\beta}_j) = \phi_i(\alpha(\overline{q}_j)) = \phi_i(\overline{q}_j) = \psi_j$$

This completes the derivation of ψ_j. Q.E.D.

Note that the number of steps in the simulation from the previous proof sometimes increases exponentially. In fact, $k_{SF}(\tau)$ may be exponentially smaller than $k_{EF}(\tau)$.

Lemma 4.5.6. *For any sufficiently large n there is tautology τ_n with an SF-proof having $O(n)$ steps but whose every EF-proof must have at least $\Omega(2^n)$ steps.*

Proof. Define $\tau_n := \neg^{(2^n)}(1)$ with $\neg^{(t)}$ standing for t occurrences of $\neg$.

Consider formulas $\beta_k = p \to (\neg)^{2^k}(p)$. Obviously SF derives β_k from β_{k-1} by $O(1)$ steps (by first substituting $(\neg)^{2^{k-1}}(p)$ for p in β_{k-1} and by modus ponens).

As β_0 is provable, every β_k has an SF-proof with $O(k)$ steps. In particular, τ_n has an SF-proof with $O(n)$ steps.

Let τ_n have an EF-proof with k steps; thus by Lemma 4.5.3 there is an F-proof π of τ_n with $O(k)$ steps.

By Lemma 4.4.4 there is an F-proof of logical depth $O(k)$ of some σ_n such that τ_n is a substitution instance of σ_n. But then $\sigma_n = \tau_n$ and hence $k = \Omega(2^n)$, which entails the lower bound.   Q.E.D.

There is a close relation between the size and the number of steps in the extended Frege system.

Lemma 4.5.7. *Any tautology τ has an EF-proof of size $O(k_{EF}(\tau) + |\tau|)$.*

Proof. Let τ be a tautology with atoms $q_1, \ldots, q_n$. For all formulas θ in atoms $q_1, \ldots, q_n$ consider the atom p_θ and a set of *defining conditions* like $\text{Ext}(\theta)$ in Section 4.2:

1. $p_{q_i} \equiv q_i$
2. $p_{\neg\theta} \equiv \neg p_\theta$
3. $p_{\theta_1 \wedge \theta_2} \equiv p_{\theta_1} \wedge p_{\theta_2}$
4. $p_{\theta_1 \vee \theta_2} \equiv p_{\theta_1} \vee p_{\theta_2}$.

Also every formula ψ built from q_i's and p_θ's is equivalent to some p_ϕ: Replace p_θ's by θ's and let ϕ be the resulting formula. We will denote such ϕ by ψ^*.

Now assume that we have an EF-proof π of τ with k steps, and assume w.l.o.g. that the first ℓ steps are the extension axioms

$$r_1 \equiv \phi_1, \ldots, r_\ell \equiv \phi_\ell$$

where ϕ_1 contains only q_i's, and ϕ_{j+1} may also contain the extension atoms $r_1, \ldots, r_j$.

Let

$$\theta_{\ell+1}, \ldots, \theta_k$$

be the remaining $k - \ell$ steps of π.

We shall construct a new EF-proof π' by the following process. First replace ϕ_1 in the first formula $r_1 \equiv \phi_1$ by p_{ϕ_1} and replace the atom r_1 by p_{ϕ_1} in the whole proof. Then replace ϕ_2 (which now has p_{ϕ_1} in place of r_1) in the second formula $r_2 \equiv \phi_2$ by $p_{\phi_2^*}$ and r_2 by $p_{\phi_2^*}$ in the whole proof. Generally in the tth step, $t \le \ell$, replace ϕ_t by $p_{\phi_t^*}$ in the tth formula and r_t everywhere in the proof also by $p_{\phi_t^*}$.

After this transformation the first ℓ formulas of π are transformed to formulas of the form

$$p_\phi \equiv p_\phi$$

which all have constant size F-proofs, and no atoms r_j's appear in the remaining $k - \ell$ steps.

Our aim is to show that for each θ_i, $\ell < i \leq k$, the atom p_{θ^*} has constant size proof from $p_{\theta^*_{\ell+1}}, \ldots, p_{\theta^*_{i-1}}$ and from some *defining conditions*. This is readily shown by induction on i, considering cases distinguished by the rule used in π to infer θ_i. As an example consider the case when θ_w was inferred from θ_u and $\theta_v (= \theta_u \to \theta_w)$ by modus ponens. Assume $p_{\theta^*_u}$, $p_{\theta^*_v}$ and use the defining conditions to infer from $p_{\theta^*_v}$ the formula $p_{\theta^*_u} \to p_{\theta^*_w}$ and apply modus ponens to get $p_{\theta^*_w}$. Finally, using the defining conditions we infer from $p_{\theta^*_k} (= p_\tau)$ the formula τ itself. The defining conditions are just the extension axioms, so the newly constructed sequence is an EF-proof π' with $O(k) + O(|\tau|)$ steps and total size $O(k) + O(|\tau|)$.   Q.E.D.

The previous proof can be used to give another proof that any two extended Frege systems polynomially simulate each other. This is because we could construct the new proof π' in another system with other language in the same way; only the *defining conditions* for p_θ (θ in the first language) would be written by using the second language.

Lemma 4.5.8. *Extended resolution ER is the system R augmented by all clauses from $Ext(\phi)$, for all formulas ϕ, as extra initial clauses (cf. Section 4.2). Then ER and EF polynomially simulate each other.*

The substitution rule allows several seemingly weaker variants that are, however, polynomially equivalent. For example, we may restrict the substitution rule by allowing substitution of only one formula at a time or, more interestingly, by allowing only *atoms* in place of formulas $\phi_1, \ldots, \phi_n$ in Definition 4.5.1. We leave the proofs of the equivalence of these versions to the reader.

4.6. Quantified propositional logic

Quantified propositional calculus is formed from a Frege system or the sequent calculus LK by introducing propositional quantifiers: $\forall x \theta(\overline{p}, x)$ whose meaning is $\theta(\overline{p}, 0) \wedge \theta(\overline{p}, 1)$, and $\exists x \theta(\overline{p}, x)$ whose meaning is $\theta(\overline{p}, 0) \vee \theta(\overline{p}, 1)$. This also defines the satisfiability of quantified propositional formulas.

In first order logic quantifiers allow one to define relations and functions not definable by quantifier-free formulas, but the propositional quantifiers do not increase the expressive power of formulas. Instead they allow us to shorten some quantifier-free propositional formulas. For example, $\bigvee_{\overline{\epsilon}} \theta(\overline{\epsilon})$ with $\overline{\epsilon}$ ranging over $\{0, 1\}^n$ has size $\Omega(2^n |\theta|)$ but an equivalent quantified formula $\exists x_1 \ldots \exists x_n \theta(\overline{x})$ has size only $O(n) + |\theta|$.

Definition 4.6.1.

1. *The class $\Sigma^q_0 = \Pi^q_0$ consists of the quantifier-free propositional formulas.*

2. *The classes Σ_{i+1}^q and Π_{i+1}^q are the smallest classes satisfying:*

    (a) $\Sigma_i^q \cup \Pi_i^q \subseteq \Sigma_{i+1}^q \cap \Pi_{i+1}^q$

    (b) *both Σ_{i+1}^q and Π_{i+1}^q are closed under $\vee$ and $\wedge$*

    (c) *if $\phi \in \Sigma_{i+1}^q$ then $\neg\phi \in \Pi_{i+1}^q$*

    (d) *if $\phi \in \Pi_{i+1}^q$ then $\neg\phi \in \Sigma_{i+1}^q$*

    (e) Σ_{i+1}^q *is closed under existential quantification*

    (f) Π_{i+1}^q *is closed under universal quantification*

3. *A quantified propositional formula is any formula appearing in one of Σ_i^q; Σ_∞^q denotes the class of all quantified propositional formulas*

We shall base our definition of the quantified propositional calculus on system LK.

Definition 4.6.2.

1. *Quantified propositional calculus G extends the system LK by allowing quantified propositional formulas in sequents and by adopting the following extra quantifier rules:*

    (a) $\forall$: **introduction**

$$\text{left} \quad \frac{A(B), \Gamma \longrightarrow \Delta}{\forall x\, A(x), \Gamma \longrightarrow \Delta} \quad and \quad \text{right} \quad \frac{\Gamma \longrightarrow \Delta, A(p)}{\Gamma \longrightarrow \Delta, \forall x\, A(x)}$$

    (b) $\exists$: **introduction**

$$\text{left} \quad \frac{A(p), \Gamma \longrightarrow \Delta}{\exists x\, A(x), \Gamma \longrightarrow \Delta} \quad and \quad \text{right} \quad \frac{\Gamma \longrightarrow \Delta, A(B)}{\Gamma \longrightarrow \Delta, \exists x\, A(x)}$$

    *where B is any formula, and with the restriction that the atom p does not occur in the lower sequents of $\forall$: **right** and $\exists$: **left**.*

2. *The system G_i is a subsystem of G that allows only Σ_i^q-formulas in sequents.*

3. *The system G_i^* is a subsystem of G_i allowing only treelike proofs.*

The definition of the systems G_i and G_i^* may seem ad hoc, but these systems naturally occur in connection with bounded arithmetic in Chapter 9.

Note that $G_0 = \text{LK}$ and hence by Lemmas 4.4.8 and 4.4.15 G_0 and G_0^* polynomially simulate each other. This is unknown for G_i and G_i^* for $i > 0$. We consider these quantified systems primarily as proof systems for quantifier-free tautologies, but later we shall also consider sets TAUT$_i$ of tautological Σ_i^q-formulas, and we shall compare the systems G_i and G_i^* as proof systems for TAUT$_i$ rather then just for TAUT.

Lemma 4.6.3. *The systems G_1^* and EF polynomially simulate each other.*

Proof. First we show that G_1^* polynomially simulates treelike SF, which implies that it also polynomially simulates EF, by Lemmas 4.4.8 and 4.5.4. Clearly it is enough to describe how G_1^* simulates an instance of the substitution rule

$$\frac{\theta(p_1, \dots, p_n)}{\theta(\phi_1, \dots, \phi_n)}$$

To $\longrightarrow \theta(p_1, \dots, p_n)$ apply n-times $\forall$: **right** to obtain

$$\longrightarrow \forall x_1 \dots \forall x_n \theta(x_1, \dots, x_n).$$

Any sequent of the form

$$\theta(\phi_1, \dots, \phi_n) \longrightarrow \theta(\phi_1, \dots, \phi_n)$$

has a short G_1^* -proof, from which follows

$$\forall x_1 \dots \forall x_n \theta(x_1, \dots, x_n) \longrightarrow \theta(\phi_1, \dots, \phi_n)$$

by n applications of $\forall$: **left**. The cut-rule infers from this sequent and the previous sequent the desired sequent

$$\longrightarrow \theta(\phi_1, \dots, \phi_n)$$

Now we want to show that EF p-simulates G_1^*. We shall first deal with a simpler case in which all formulas in a G_1^*-proof either are quantifier free or begin with a block of existential quantifiers followed by a quantifier-free kernel; we shall call such formulas *strict* Σ_1^q.

In this case every sequent in a proof looks like

$$\dots, \alpha_i(\overline{p}), \dots, \exists \overline{x}_j \alpha_j'(\overline{p}, \overline{x}_j), \dots \longrightarrow \dots, \beta_s(\overline{p}), \dots, \exists \overline{y}_t \beta_t'(\overline{p}, \overline{y}_t), \dots$$

where α_i, β_s range over the quantifier-free formulas in the sequent and $\exists \overline{x}_j \alpha_j'(\overline{p}, \overline{x}_j), \exists \overline{y}_t \beta_t'(\overline{p}, \overline{y}_t)$ over formulas with a block of existential quantifiers in front of quantifier-free kernels α_j', β_t''s.

Call a sequence of formulas satisfying the conditions of Definition 4.5.2 with the requirement that the extension atoms cannot appear in the last formula dropped an *EF-sequence*.

Claim. *Assume that the sequent S of the form given previously has a G_1^*-proof in which all formulas are either quantifier free or strict Σ_1^q, with k sequents and of size m.*

Then there is an EF-sequence from $\{\dots, \alpha_i(\overline{p}), \dots, \alpha_j'(\overline{p}, \overline{q}_j), \dots\}$ with extension atoms $\overline{r}_1, \dots$, with the last formula

$$\bigvee_s \beta_s(\overline{p}) \vee \bigvee_t \beta_t'(\overline{p}, \overline{r}_t)$$

and with O(k) steps and size O(m).

The claim is proved readily by induction on k and clearly implies the lemma. The case when not all formulas are strict Σ_1^q is treated similarly, only the definition of EF-sequences is slightly more complicated. We leave it to the reader.

Q.E.D.

The following lemma is proved completely analogously.

Lemma 4.6.4. *For any $i > 0$, the system G_i polynomially simulates G_{i+1}^*-proofs of Σ_i^q-formulas.*

The system G is akin to sequent predicate calculus and shares some analogous properties. We mention only one, the *midsequent theorem* (see Takeuti 1975), but we will not prove it (the reader can follow the idea of the proof of Theorem 4.3.3).

Lemma 4.6.5. *Let $\Gamma \longrightarrow \Delta$ be a valid sequent consisting of quantified propositional formulas in a prenex normal form.*

Then there is a treelike, cut-free G-proof of $\Gamma \longrightarrow \Delta$ in which there is a sequent S (the midsequent) such that

1. no quantifier inferences occur in the proof above S

2. no propositional inferences occur in the proof below S

There is an obvious generalization of the systems LK and G to the "higher type" propositional calculi. We shall not pursue this generalization as any available information about such systems is only a trivial generalization of facts about LK and G.

4.7. Bibliographical and other remarks

The notions of a propositional proof system and of a polynomial simulation (Definitions 4.1.1 and 4.1.2(b)) are from Cook and Reckhow (1979).

Theorem 4.2.1 is from Davis and Putnam (1960).

Theorem 4.2.3 was used implicitly by various authors and explicitly noted in Lovász et al. (1991). The lower bound for resolution in Theorem 4.2.4 was preceded by a lower bound for *regular resolution* in Tseitin (1968). Urquhart (1987,1992) investigated the complexity of resolution further, in relation to cut-free Gentzen's systems and distinguishing between sequencelike and treelike resolution refutations (for regular resolution treelike and sequencelike systems are equally efficient). The proof of Theorem 4.2.4 follows Buss and Turán (1988) closely. See Krajíček (1995b) for another proof.

As we excluded the implication from the language of LK, our system has an additional structural property: All weakening inferences can be postponed till the end of any proof. Interpolation theorems and their relevance to lower bounds are studied in Krajíček (1995b); see also Krajíček and Pudlák (1995). Theorem 4.3.9 was originally proved differently by Statman (1978). The proof of Theorem

4.3.10 follows Krajíček (1994). Notions of *Frege* and *extended Frege* systems were defined in Cook and Reckhow (1979). Gentzen refers to such systems as Hilbert-style systems; see Gentzen (1969). Parikh (1973) was the first to observe that a bound to the number of steps implies a bound to the logical depth of a proof. His proof implicitly gave an exponential upper bound. The *optimal* bound of Lemma 4.4.4 was proved in Krajíček (1989a) for the predicate calculus of arbitrary order. The corollary for the propositional calculus follows also from Lemma 4.5.7: The proof constructed there has formulas of constant logical depth, and thus replacing the extension atoms by their definitions increases the logical depth only proportionally to the number of steps. Lemma 4.4.5 is taken from Krajíček (1989a). Lemma 4.4.6 was noted in Buss (1993a) and Krajíček (1995a). Lemma 4.4.8 is from Krajíček (1989a). A detailed study of the deduction lemma, 4.4.10, is in Bonet (1993).

Lemma 4.4.14 is implicitly contained in Reckhow (1976).

Definitions 4.5.1 and 4.5.2 are from Cook and Reckhow (1979).

Dowd (unpublished) observed that Lemma 4.5.5 is a corollary of a relation of EF to a bounded arithmetic theory PV established by Cook (1975); see Sections 5.3 and 9.3. This was independently observed in Krajíček and Pudlák (1989a), where the explicit polynomial simulation was constructed.

Lemma 4.5.6 is due to Tseitin and Choubarian (1975); the proof is from Krajíček (1989b). Lemma 4.5.7 is from Statman (1977).

Definitions 4.6.1 and 4.6.2 are from Krajíček and Pudlák (1990a).

A system that I did not include into this chapter is the system of *cutting planes* of Cook, Coulard, and Turán (1987). It is defined in Section 13.1. For *natural deduction* systems I refer the reader to Smullyan (1968).

The measure ℓ_P of Definition 4.4.3 appears a bit artificial. But for systems $P \geq_p EF$ it is polynomially related to the size (the most important measure by Theorem 4.1.2) and, in fact, all important lower bounds are actually lower bounds to the measure ℓ_P.

5

Basic bounded arithmetic

Bounded arithmetic was proposed in Parikh (1971), in connection with length-of-proofs questions. He called his system PB, presumably as the alphabetical successor to PA, but we shall stay with the established name $I\Delta_0$ (for "induction for Δ_0 formulas"). This theory and its extensions by Π_2^0 axioms saying that some particular recursive function is total were studied and developed in the fundamental work of J. Paris and A. Wilkie, and their students C. Dimitracopoulos, R. Kaye, and A. Woods.

They studied this theory both from the logical point of view, in connections with models of arithmetic, and in connection with computational complexity theory, mostly reflected by the definability of various complexity classes by subclasses of bounded formulas. They also investigated the relevance of Gödel's theorem to these weak subtheories of PA and closely related interpretability questions.

Further impetus to the development of bounded arithmetic came with Buss (1986), who formulated a bounded arithmetic system S_2, a conservative extension of the system $I\Delta_0 + \Omega_1$ investigated earlier by J. Paris and A. Wilkie, and its various subsystems and second order extensions. The particular choice of the language and the definition of suitable subtheories of S_2 allowed him to formulate a very precise relation between the quantifier complexity of a bounded formula and the complexity of the relation it defines, measured in terms of the levels of the *polynomial time hierarchy* PH. He also proved a first witnessing theorem for bounded arithmetic precisely characterizing the computational complexity of a class of functions definable in certain subtheories of S_2.

Later developments, which established some deeper connections between bounded arithmetic and the complexity theory, built on all this foundational work.

Paris and Wilkie also considered the relevance of the length-of-proof questions in propositional logic to independence questions in bounded arithmetic (cf. Paris and Wilkie 1985). Earlier Cook (1975) constructed an equational theory PV (a subsystem of bounded arithmetic, as we shall see later) and proved another relation of PV to propositional logic.

In this chapter we first define and study Parikh's system $I\Delta_0$ and its extensions. Then we introduce Buss's first order systems S_2 and T_2, Cook's PV, and Buss's second order systems U_2 and V_2, and we prove basic relations among various subsystems of these theories.

5.1. Theory $I\Delta_0$

Recall the language L_{PA} from Section 3.2 and Definition 3.2.1 of Δ_0-formulas.

Definition 5.1.1. *Bounded arithmetic theory $I\Delta_0$ is a first order theory of arithmetic in the language L_{PA}, a subtheory of* Peano arithmetic *PA. It is axiomatized by the following axioms called PA$^-$:*

1. $a + 0 = a$
2. $(a + b) + c = a + (b + c)$
3. $a + b = b + a$
4. $a < b \to \exists x, a + x = b$
5. $0 = a \lor 0 < a$
6. $0 < 1$
7. $0 < a \to 1 \le a$
8. $a < b \to a + c < b + c$
9. $a \cdot 0 = 0$
10. $a \cdot 1 = a$
11. $(a \cdot b) \cdot c = a \cdot (b \cdot c)$
12. $a \cdot b = b \cdot a$
13. $(a < b \land c \ne 0) \to a \cdot c < b \cdot c$
14. $a \cdot (b + c) = (a \cdot b) + (a \cdot c)$

and by the Δ_0-induction scheme IND:

$$(\phi(0) \land \forall x(\phi(x) \to \phi(x + 1))) \to \forall x \phi(x)$$

where ϕ is a bounded *formula (= Δ_0-formula), which may have other free variables besides x.*

Note that the theory PA$^-$ extends Robinson's arithmetic Q, originally considered by Parikh. Its models are exactly nonnegative parts of discretely ordered commutative rings.

It is often easier to work with basic theory PA$^-$ instead of Q. Then notions like order and cut make sense; this is not true in every model of Q: A model of Q is, for example, a copy of N followed by one nonstandard element e in which we define $e + x = e$ all x, $e \cdot 0 = 0$, and $e \cdot x = e$ for $x \ne 0$.

Natural models for $I\Delta_0$ are cuts in models of PA. We shall define the general notion of a cut in a model of PA$^-$.

Definition 5.1.2. *Let M be a model of PA^-. A* cut *in M is any subset $I \subseteq M$ closed under $\leq$ and $x + 1$*

$$\forall x, y \in M, (x \in I \wedge y \leq x) \rightarrow (y \in I \wedge x + 1 \in I)$$

We denote this by $I \subseteq_e M$ (e is for "end-extension").

Note that a cut is not necessarily closed under the addition and the multiplication in M. Examples of cuts are the model itself and N, the initial segment of M isomorphic to the natural numbers.

Lemma 5.1.3. *Let M be a model of $I\Delta_0$ and $I \subseteq_e M$ any cut in M closed under the addition and the multiplication of M.*
   Then I is a model of $I\Delta_0$.

Proof. This lemma is simple but it rests on an important property of bounded formulas which is worthwhile to state explicitly:

Claim. *Let $\phi(a, \overline{b})$ be a bounded formula with all free variables displayed. Assume that I is a cut in M closed under the addition and the multiplication, and $\overline{v}$ are some elements of I. Then for every $u \in I$*

$$I \models \phi(u, \overline{v}) \quad \text{iff} \quad M \models \phi(u, \overline{v})$$

The claim is easily established by induction on the quantifier complexity of ϕ, using the fact that I is closed under $\leq$.

To prove the lemma it is enough to establish that for any bounded formula $\phi(a)$ with parameters from I the induction for $\phi(a)$ holds in I, as the axioms of PA^- are clearly preserved from M to I.

Assume

$$I \models \phi(0) \wedge \forall x(\phi(x) \rightarrow \phi(x + 1)) \wedge \neg\phi(v)$$

for some $v \in I$. By the *Claim* also

$$M \models \phi(0) \wedge (\phi(u) \rightarrow \phi(u + 1))$$

for all $u \in I$ and also $M \models \neg\phi(v)$. Hence in M the induction fails for the bounded formula

$$x > v \vee \phi(x)$$

which contradicts the assumption that $M \models I\Delta_0$.   Q.E.D.

If M is not a model of $I\Delta_0$, not all cuts in M closed under $+$ and $\cdot$ are models of $I\Delta_0$ (e.g., M itself) but provided M satisfies at least theory Q, a cut that is a model of $I\Delta_0$ is actually *definable* in M. This is a nontrivial result of A. Wilkie (unpublished); see also Hájek and Pudlák (1993) and Section 10.6.

Generally, to construct a model of $I\Delta_0$ is difficult: There are no countable recursive models of $I\Delta_0$ (a theorem of Tennenbaum 1959) and, in fact, not even of a weak subtheory IE_1 of $I\Delta_0$ with the induction scheme restricted only to E_1-formulas, that is, bounded existential formulas (Paris 1984). A subtheory of $I\Delta_0$ with the induction restricted to *open* formulas only, called IOpen, does have recursive nonstandard models (Sheperdson 1964). The construction of such models and the question of deciding whether a *Diophantine* equation has a solution in a model of IOpen is of relevance to criteria for solvability of such equations in number theory, but less to issues treated here. A curious reader is advised to consult van den Dries (1980, 1990) and Otero (1991).

Theorem 5.1.4 (Parikh 1971). *Assume that $\theta(\overline{a}, b)$ is a Δ_0-formula and that*

$$I\Delta_0 \vdash \forall \overline{x}\, \exists y \theta(\overline{x}, y)$$

Then there is a term $t(\overline{x})$ such that

$$I\Delta_0 \vdash \forall \overline{x}\, \exists y < t(\overline{x}), \theta(\overline{x}, y)$$

Proof. We prove the theorem by a simple compactness argument.

Assume that $I\Delta_0$ proves $\forall \overline{x}\, \exists y \theta(\overline{x}, y)$, but it does not prove $\forall \overline{x}\, \exists y < t(\overline{x})$, $\theta(\overline{x}, y)$, for any term t. This means that it also does not prove any disjunction of the form

$$\bigvee_i \forall \overline{x}\, \exists y < t_i, \theta(\overline{x}, y)$$

as otherwise it would prove

$$\forall \overline{x}\, \exists y < t, \theta(\overline{x}, y)$$

for $t := t_1 + t_2 + \ldots$.

By compactness then the theory

$$I\Delta_0 + \{\forall y < t(\overline{c})\neg\theta(\overline{c}, y) \mid t \text{ a term}\}$$

is consistent where $\overline{c}$ are new constants.

Let M be a model of this theory and define a cut I in M by

$$b \in I \quad \text{iff} \quad M \models b < t(\overline{c}), \quad \text{some term } t$$

Then I is a model of $I\Delta_0$ (by Lemma 5.1.3) but also $I \models \exists \overline{x}\, \forall y \neg\theta(\overline{x}, y)$, contradicting the hypothesis of the theorem. Q.E.D.

It follows from Parikh's theorem that $I\Delta_0$ cannot Δ_0-define a function that eventually majorizes all polynomials, for example, the exponentiation $x^y = z$. Consequently, $I\Delta_0$ cannot directly formalize constructions requiring exponentiation (such as the cut-elimination).

In fact, it is far from obvious that there is even a Δ_0-definition of the *graph* of the exponentiation $\{(x, y, z) \mid x^y = z\}$.

The existence of such a definition follows from Bennett's Lemma 3.2.6 and Theorem 3.2.3; see also Bennett (1962). Paris (described in Dimitracopoulos 1980) and later Pudlák (1983) constructed a definition of the graph of the exponentiation about which $I\Delta_0$ could prove the recursive properties of exp:

$$x^0 = 1 \quad \text{and} \quad x^{(y+1)} = x^y \cdot x$$

With such a definition we can define other functions in $I\Delta_0$

1. $\lceil \log_2(x) \rceil = y$ iff $2^{(y-1)} < x \leq 2^y$
2. $|x| := \lceil \log_2(x+1) \rceil$ if $x > 0$, and $|0| := 0$.

and, more generally, one can define in $I\Delta_0$ the basic relations and operations of the rudimentary sets (i.e., the concatenation, the part-of quantification) and prove again the basic closure properties.

Of particular importance are functions helping to formalize syntactic constructions so that $I\Delta_0$ can speak about proofs or computations. The existence of Δ_0-definitions of these basic notions follows from Theorem 3.2.7.

Coding of finite sequences is the most important function of this type and we state its properties in a separate lemma. Section 5.4 is devoted to constructions of such a definition.

Lemma 5.1.5. *There is a Δ_0-formula $\theta(w, i, x)$ that we shall write as*

$$(w)_i = x$$

such that $I\Delta_0$ proves:

1. $(w)_i = x \wedge (w)_i = y \rightarrow x = y$
2. $(w)_i = x \wedge j < i \rightarrow \exists y \leq w, (w)_j = y$
3. $\exists w \forall i \forall x, (w)_i \neq x$
4. $\forall w \forall i \forall x, z \exists w' \forall j \neq i, (w)_j = (w')_j \wedge ((w)_i = z \rightarrow (w')_i = x)$

Theories that can interpret a theory of a binary operation satisfying conditions 1–4 of the lemma are called *sequential* in Pudlák (1985). They are very important as they can interpret theory Q and it is often easier to interpret a sequential theory in a theory first than some fragment of arithmetic directly.

What one cannot prove in $I\Delta_0$, however, is that there is a sequence of all numbers smaller than x, as its code would be exponentially large in contradiction to Parikh's theorem.

One also cannot prove in $I\Delta_0$ that in a word a letter can be substituted for by another word

$$\forall w, x, u \exists w', w' = w(x/u)$$

as the code of w' would have length bounded by $|w| \cdot |u|$ only: That is, its size would be bounded by $2^{|w| \cdot |u|}$, which is superpolynomial. The *substitution* plays, however, an essential role in basic logical notions, including the definitions of predicate calculus, and to avoid its use one would be forced into unnatural definitions of formulas with only one explicit occurrence of a variable, and so forth.

Perhaps more importantly, coding of polynomial length proofs and polynomial time computations requires totality of functions $2^{|x|^k}$, all k.

For this reason Paris and Wilkie (1981a,b; 1987b) studied the extension $I\Delta_0 + \Omega_1$ obtained by adding to $I\Delta_0$ a Π_2^0-sentence

$$\Omega_1 \ : \ \forall x \exists y, x^{|x|} = y$$

Function $x^{|x|}$ is denoted by $\omega(x)$.

We shall treat this in greater detail in the next section, but first we examine the question whether one could at least *interpret* (cf. Section 10.6) theories like $I\Delta_0 + \Omega_1$ or $I\Delta_0 + \text{Exp}$ in $I\Delta_0$ and, in particular, whether one can define in $I\Delta_0$ cuts closed under $\omega(x)$ or exp. A positive result is due to Solovay (unpublished); see Hájek and Pudlák (1993).

Theorem 5.1.6. *Define functions $\omega_k(x)$ by*

$$\omega_1(x) := \omega(x), \ \omega_{k+1}(x) := 2^{\omega_k(|x|)}$$

Then for every definable cut $J(x)$ and every k there is formula $I_k(x)$ such that $I\Delta_0$ proves three properties:

1. I_k is closed under ω_k

$$\forall x, (I_k(x) \to \exists y, I_k(y) \wedge \omega_k(x) = y)$$

2. I_k defines a cut

$$I_k(0) \wedge \forall x, y((I_k(x) \wedge y < x) \to (I_k(x+1) \wedge I_k(y)))$$

3.

$$\forall x, I_k(x) \to J(x)$$

Proof. Note that $\omega_k(x)$ majorizes all $\omega_\ell(x)$ for $\ell < k$ and is majorized by 2^x.

First define $J^1(x)$ by

$$J^1(x) := \forall y, J(y) \to J(y+x)$$

and $J^2(x)$ by

$$J^2(x) := \forall y, J^1(y) \to J^1(y \cdot x)$$

$J^2(x)$ defines a cut closed under addition and multiplication.

Then define $I_1(x)$ by

$$I_1(x) := \forall y, J^2(y) \to \exists z, 2^{|y||x|} = z \wedge J^2(z)$$

Then $I\Delta_0$ proves that I_1 is a nonempty cut closed under ω_1.

Having the formula $I_k(x)$ about which $I\Delta_0$ proves that it defines a nonempty cut closed under $\omega_k(x)$ define a formula $I_{k+1}(x)$ by

$$I_{k+1}(x) := \forall y, I_k(y) \rightarrow \exists z, 2_k^{|y|^{(k)}|x|^{(k)}} = z \wedge I_k(z)$$

where 2_k^x is the k-times iterated function 2^x. It is easy to verify that this formula is in $I\Delta_0$ a nonempty cut closed under $\omega_{k+1}(x)$.   Q.E.D.

The following negative result is interesting.

Theorem 5.1.7 (Wilkie 1986, Paris and Wilkie 1987a). *There is no definable cut that would be provably in $I\Delta_0$ closed under all $\omega_k(x)$, $k \geq 1$.*

In particular, no definable cut is provably closed under 2^x.

5.2. Theories S_2 and T_2

The language L of the systems to be defined in this section extends the language L_{PA}. The idea is that one adds enough function symbols to spare the tedious introduction of coding of syntactical objects within the systems.

The language L extends L_{PA} by three new function symbols

$$L = L_{PA} \cup \left\{ \left\lfloor \frac{x}{2} \right\rfloor, |x|, x\#y \right\},$$

where the last symbol has the meaning $2^{|x| \cdot |y|}$.

For a convenience we shall sometimes also add the symbol $(w)_i$, for coding sequences, into the language, so

$$L^+ := L \cup \{(w)_i\}$$

The basic theory that will be included in all our systems is denoted *BASIC* and it extends Robinson's Q.

Definition 5.2.1. *The theory BASIC consists of the following 32 axioms in the language L:*
  1. $a \leq b \rightarrow a \leq b+1$
  2. $a \neq a+1$
  3. $0 \leq a$
  4. $(a \leq b \wedge a \neq b) \rightarrow a+1 \leq b$
  5. $a \neq 0 \rightarrow 2a \neq 0$
  6. $a \leq b \vee b \leq a$
  7. $(a \leq b \wedge b \leq a) \rightarrow a = b$
  8. $(a \leq b \wedge b \leq c) \rightarrow a \leq c$
  9. $|0| = 0$
  10. $a \neq 0 \rightarrow (|2a| = |a| + 1 \wedge |2a+1| = |a| + 1)$

11. $|1| = 1$
12. $a \leq b \rightarrow |a| \leq |b|$
13. $|a\#b| = |a| \cdot |b| + 1$
14. $0\#a = 1$
15. $a \neq 0 \rightarrow (1\#(2a) = 2(1\#a) \wedge 1\#(2a + 1) = 2(1\#a))$
16. $a\#b = b\#a$
17. $|a| = |b| \rightarrow a\#c = b\#c$
18. $|a| = |b| + |c| \rightarrow a\#d = (b\#d) \cdot (c\#d)$
19. $a \leq a + b$
20. $(a \leq b \wedge a \neq b) \rightarrow (2a + 1 \leq 2b \wedge 2a + 1 \neq 2b)$
21. $a + b = b + a$
22. $a + 0 = a$
23. $a + (b + 1) = (a + b) + 1$
24. $(a + b) + c = a + (b + c)$
25. $a + b \leq a + c \rightarrow b \leq c$
26. $a \cdot 0 = 0$
27. $a \cdot (b + 1) = a \cdot b + a$
28. $a \cdot b = b \cdot a$
29. $a \cdot (b + c) = (a \cdot b) + (a \cdot c)$
30. $1 \leq a \rightarrow ((a \cdot b \leq a \cdot c) \equiv (b \leq c))$
31. $a \neq 0 \rightarrow |a| = |\lfloor (a/2) \rfloor| + 1$
32. $a = \lfloor (b/2) \rfloor \equiv (2a = b \vee 2a + 1 = b)$

For future theories, when L is replaced by L^+ the common theory *BASIC* will automatically be extended to *BASIC$^+$* by adding the four conditions from Lemma 5.1.5.

The next two definitions introduce the basic systems of bounded arithmetic.

Definition 5.2.2. T_2^i *is a theory in the language L extending BASIC by the* induction axiom IND

$$\phi(0) \wedge \forall x(\phi(x) \rightarrow \phi(x + 1)) \rightarrow \forall x \phi(x)$$

for all Σ_i^b-formulas $\phi(a)$. The formula $\phi(a)$ may have other free variables than a. The theory T_2 is the union of all theories T_2^i.

Note that $IE_i \subseteq T_2^i$.

Definition 5.2.3. S_2^i *is a theory in language L extending BASIC by the* polynomial induction axiom PIND

$$\phi(0) \wedge \forall x(\phi(\lfloor \frac{x}{2} \rfloor) \rightarrow \phi(x)) \rightarrow \forall x \phi(x)$$

for all Σ_i^b-formulas $\phi(a)$. The formula $\phi(a)$ may have other free variables than a. The theory S_2 is the union of all theories S_2^i.

Finally we introduce a third version of induction.

Definition 5.2.4. *The scheme of the* length induction axioms LIND *is*

$$\phi(0) \wedge \forall x (\phi(x) \rightarrow \phi(x+1)) \rightarrow \forall x \phi(|x|)$$

The formula $\phi(a)$ *may have other free variables than* a.

The theory $\Sigma_i^b -$ LIND *is BASIC augmented by the axioms of* LIND *for all* Σ_i^b-*formulas* $\phi(a)$.

Letters S, T are next to letters P, Q, R taken previously in Tarski, Mostowski, and Robinson (1953) to denote PA, Robinson's Q, and a theory R defined there. The superscript i in T_2^i and S_2^i refers to the restriction of the induction scheme to Σ_i^b-formulas, while the index 2 refers to the presence of function # in L. An index k would refer to the presence of a function symbol $\#_k$ in L, a function of the growth rate of approximately ω_{k+1}. We shall consider such systems only exceptionally (see, e.g., Corollary 10.5.4).

Lemma 5.2.5. *For any* $i \geq 1$:

$$S_2^i = \Sigma_i^b - LIND = \Pi_i^b - PIND = \Pi_i^b - LIND$$

and

$$T_2^i = \Sigma_i^b - IND = \Pi_i^b - IND$$

Proof. Let ϕ be a Σ_i^b-formula. Assume first that

$$\phi(0) \wedge \forall x \left(\phi\left(\left\lfloor \frac{x}{2} \right\rfloor\right) \rightarrow \phi(x) \right) \wedge \neg\phi(a)$$

Then

$$\psi(0) \wedge \forall x \leq |a|(\psi(x) \rightarrow \psi(x+1)) \wedge \neg\psi(|a|)$$

where $\psi(x) := \phi(a_x)$ for a_x denoting the number consisting of the first x bits of a. Formula $\psi(x)$ is Σ_i^b as $a_x = y$ is Σ_1^b-definable

$$a_x = y \quad \text{iff} \quad |y| = x \wedge \exists z \leq a, a = y + 2^{|y|} \cdot z$$

where $2^{|y|}$ is just $y\#1$.

That contradicts $\Sigma_i^b -$ LIND.

Analogous reasoning gives the rest of the lemma.   Q.E.D.

Definition 5.2.6. *The following are four minimization/maximization principles:*
   (a) *MIN*

$$\phi(a) \rightarrow \exists x \leq a \forall y < x, \phi(x) \wedge \neg\phi(y)$$

(b) *LENGTH-MIN*

$$\phi(a) \to \exists x \le a \forall y \le a, \phi(x) \wedge (|y| < |x| \to \neg\phi(y))$$

(c) *MAX*

$$\phi(0) \to \exists x \le a \forall y \le a, \phi(x) \wedge (x < y \to \neg\phi(y))$$

(d) *LENGTH-MAX*

$$\phi(0) \to \exists x \le a \forall y \le a, \phi(x) \wedge (|x| < |y| \to \neg\phi(y))$$

Recall the definition of Δ_i^b-formulas (cf. Definition 3.2.11, part 4).

Lemma 5.2.7. *For all $i \ge 1$:*
(a) $T_2^i = \Sigma_i^b\text{-}MAX = \Sigma_i^b\text{-}MIN = \Pi_{i-1}^b\text{-}MAX = \Pi_{i-1}^b\text{-}MIN$
(b) $S_2^i = \Sigma_i^b\text{-}LENGTH\text{-}MAX = \Sigma_i^b\text{-}LENGTH\text{-}MIN$
$\qquad = \Pi_{i-1}^b\text{-}LENGTH\text{-}MAX = \Pi_{i-1}^b\text{-}LENGTH\text{-}MIN$
where for $i = 1$ the class Π_{i-1}^b is replaced by the class of formulas Δ_1^b in S_2^1.

Proof.

(a) Let $\phi(x)$ be a Σ_i^b-formula satisfying

$$\phi(0) \wedge \neg\phi(a)$$

By Σ_i^b-MAX there is maximal $b < a$ such that $\phi(b)$, that is, also $\neg\phi(b+1)$. Hence $\phi(x)$ does not satisfy the induction hypothesis. This shows that Σ_i^b-MAX implies T_2^i. For the opposite direction let $\phi(x)$ be a Σ_i^b-formula and define another Σ_i^b-formula $\psi(x)$ by

$$\psi(x) := \exists y \le a, x \le y \wedge \phi(y)$$

Condition $\phi(0)$ implies $\psi(0)$. By Σ_i^b-IND either $\psi(a)$ holds, in which case $\phi(a)$ holds too, or $\psi(b) \wedge \neg\psi(b+1)$ holds for some $b < a$, in which case b is the maximal element $\le a$ satisfying $\phi(x)$.
The MAX and MIN principles for Σ_i^b (resp. for Π_{i-1}^b) are equivalent as x is the maximal element $\le a$ satisfying $\phi(x)$ iff $y = a - x$ is the minimal element $\le a$ satisfying $\phi(a - y)$. Note that $a - x$ is Δ_1^b-definable in S_2^1.
Now we show that Σ_i^b-MAX implies Π_{i-1}^b-MIN: Take a Π_{i-1}^b-formula $\phi(x)$ and define Σ_i^b-formula $\psi(x)$ by:

$$\psi(x) := \exists y \le a, x + y = a \wedge \phi(y)$$

Formula $\phi(a)$ implies $\psi(0)$, and if b is the maximal element $\le a$ satisfying ψ, then $a - b$ is the minimal element satisfying ϕ.

Finally assume Π^b_{i-1}-MIN and let $\phi(x) = \exists y \le t(x,a)\psi(x,y)$ be a Σ^b_i-formula with ψ being Π^b_{i-1} (or Δ^b_1 if $i = 1$). Define a formula θ by

$$\theta(\langle z, y\rangle) := \forall x \le a, x + z = a \to (y \le t(x,a) \wedge \psi(x,y))$$

where $\langle z, y\rangle$ is the pairing function, which is increasing w.r.t the lexicographic ordering. Formula $\phi(0)$ implies $\theta(\langle a, y_0\rangle)$ for some $y_0 \le t(0,a)$. By the MIN axiom there is minimal $\langle b, c\rangle$ for which θ holds; clearly then $a - b$ is the maximal element $\le a$ for which ϕ holds.

(b) The proofs of Part **(b)** are completely analogous.

Q.E.D.

Lemma 5.2.8. *For all $i \ge 1$: $S^i_2 \subseteq T^i_2 \subseteq S^{i+1}_2$ and thus: $T_2 = S_2$.*

Proof. $S^i_2 \subseteq T^i_2$ is easier: Let ϕ be a Σ^b_i-formula and assume

$$\phi(0) \wedge \forall x(\phi(x) \to \phi(x+1))$$

Hence by T^i_2 also $\forall x \phi(x)$ and, in particular, also $\forall x \phi(|x|)$. Hence T^i_2 implies the Σ^b_i-LIND which is by Lemma 5.2.5 equivalent to S^i_2.

For $T^i_2 \subseteq S^{i+1}_2$ we use the idea of shortening cuts from Theorem 5.1.6. Let ϕ be a Σ^b_i-formula satisfying

$$\phi(0) \wedge \forall x(\phi(x) \to \phi(x+1)) \wedge \neg\phi(a)$$

Define formula ψ by

$$\psi(x) := \forall y \le a, \phi(y) \to \phi(x+y).$$

Then ψ satisfies the assumptions of the PIND scheme and hence $\psi(a)$ follows from S^{i+1}_2. But $\psi(a)$ implies $\phi(a)$.

Q.E.D.

The next lemma is proved by the same shortening of cuts as in the second part of the previous lemma; we leave it to the reader.

Lemma 5.2.9. *For all $i \ge 1$ and for arbitrary Σ^b_i-formula ϕ and Π^b_i-formula ψ, if the theory S^i_2 proves the equivalence*

$$\forall x \le a, \phi(x) \equiv \psi(x)$$

then S^i_2 proves the formula

$$\phi(0) \wedge \forall x \le a(\phi(x) \to \phi(x+1)) \to \phi(a)$$

Such a scheme is called Δ^b_i-IND.

The lemma can be strengthened (Corollary 8.2.7) but that requires first some other nontrivial results.

The class of *Boolean combinations* of formulas from a class Γ is denoted by $\mathcal{B}(\Gamma)$.

Lemma 5.2.10. *For all $i \geq 1$, the theory T_2^i proves the induction scheme IND for all $\mathcal{B}(\Sigma_i^b)$-formulas.*

Proof. Every $\mathcal{B}(\Sigma_i^b)$-formula $\psi(x)$ is logically equivalent to a formula of the form

$$\phi_k(x) \wedge \neg(\phi_{k-1}(x) \wedge \neg(\ldots \neg(\phi_2(x) \wedge \neg\phi_1(x))\ldots)$$

with all $\phi_j \in \Pi_i^b$ (analogously with the difference hierarchy of Hausdorf 1978). We shall call a formula ψ such that ψ or $\neg\psi$ can be expressed in this form a *level k formula*.

We shall show by induction on ℓ that T_2^i proves the induction axiom for every level ℓ formula. Denote

$$\psi_\ell(x) := \phi_\ell(x) \wedge \neg(\phi_{\ell-1}(x) \wedge \neg(\ldots \neg(\phi_2(x) \wedge \neg\phi_1(x))\ldots)$$

For $\ell = 1$ this follows from Lemma 5.2.5. Let $\ell > 1$ and let a be arbitrary, and assume that the induction assumption for ψ_ℓ holds on the interval $[0, a]$.

If $\forall x \leq a\phi_\ell(x)$ holds, then the induction for ψ_ℓ on $[0, a]$ is equivalent to the induction for $\neg\psi_{\ell-1}$, which is a level $\ell - 1$ formula. If $\forall x \leq a\phi_\ell(x)$ fails, take $b \leq a$ the least number such that $\neg\phi_\ell(b)$; it exists provably in T_2^i by Lemma 5.2.7 as $\neg\phi_\ell \in \Sigma_i^b$. Then the induction hypothesis for ψ_ℓ on $[0, b - 1]$ implies the induction hypothesis for the level $\ell - 1$ formula $\neg\psi_{\ell-1}$ and hence $\neg\psi_{\ell-1}(b - 1)$ holds. But that means that the implication

$$\psi_\ell(b - 1) \rightarrow \psi_\ell(b)$$

fails, contradicting the induction assumption for ψ_ℓ.

This establishes the induction for level ℓ formula ψ, which can be represented as ψ_ℓ. If ψ is a level ℓ formula such that $\neg\psi$ can be represented as ψ_ℓ, the induction for ψ follows from the induction for $\neg\psi(a - x)$, which can be represented as ψ_ℓ. Q.E.D.

Definition 5.2.11.

(a) Bounded collection *scheme B is the scheme*

$$\forall x \leq a\exists y, \phi(x, y) \rightarrow \exists z\forall x \leq a\exists y \leq z, \phi(x, y)$$

where ϕ is a bounded formula. Symbol $B\Sigma_i^b$ denotes the scheme restricted to Σ_i^b-formulas only.

(b) Sharply bounded collection *scheme BB is the scheme*

$$\forall i \leq |a|\exists y \leq b, \phi(i, y) \rightarrow \exists w\forall i \leq |a|, \phi(i, (w)_i)$$

Symbol $BB\Sigma_i^b$ denotes the scheme restricted to Σ_i^b-formulas only.

The collection scheme is sometimes also called the *replacement scheme* or, in the context of second order systems, the *choice scheme*. We only remark that the bounded collection scheme is as strong as the induction scheme when accepted for all formulas (cf. Hájek and Pudlák 1993).

Lemma 5.2.12. *For all* $i \geq 1$

$$S_2^i \vdash BB\Sigma_i^b$$

Proof. Note that the formula

$$\psi(x) := \exists w \leq a\#b \, \forall i \leq |a|, i \leq x \rightarrow \phi(i, (w)_i)$$

is Σ_i^b whenever ϕ is, and satisfies the induction assumption

$$\psi(0) \wedge \forall x \leq |a|(\psi(x) \rightarrow \psi(x+1))$$

and hence also, by S_2^i, $\psi(|a|)$ holds.

The bound $a\#b$ to w follows as w codes a sequence of at most $|a|$ numbers each of length at most $|b|$, hence itself has length at most $|a| \cdot |b|$. That is, its code is at most $2^{|a| \cdot |b|} \leq a\#b$. (In fact, the bound $a\#b$ depends on a particular way of coding sequences and could be replaced by another term; see Section 5.4.)   Q.E.D.

Note that the same proof actually shows that S_2^i proves the scheme of the *strong sharply bounded collection* scheme

$$\exists w \forall i \leq |a|, (\exists y \leq b\phi(i, y) \rightarrow \phi(i, (w)_i))$$

A conservation result for the sharply bounded collection scheme was obtained by Ressayre (1986).

Lemma 5.2.13. *Let* $i \geq 1$. *Then the theory*

$$S_2^i + BB\Sigma_{i+1}^b$$

is $\forall \Sigma_{i+1}^b$-*conservative over theory* S_2^i.

The proof follows the idea of the proof of $\forall \Sigma_{n+1}^0$-conservativity of $B\Sigma_{n+1}^0$ over theory $I\Sigma_n^0$ (cf. Hájek and Pudlák 1993).

Lemma 5.2.14. *Denote by* $S_2^i(L^+)$ *a theory defined as* S_2^i *but in the language* L^+ *and with BASIC$^+$ instead of BASIC. Call a bounded formula in the language* L^+ *strictΣ_i^b if it has the form*

$$\exists x_1 \leq t_1 \forall x_2 \leq t_2 \ldots \phi(\overline{a}, x_1, \ldots, x_i)$$

with i *alternating bounded quantifiers and a sharply bounded kernel* ϕ.
Then any Σ_i^b-*formula is in* $S_2^i(L^+)$ *equivalent to a strict* Σ_i^b-*formula.*

Proof. Let $\phi(x)$ be a Σ_i^b-formula in prenex normal form. We want to show that ϕ is in S_2^i equivalent to a formula ψ in prenex form, in which all sharply bounded quantifiers follow after all bounded but not sharply bounded ones. Let

$$\forall i \leq |t| \exists y \leq s, \theta(i, y)$$

be a subformula of ϕ: That is, θ is a Σ_i^b-formula too. By the sharply bounded collection $BB\Sigma_i^b$ available through Lemma 5.2.12 in S_2^i this subformula is equivalent to

$$\exists w \leq t\#s \forall i \leq |t|, \theta(i, (w)_i)$$

This demonstrates how to switch a pair of sharply bounded/bounded quantifiers. Repeating this yields ψ.

To get from ψ a strict Σ_i^b-formula we only have to replace two consecutive occurrences of the same bounded, but not sharply bounded quantifier by one; but this is easily achieved by a pairing function that can be defined by using the function $(w)_i$.   Q.E.D.

Lemma 5.2.15. *The theory S_2 is a conservative extension of the theory $I\Delta_0 + \Omega_1$. That is: Any formula in language L_{PA} provable in S_2 is also provable in $I\Delta_0 + \Omega_1$.*

In fact, every model M of $I\Delta_0 + \Omega_1$ can be expanded to a model of S_2.

Proof. The theory S_2 defines the function $\omega_1(x)$ and proves the axiom Ω_1: This follows from a trivial bound

$$\omega_1(x) \leq ((x\#x) + 2)^2$$

This shows that $I\Delta_0 + \Omega_1 \subseteq S_2$.

Let M be a model of $I\Delta_0 + \Omega_1$. By Theorem 3.2.6 there is a Δ_0-definition of the graph of exponentiation and by the remark before Lemma 5.1.5 there is such definition for which $I\Delta_0$ can prove the recursive equations. It follows that $I\Delta_0$ can also Δ_0-define the graph of the function $a\#b$, and from the bound

$$a\#b \leq \omega_1(a \cdot b + 2)$$

it follows that $I\Delta_0 + \Omega_1$ proves the totality of $a\#b$. The axioms of BASIC pose no problem; nor do the Δ_0-definitions of $|x|$ and $\lfloor (x/2) \rfloor$. This demonstrates that M can be extended by functions to obey BASIC, and from the fact that they are Δ_0- definable in M it follows that induction will hold for all bounded formulas in the expanded language.   Q.E.D.

5.3. Theory PV

Building on an earlier work of Bennett (1962), Cobham (1965) characterized the class of polynomial time functions in the following "machine independent" way.

We say that a function f is defined from functions g, h_0, h_1, and ℓ by *limited recursion on notation* if:

1. $f(\overline{x}, 0) = g(\overline{x})$,
2. $f(\overline{x}, s_i(y)) = h_i(\overline{x}, y, f(\overline{x}, y))$, for $i = 0, 1$,
3. $f(\overline{x}, y) \le \ell(\overline{x}, y)$,

where $s_0(y)$ and $s_1(y)$ are two functions adding 0, respectively 1, to the right of the binary representation of y

$$s_0(y) := 2y$$

$$s_1(y) := 2y + 1$$

Theorem 5.3.1 (Cobham 1965). *The class of polynomial time functions is the smallest class of functions containing constant* 0, *functions* $s_0(y)$, $s_1(y)$ *and* $x\#y$, *and closed under:*

1. *permutation and renaming of variables*
2. *composition of functions*
3. *limited recursion on notation*

We might note at this point that it is possible to enlarge basic functions by finitely many polynomial time functions such that requirement 3 becomes redundant in the theorem: That is, the class of polynomial time functions has a *finite basis* (cf. Muchnik 1970).

Building on this theorem Cook (1975) defined formal system PV (for *polynomially verifiable*). There are two motivations for considering a system like that: One is its relation to the *extended Frege* system (Corollary 9.2.4); another is more philosophical, to define a system in which instances of general proofs can be verified by constructive, computationally feasible procedures.

Definition 5.3.2. *We simultaneously define function symbols of rank k and PV-derivations of rank k, $k = 0, 1, \ldots$. The language of PV will then consist of all function symbols of any rank, and a PV-derivation will be a PV-derivation of any rank.*

(a) *Function symbols of rank* 0 *are constant* 0; *unary* $s_0(y)$, $s_1(y)$, *and* $Tr(x)$; *and binary* $x \frown y$, $x\#y$, *and* $Less(x, y)$.

(b) *Defining equations of rank* 0 *are:*

$$Tr(0) = 0$$

$$Tr(s_i(x)) = x, \qquad i = 0, 1$$

$$x \frown 0 = x$$

$$x \frown (s_i(y)) = s_i(x \frown y), \qquad i = 0, 1$$

$$x\#0 = 0$$

$$x \# s_i(y) = x \frown (x \# y), \qquad i = 0, 1$$

$$Less(x, 0) = x$$

$$Less(x, s_i(y)) = Tr(Less(x, y)), \qquad i = 0, 1$$

$Tr(x)$ *truncates* x *(i.e., deletes the rightmost bit),* $x \frown y$ *is the concatenation,* $x \# y$ *is* $|y|$ *concatenated copies of* x*, and* $Less(x, y)$ *is* x *with* $|y|$ *right bits deleted.*

(c) *PV rules are as follows*

   R1

$$\frac{t = u}{u = t}$$

   R2

$$\frac{t = u \quad u = v}{t = v}$$

   R3

$$\frac{t_1 = u_1, \ldots, t_k = u_k}{f(t_1, \ldots, t_k) = f(u_1, \ldots, u_k)}$$

   R4

$$\frac{t = u}{t(x/v) = u(x/v)}$$

   R5 *Let* $E_1, \ldots, E_6$ *be the equations (1–3) from the definition of the limited recursion on notation: three for* f_1 *and three for* f_2 *in place of* f*. Then*

$$\frac{E_1, \ldots, E_6}{f_1(\overline{x}, y) = f_2(\overline{x}, y)}$$

(d) *PV-derivations of rank* k *are sequences of equalities* $E_1, \ldots, E_t$ *in which every function symbol is of rank* $\leq k$ *and every* E_i *is either a defining equation of rank* $\leq k$ *or derived from some earlier equations by one of the PV-rules.*

(e) *Let* t *be a term consisting of function symbols of rank* $\leq k$*. Then* f_t *is a function symbol of rank* $k + 1$ *and* $f_t = t$ *is a defining equation of rank* $k + 1$*.*

(f) *Other function symbols of rank* $k + 1$ *are obtained as follows. Whenever* $g, h_0, h_1, \ell_0, \ell_1$ *are function symbols of maximum rank* k *and* $\pi_i, i = 0, 1,$ *are PV-derivations of rank* k *of equality*

$$Less(h_i(\overline{x}, y, z), z \frown \ell_i(\overline{x}, y)) = 0$$

   then

$$f = f_{\langle g, h_0, h_1, \ell_0, \ell_1, \pi_0, \pi_1 \rangle}$$

is a function symbol of rank $k+1$, and the two equations (1) and (2) defining
f from g, h_i by limited recursion on notation are defining equations of rank
$k + 1$.

PV has a function symbol f for every function introduced from earlier ones by
limited recursion on notation, provided one can first prove (an equivalent statement
to) that f is bounded by a function obtained earlier.

Lemma 5.3.3. *Let $\underline{n}$ denote the dyadic numeral of number n:*

$$\underline{0} := 0, \quad \underline{2n} := s_0(\underline{n}), \qquad \underline{2n+1} := s_1(\underline{n})$$

and let $f(x_1, \ldots, x_k)$ be any polynomial time function.
*Then there is a PV-function symbol $f_{PV}(x_1, \ldots, x_k)$ such that for every $m_1, \ldots,$
m_k and $n = f(m_1, \ldots, m_k)$ the theory PV proves the equation*

$$f_{PV}(\underline{m}_1, \ldots, \underline{m}_k) = \underline{n} \,.$$

*On the other hand, every PV-function symbol defines in $\mathbf{N}$ a polynomial time
function.*

Proof. The last part follows from Cobham's Theorem 5.3.1. The first part is not
obvious because of the requirement in part (f) of the definition of PV that before we may introduce a symbol for the function defined by limited recursion on
notation we must have a proof that it is bounded by some previously introduced
function. An inspection of the proof of Theorem 5.3.1 shows that as one constructs
polynomial time functions, one also has obvious PV-proofs of their boundedness.
                                                                            Q.E.D.

Now we state a relation between the systems S_2^1 and PV. Denote by $S_2^1(PV)$ the
theory defined as S_2^1 is in the language L augmented by all PV-function symbols,
with all PV defining equations as new axioms and with the PIND rule extended to
all Σ_1^b-formulas in the new language.

Theorem 5.3.4 (Buss 1986). *The theory $S_2^1(PV)$ is conservative over PV. That
is: Whenever an equation $t = u$ between PV-terms is provable in $S_2^1(PV)$, then it
is provable already in PV.*

The proof requires some results from Chapter 7 and we postpone it till Section
7.2 (cf. Corollary 7.2.4 and the text after it).

Cook (1975) also defines an extension PV1 of PV, allowing open formulas and
propositional reasoning instead of only equations, and adding PIND scheme for
all open formulas. It holds then (and follows immediately from Theorem 5.3.4)
that PV1 is also conservative over PV.

Instead we shall define theories PV_i, $i = 1, 2, \ldots$, which will act as a universal
axiomatization of a conservative extension of theories T_2^{i-1}.

PV_1 consists of all equations $t = u$ provable in PV but has also a form of induction axiom: For an open formula $\psi(x)$ define a function $h(b, u)$ by

(a) $h(b, 0) = (0, b)$

(b) if $h(b, \lfloor (u/2) \rfloor) = (x, y)$ and $u > 0$ then set

$$
h(b, u) := \begin{cases} (\lceil (x + y/2) \rceil, y) & \text{if } \lceil (x + y/2) \rceil < y \wedge \psi(\lceil (x + y/2) \rceil) \\ (x, \lceil (x + y/2) \rceil) & \text{if } x < \lceil (x + y/2) \rceil \wedge \neg\psi(\lceil (x + y/2) \rceil) \\ (x, y) & \text{otherwise} \end{cases}
$$

Then PV_1 contains the axiom

$$(\psi(0) \wedge \neg\psi(b) \wedge h(b, b) = (x, y)) \rightarrow (x + 1 = y \wedge \psi(x) \wedge \neg\psi(y))$$

This axiom simulates the binary search and is thus related to the PIND scheme. The logic of PV_1 is the usual first order predicate calculus.

Theory PV_{i+1} contains PV_1 and has inductively defined characteristic functions of all Σ_i^b- predicates in its language, in particular universal axioms of the form

$$\exists y \le t(\overline{x}), f(\overline{x}, y) = 1 \rightarrow g(\overline{x}) \le t(\overline{x}) \wedge f(\overline{x}, g(\overline{x})) = 1$$

where g are new function symbols introduced inductively for all formulas of the form

$$\exists y \le t(\overline{x}), f(\overline{x}, y) = 1$$

Furthermore, PV_{i+1} is closed under definition by cases, composition, and limited recursion on notation (i.e., has function symbols for all functions introduced by these processes) and also contains the preceding axiom for every open ψ in its language. The logic of PV_{i+1} is also the first order predicate calculus.

Note that each PV_i is a universal theory. We state a theorem that will be proved in Section 6.1 (cf. Corollary 6.1.3).

Theorem 5.3.5. *For every $i \ge 1$, PV_{i+1} is fully conservative over the theory T_2^i. The theory PV_1 is conservative over PV.*

5.4. Coding of sequences

In this section we sketch a way to code finite sets and sequences in S_2^1. This is necessary in order to be able to formalize various syntactic and logical notions and computations of machines in subsystems of S_2.

For the language L_{PA} (and $I\Delta_0 + \Omega_1$) this is quite nontrivial as one must first find a well-behaved Δ_0-definition of the graph of exponentiation, in order to speak about lengths of numbers and their bits. The existence of a Δ_0-definition of the graph of exponentiation follows from Bennett (1962) (cf. Theorem 3.2.6), but the theorem does not imply that there is such a definition about which $I\Delta_0$ or $I\Delta_0 + \Omega_1$

could prove the basic recursive properties

   1. $x^0 = 1$

   2. $x^{y+1} = x^y \cdot x$

Such a bounded definition was constructed by J. Paris (in Dimitracopulos 1980) (see also Pudlák 1983).

Another crucial function whose well-behaved bounded definition is needed is Numones(x): the number of ones in the binary expansion of x. The function is clearly computable in logarithmic space, and thus its graph is Δ_0-definable by Corollary 3.2.9, but proving the basic recurrence properties

   1. Numones$(0) = 0$

   2. Numones$(2x) =$ Numones(x)

   3. Numones$(2x + 1) =$ Numones$(x) + 1$

again requires some work. In $I\Delta_0 + \Omega_1$ this is easier. Theorem 3.2.7 is a general tool showing that all usual concepts defined by inductive properties can be defined in a well-behaved way in $I\Delta_0 + \Omega_1$.

With these two definitions in hand one defines the basic relations and functions on finite words, identifying a number with its dyadic representation and the coding of sequences then follows the development of rudimentary sets in Bennett (1962). In $I\Delta_0 + \Omega_1$ the formalization of syntax and logic is then smooth.

If one wants to formalize logical notions in $I\Delta_0$ only, there are other complications. For example, the term resulting from substitution of a term u into a term v for a variable x will not in general have length proportional to $|u| + |v|$; hence its code will not be bounded by a polynomial in u, v and by Theorem 5.1.4 $I\Delta_0$ cannot prove that the substitution is always defined.

In some situations one can restrict the syntax, for example, to terms and formulas with only one occurrence of each variable (or to their representation with this property), but the formalizations obtained in this way are unnatural.

We refer the reader to Paris and Wilkie (1987b) or Hájek and Pudlák (1993) for detailed development of coding, sequences, and syntax in $I\Delta_0$ and $I\Delta_0 + \Omega_1$.

With the language L of S_2 the situation is much simpler because we have the length function $|x|$ in language allowing us to define the graph of exponentiation immediately by

$$2^x = y \equiv \exists z < y, z + 1 = y \wedge |z| = x \wedge |y| = x + 1$$

which is equivalent to

$$2^x = y \equiv \forall z < y, z + 1 = y \rightarrow |z| = x \wedge |y| = x + 1$$

We want to define the basic notions of rudimentary sets and of coding of sequences by means of Δ_1^b-formulas in S_2^1, in such a way that S_2^1 can prove the basic properties and, in particular, the properties of Lemma 5.1.5. This is done in great detail in Buss (1986). Another approach is to follow the development of Paris and Wilkie (1987b) and Hájek and Pudlák (1993) in S_2^1 and to verify that all notions that are only Δ_0 there are Δ_1^b in S_2^1. To illustrate these topics we outline a way of coding sequences, but we shall proceed rather swiftly, leaving details to the reader.

First we define *the pairing function*

$$\langle a, b \rangle := \left\lfloor \frac{(a+b)(a+b+1)}{2} \right\rfloor + a$$

It is defined by a term (hence is Δ_1^b) and S_2^1 can prove the basic property

$$\langle a, b \rangle = \langle u, v \rangle \equiv (a = u \wedge b = v)$$

Then we define the predicate "*a is a power of* 2"

$$\mathrm{Pow}(a) \equiv \exists x \leq a, x + 1 = a \wedge |x| + 1 = |a|$$

which is provably in S_2^1 equivalent to

$$\forall x \leq a, x + 1 = a \rightarrow |x| + 1 = |a|$$

Next define the function *the ith bit of a*

$$\mathrm{bit}(a, i) := \begin{cases} 1 & \text{if } \exists u, v, w \leq a, u + v + 2vw = a \wedge \mathrm{Pow}(v) \wedge |v| = i + 1 \\ & \qquad\qquad \wedge\, u < v \\ 0 & \text{otherwise} \end{cases}$$

which is also Δ_1^b as $\mathrm{bit}(a, i) = 1$ is also equivalent to

$$\forall u, v, w \leq a, u + v + 2vw = a \wedge \mathrm{Pow}(v) \wedge |u| \leq i \rightarrow |v| = i + 1$$

Using this function we define *the elementhood predicate*

$$i \in a \equiv \mathrm{bit}(a, i) = 1$$

Claim 1. *Functions and predicates* $\langle a, b \rangle$, $\mathrm{Pow}(a)$, $\mathrm{bit}(a, i)$, *and* $i \in a$ *are* Δ_1^b *in* S_2^1.

We want to code arbitrary 0–1 words. This cannot be done just by binary expansions of numbers that always start with 1. So we think of a *word* as pair $\langle u, v \rangle$, coding the word consisting of first right $|v|$ bits of u. With this interpretation in mind define the *equality of words* $a =_w b$ by

$$\exists x, y \leq a \exists u, v \leq b, \langle x, y \rangle = a \,\wedge\, \langle u, v \rangle = b \,\wedge$$

$$\wedge\, (\forall i \leq |y|, i \in x \equiv (i \in u \wedge i \leq |v|)) \wedge (\forall i \leq |v|, i \in u \equiv (i \in x \wedge i \leq |y|))$$

which is again Δ_1^b as x, y and u, v are unique. We also define the function the ith *letter in word a*

$$\mathrm{Letter}(a, i) := \begin{cases} 1 & \text{if } \exists u, v \leq a, \langle u, v \rangle = a \wedge i \in u \wedge i \leq |v| \\ 0 & \text{otherwise} \end{cases}$$

The idea of coding sequences of words is that a sequence will be coded by a pair $\langle a, b \rangle$, where the ith bit 1 in b marks the end of the ith subword of a: That is, a sequence $w_1, \ldots, w_t$ of t words will be coded by number a whose binary expansion is $w_t \frown \ldots \frown w_1$ and number b, which has bit 1 in positions $|w_1|$, $|w_1| + |w_2|, \ldots, |w_1| + \cdots + |w_t| = |a|$.

This idea requires that we must be able to define the function *counting the number of ones among the first i bits of a*

$$\text{Numones}(a, i) := |\{j \mid j \le i \wedge j \in a\}|$$

Define

Numones$(a, i) = k$ iff

$$\exists x \le (a\#a)\#a \forall u \le |a|, \langle 1, u \rangle \in x \equiv (u \in a \wedge \forall v < u, v \notin a)$$
$$\wedge \, \forall t, u, v \le |a|, \langle t, u \rangle \in x \wedge u < v \wedge v \in a \wedge (\forall s < v, u < s \to s \notin a)$$
$$\to \, \langle t + 1, v \rangle \in x \wedge \exists u \le i, \langle k, u \rangle \in x \wedge \forall u \le i, \langle k + 1, u \rangle \notin x$$

In words: x codes an increasing map from $\{1, \ldots, k\}$ onto the 1's of a. Such an x is unique; hence the definition is Δ_1^b, and the inductive character of the definition of x allows us to prove basic inductive properties of Numones(a) (see previous discussion).

We are ready to define *sequences* and the function $(w)_i$

$$\text{Seq}(w) \equiv \exists x, y \le w, \langle x, y \rangle = w \wedge |x| = |y|$$

and for w a sequence

$$(w)_i = u \equiv \exists x, y \le w, \langle x, y \rangle = w \wedge \forall t \le |u| \forall j, \qquad k \le |x|,$$

$$\text{Numones}(y, j) = i - 1 \wedge \text{Numones}(y, k) = i \to k = j + |u|$$
$$\wedge \, t \in u \equiv (j + t) \in x$$

Lemma 5.4.1. *The function* $(w)_i = u$ *is* Δ_1^b*-definable in* S_2^1 *and* S_2^1 *proves the conditions of Lemma 5.1.5.*

We shall conclude this section by a lemma stating that some predicates can be in a sense coded in S_2^i. It extends Lemma 5.2.12.

Lemma 5.4.2. *Let* $A(a)$ *be a* $\Sigma_0^b(\Sigma_i^b)$ *-formula, that is, a formula obtained from* Σ_i^b*-formulas by logical connectives and sharply bounded quantification. Let* $i \ge 1$. *The theory* S_2^i *proves*

$$\forall x \exists y \forall t \le |x|, A(t) \equiv (t \in y)$$

That is: Any bounded set of lengths defined by a $\Sigma_0^b(\Sigma_i^b)$*-formula can be coded by a number.*

Proof. It is enough to show that for $A \in \Sigma_i^b$, which also implies that any $\Sigma_0^b(\Sigma_i^b)$-predicate can be (on any interval $[0, |x|]$) expressed as Δ_1^b whose coding follows from the case $i = 1$.

Consider a Σ_i^b-formula $B(s)$ with parameter x

$$\exists y \leq x \forall t \leq |x|, (t \in y \rightarrow A(t)) \wedge |y| = s$$

Clearly $B(0)$ holds as y corresponds to the empty set, and so by the Σ_i^b-LENGTH-MAX principle available in S_2^i by Lemma 5.2.7 there is maximal $s \leq |x|$ satisfying B. It is straightforward to verify that y corresponding to this s codes A on interval $[0, |x|]$.   Q.E.D.

Corollary 5.4.3. *For $i \geq 1$ the theory S_2^i proves the $\Sigma_0^b(\Sigma_i^b)$-PIND scheme.*

An interesting topic related to coding are partial truth definitions; see Paris and Dimitracopoulos (1982).

5.5. Second order systems

In this section we shall introduce some second order systems of bounded arithmetic, most from Buss (1986). We shall, however, proceed by model-theoretic reasoning rather than by direct proof-theoretic investigations. This will allow us to give simple model-theoretic proofs for the so-called RSUV isomorphism and translate several results from the previous section directly to these systems. It also allows us to relate the use of a second order object to the limited use of exponentiation.

Consider M a nonstandard model of S_2 and define a particular cut $I \subseteq_e M$ by

$$\text{for } a \in M : a \in I \quad \text{iff} \quad M \models \exists x, a = |x|$$

For the obvious reason we shall denote this cut by $\text{Log}(M)$. This cut is closed under addition and multiplication (as $|a| \cdot |b| + 1 = |a\#b|$), but it is not necessarily closed under $\#$ (that would require that M is closed under $\omega_2(x)$, which we do not assume).

Take a collection $\mathcal{X}_M$ of those subsets of $\text{Log}(M)$ *coded* in model M, that is, those $\alpha \subseteq \text{Log}(M)$ such that for some $a \in M$

$$\forall i \in \text{Log}(M), (M \models i \in \alpha) \equiv (\text{bit}(a, i) = 1)$$

We shall denote such an α by $\tilde{a}$.

Consider now the two-sorted first order structure $(\text{Log}(M), \mathcal{X}_M)$ with all symbols of $L \setminus \{x\#y\}$ defined for elements of $\text{Log}(M)$ by restricting the operations and relations from M, with $=$ defined on $\mathcal{X}_M$ and with the relation $i \in \alpha$ defined for pairs from $\text{Log}(M) \times \mathcal{X}_M$ by the preceding condition. We call elements of the second sort $\mathcal{X}_M$ sets.

We call our systems *second order* although we treat them only as two-sorted first order structures (as the underlying logic always remains just the first order predicate calculus). In second order logic the underlying logic has to assume various principles about sets and some set theory. In fact, there is no specific second order logic. We shall, however, call our systems second order, to honor the established terminology.

Second order bounded arithmetic theories U_j^i and V_j^i were introduced in Buss (1986). In this section we define these theories and obtain some basic information about their strength. L_2 is a second order language whose first order part is L, with second order variables $\alpha^t, \beta^s, \ldots$ ranging over finite sets of numbers, where $t, s, \ldots$ are first order terms, and with a membership relation $x \in \alpha^t$. The superscript t in α^t is introduced for technical reasons as an explicit upper bound to elements in α^t; we will mostly omit the superscript as such upper bounds are implicit in (proofs of) bounded formulas, and we shall display it only to simplify the presentation. More systems are introduced in Buss (1986), using varying language (whether variables for functions are included, or variables for unbounded sets, etc). In accordance with the notation of Buss (1986) the systems we shall introduce should be called $\tilde{U}_j^i(\text{BD})$ or $\tilde{V}_j^i(\text{BD})$, but we shall abuse the original notation slightly and adopt the simplest notation. In any case, all these theories prove the same bounded formulas.

Definition 5.5.1. Bounded second order formulas *are formulas of L_2 all of whose first order quantifiers are bounded.*

$\Sigma_0^{1,b}$-*formulas are bounded formulas without second order quantifiers.*

The classes of $\Sigma_i^{1,b}$-formulas and $\Pi_i^{1,b}$-formulas are classes of bounded formulas defined analogously to classes Σ_i^b and Π_i^b, counting the number of alternations of second order quantifiers and not counting the first order quantifiers.

A $\Sigma_i^{1,b}$-formula is $\Delta_i^{1,b}$ (resp. $\Delta_i^{1,b}$ in a theory T) if it is equivalent to a $\Pi_i^{1,b}$-formula in predicate calculus (resp. equivalent to it in T). In particular, all $\Sigma_0^{1,b}$-formulas are $\Delta_1^{1,b}$.

Definition 5.5.2. *The theory $I\Sigma_0^{1,b}$ is a theory in language L_2 with the following axioms:*

1. *BASIC*

2. *the extensionality axiom*

$$\forall \alpha^{t(x)}, \beta^{s(x)}, (\forall y \le t(x) + s(x), y \in \alpha \equiv y \in \beta) \to \alpha = \beta$$

3. *axioms stating that all sets are bounded*

$$\forall \alpha^{t(\overline{x})} \forall \overline{x}, y;\ y \in \alpha^t \to y \le t(\overline{x})$$

4. $\Sigma_0^{1,b}$-*IND scheme*

5. bounded comprehension scheme *for $\Sigma_0^{1,b}$-formulas*

$$\Sigma_0^{1,b} - CA$$

$$\forall x \exists \psi^x \forall y < x, y \in \psi^x \equiv A(y), \qquad \text{for all } A \in \Sigma_0^{1,b}$$

Definition 5.5.3. *The theory V_1^i is the theory of all structures $(\mathrm{Log}(M), \mathcal{X}_M)$ obtained from all models M of the theory S_2^i.*

Lemma 5.5.4. *The theory V_1^i is equivalent to the theory $I\Sigma_0^{1,b}$ extended by the IND scheme for all #-free $\Sigma_i^{1,b}$-formulas.*

Proof. Let M be a model of S_2^i and $(\mathrm{Log}(M), \mathcal{X}_M)$ be a model of V_1^i, defined earlier. We want to show first that in all such structures the theory $I\Sigma_0^b$ holds and $\Sigma_1^{1,b}$-IND holds (for formulas in the language of V_2^i).

Note that statement like $u \in \alpha^t$ or $\alpha^t = \beta^s$ in $(\mathrm{Log}(M), \mathcal{X}_M)$ can be equivalently stated in M by Δ_1^b-formulas

$$\mathrm{bit}(a, u) = 1 \wedge u \leq t \quad \text{or} \quad a = b$$

for $\alpha = \tilde{a}$ and $\beta = \tilde{b}$, first order bounded quantifiers $\exists u \leq v$ and $\forall u \leq v$ in $(\mathrm{Log}(M), \mathcal{X}_M)$ are equivalent to sharply bounded ones $\exists u \leq |2^v|$ and $\forall u \leq |2^v|$ in M, as 2^v exists in M, and second order quantifiers $\exists \alpha^t$ and $\forall \beta^s$ translate to first order bounded quantifiers in M: $\exists a \leq 2^t$ and $\forall b \leq 2^s$.

In this view the first three axioms of $I\Sigma_0^{1,b}$ are trivially satisfied. To see that bounded comprehension scheme 5 holds, let $A(x)$ be a $\Sigma_0^{1,b}$ with x the only free variable and with possibly some other first or second order parameters from $(\mathrm{Log}(M), \mathcal{X}_M)$, and let $u \in \mathrm{Log}(M)$. Then the formula

$$\exists \psi^a \forall y < u, y \in \psi^a \equiv A(y)$$

translates in M into a Σ_1^b-formula (as A translates into a Δ_1^b formula) that clearly holds for $a = 0$ and satisfies the induction assumption. By Σ_1^b-LIND in M then holds for any length, in particular for $u = |2^u| \in \mathrm{Log}(M)$.

To verify the IND scheme for $\Sigma_1^{1,b}$-formulas let $A(y)$ be a $\Sigma_1^{1,b}$-formula, with y the only free variable and with parameters. The formula $A(y)$ translates in M into a Σ_1^b-formula $A^*(y)$, and the assumption

$$(\mathrm{Log}(M), \mathcal{X}_M) \models A(0) \wedge \forall x, A(x) \rightarrow A(x+1)$$

implies that

$$M \models A^*(0) \wedge \forall x \leq |2^u|, A^*(x) \rightarrow A^*(x+1)$$

holds for all $u \in \mathrm{Log}(M)$.

By Σ_1^b-LIND in M there follows

$$M \models A^*(u)$$

which gives back

$$(\mathrm{Log}(M), \mathcal{X}_M) \models A(u)$$

This proves that V_1^i contains $I\Sigma_0^{1,b} + \Sigma_1^{1,b}$-IND.

For the opposite inclusion let $(K, \mathcal{X})$ be a model of $I\Sigma_0^{1,b} + \Sigma_1^{1,b}$-IND. We want to show that $(K, \mathcal{X})$ also satisfies V_1^i. To this end we shall define a model M of S_2^i such that

$$(K, \mathcal{X}) = (M, \mathcal{X}_M)$$

The idea of the construction is to use pairs $(a, \alpha^b) \in K \times \mathcal{X}$ to code elements of M; specifically, (a, α^b) would code an element with the value $\sum_{i < a, i \in \alpha^b} 2^i$. We shall omit the superscripts of the second order variables.

Claim 1. *There are relations $\Delta_1^{1,b}$-definable in V_1^i*

$$R_=((a_1, \alpha_1), (a_2, \alpha_2)), \qquad R_\le ((a_1, \alpha_1), (a_2, \alpha_2))$$

and

$$R_f ((a_1, \alpha_1), \ldots, (a_{k+1}, \alpha_{k+1}))$$

one for each k-ary function symbol f of the language L, such that V_1^1 proves the translations of all axioms of BASIC obtained by replacing $=$ and $\le$ by $R_=$ and $R_\le$, and $f(a_1, \ldots, a_k) = a_{k+1}$ by R_f. V_1^1 also proves all translations of the equality axioms.

The relation $R_=$ is $\Sigma_0^{1,b}$-definable by utilizing the extensionality axiom, and the relation $R_\le$ is also $\Sigma_0^{1,b}$-definable by formalizing the lexicographic ordering of sets. The relation R_f is defined as a formalization of computations of bits of $f(x_1, \ldots, x_k)$ from bits of $x_1, \ldots, x_k$. In the case of $\lfloor (x/2) \rfloor$ this is trivial, in the case of addition and multiplication this is a formalization of the table of computations, and in the cases of $|x|$ and $x \# y$ this involves formalizing computations of the values of the functions via the basic recurrence properties (in fact, the same applies to all polynomial time functions introduced by Cobham definition 5.3.1). These computations are unique and hence the defined relations will be $\Delta_1^{1,b}$, but then we need bounded comprehension for $\Delta_1^{1,b}$-formulas.

The formula

$$\exists \psi^a \forall y < a, \; y \in \psi^a \equiv A(y)$$

is $\Sigma_1^{1,b}$ if A is $\Delta_1^{1,b}$ and is easily proved by induction on a; hence bounded $\Delta_1^{1,b}$-CA is provable in V_1^1. The details are left to the reader.

Let M be the structure $(K, \mathcal{X})/R_=$ with the symbols of L interpreted by the relations of the claim. Numbers $u \in K$ can be best represented by pairs $(|u|, \alpha_u)$ with $\alpha_u := \{i_0 < \cdots < i_k\}$ for which $K \models u = 2^{i_0} + \ldots 2^{i_k}$.

Claim 2. *K is isomorphic to* $\text{Log}(M)$.

To see this one needs to show in $(K, \mathcal{X})$

$$(b, \beta)R_=(|a|, \alpha) \rightarrow \exists c \le a, (b, \beta)R_=(|c|, \alpha_c)$$

and

$$R_f\left((|a_1|, \alpha_{a_1}), \ldots, (|a_{k+1}|, \alpha_{a_{k+1}})\right) \rightarrow f(a_1, \ldots, a_k) = a_{k+1}$$

Both implications are readily proved by $\Delta_1^{1,b}$-IND on a (resp. on the sum $a_1 + \ldots \alpha_{k+1}$), utilizing *Claim 1*. Moreover, the length of every $(a, \alpha)/R_= \in M$ is a (i.e., $\text{Log}(M) \subseteq K$), and for every $a \in K$, $M \models a = |(a + 1, \{a\})/R_=|$. That is, $K \subseteq \text{Log}(M)$.

Claim 3. *The model M satisfies* S_2^i.

Let A be a Σ_1^b-formula with parameters from M. Using the definition of M it can be translated into a $\Sigma_1^{1,b}$-formula $A'((a, \alpha))$ with parameters from $(K, \mathcal{X})$. Assume that A satisfies in M

$$A(0) \wedge \left(\forall x < a, A\left(\left\lfloor \frac{x}{2} \right\rfloor\right) \rightarrow A(x)\right) \wedge \neg A((a, \alpha)/R_=)$$

Then $(K, \mathcal{X})$ satisfies

$$A'((1, \{0\})) \wedge \left(\forall x < a \forall \phi, A'("\lfloor(x, \phi)/2\rfloor") \rightarrow A'((x, \phi))\right) \wedge \neg A'((a, \alpha))$$

Now "$\lfloor(x, \phi)/2\rfloor$" is just $(x - 1, \phi)$ and w.l.o.g. $(1, \alpha)R_=(1, \{0\})$, so specifically we have

$$(K, \mathcal{X}) \models B(1) \wedge (\forall x \le a, B(x) \rightarrow B(x + 1)) \wedge \neg B(a)$$

for a $\Sigma_1^{1,b}$-formula $B(x)$

$$B(x) := A'((x, \alpha))$$

But this contradicts $\Sigma_1^{1,b}$-IND in $(K, \mathcal{X})$. This proves the claim.　　　Q.E.D.

Definition 5.5.5.
(a) *The theory V_2^i extends V_1^i by augmenting the language by the function symbol # and accepting the axiom scheme IND for $\Sigma_i^{1,b}$-formulas in the extended language.*
(b) *The theory U_j^i, $j = 1, 2$, extends the theory $I\Sigma_0^{1,b}$ by $\Sigma_1^{1,b}$-PIND scheme in the language $L \setminus \{\#\}$ for $j = 1$ and L for $j = 2$.*

The next lemma is entirely analogous to Lemma 5.2.8.

Lemma 5.5.6. *The theory V_2^i contains U_2^i and U_2^{i+1} contains V_2^i.*

Lemma 5.5.7. *The theory* U_1^i *proves the* separation scheme $\Pi_i^{1,b}$-SEP *scheme*

$$(\forall x, \neg A(x) \vee \neg B(x)) \rightarrow \forall x \exists \psi^x \forall y < x,$$

$$(A(y) \rightarrow y \in \psi) \wedge (y \in \psi \rightarrow \neg B(y))$$

with $A, B \in \Pi_i^{1,b}$.

Proof. Consider a formula $C(a, b)$.

$$\forall s \leq b \exists \psi^b \forall y \leq b; s \leq y \leq s+a \rightarrow \left((A(y) \rightarrow y \in \psi^b) \wedge (y \in \psi^b \rightarrow \neg B(y))\right)$$

Assuming $\forall x, \neg A(x) \vee \neg B(x)$, the theory U_1^1 proves

$$C\left(\left\lfloor \frac{a}{2} \right\rfloor, b\right) \rightarrow C(a, b)$$

as it can prove that two sets have a union.

$C(0, b)$ is trivial. By $\Sigma_1^{1,b}$-PIND then there follows the formula $C(b, b)$, and the instance of $\Pi_i^{1,b}$-SEP scheme follows.   Q.E.D.

The separation scheme immediately implies a form of comprehension scheme.

Lemma 5.5.8. *The theory* U_1^i *proves the* $\Delta_i^{1,b}$-CA *scheme*

$$(\forall x, A(x) \equiv B(x)) \rightarrow \forall x \exists \psi \forall y < x, y \in \psi \equiv A(y)$$

with $A \in \Sigma_i^{1,b}$ *and* $B \in \Pi_i^{1,b}$.

$\Delta_1^{1,b}$-CA over $I\Sigma_0^{1,b}$ readily gives the $\Delta_1^{1,b}$-IND scheme.

Lemma 5.5.9. *The theory* U_1^i *proves* $\Delta_i^{1,b}$-IND.

Lemma 5.5.10. *The theory* U_1^i *proves the* choice scheme $\Sigma_i^{1,b}$-AC:

$$\forall x < a \exists \psi A(x, \psi) \rightarrow \exists \varphi \forall x < a, A(x, (\varphi)x)$$

with $A \in \Sigma_i^{1,b}$ *and* $y \in (\varphi)_x$ *defined as* $[x, y] \in \varphi$, $[x, y]$ *a pairing function.*

Proof. The lemma is proved analogously as Lemma 5.5.7: show that the required φ exists when x is drawn from subintervals of $[0, a]$ of length $1, 2, 4, \ldots$. Hence $\Sigma_i^{1,b}$-LIND, which is available in U_1^i analogously with Lemma 5.2.5, is sufficient to obtain the statement.   Q.E.D.

A modified choice scheme is the scheme of the *dependent choice* $\Sigma_i^{1,b}$-DC:

$$\forall x < a \forall \varphi \exists \psi A(x, \varphi, \psi) \rightarrow \forall \varphi \exists \theta, (\theta)_0 = \varphi \wedge \forall x < a, A(x, (\theta)_x, (\theta)_{x+1})$$

with $A \in \Sigma_i^{1,b}$.

Lemma 5.5.11. $\Sigma_i^{1,b}$-DC *is provable in* V_2^i *and* $\Sigma_i^{1,b}$-DC $+I\Sigma_0^{1,b}$ *proves* V_2^i.

Proof. It is clear that $\Sigma_i^{1,b}$-DC is provable in V_2^i by induction on a in a $\Sigma_i^{1,b}$-formula with parameter φ

$$\exists \theta, (\theta)_0 = \varphi \wedge \forall x < a, A(x, (\theta)_x, (\theta)_{x+1})$$

To see the opposite direction assume that a $\Sigma_i^{1,b}$-formula $\exists \psi A(x, \psi)$, with $A \in \Pi_{i-1}^{1,b}$, satisfies

$$A(0, \psi_0) \wedge \forall x < a, (\exists \psi_x A(x, \psi_x)) \rightarrow (\exists \psi_{x+1} A(x+1, \psi_{x+1}))$$

The second conjuct can be written as

$$\forall x < a \forall \psi_x \exists \psi_{x+1}, A(x, \psi_x) \rightarrow A(x+1, \psi_{x+1})$$

which is the antecedent of an instance of DC scheme for the $\Delta_i^{1,b}$-formula $A(x, \psi_x) \rightarrow A(x+1, \psi_{x+1})$. By the DC scheme with $\varphi = \psi_0$ we get

$$\exists \theta, (\theta)_0 = \psi_0 \wedge \forall x < a, A(x, (\theta)_x) \rightarrow A(x+1, (\theta)_{x+1})$$

By $\Pi_{i-1}^{1,b}$-IND on the formula $A(x, (\theta)_x)$ then $\exists \psi A(a, \psi)$ follows.

The argument so far shows that $\Sigma_i^{1,b}$-DC $+\Pi_{i-1}^{1,b}$-IND proves $\Sigma_i^{1,b}$-IND. But analogously with Lemma 5.2.5 $\Pi_{i-1}^{1,b}$-IND is implied by $\Sigma_{i-1}^{1,b}$-IND. We may then repeat the same argument for $i-1$ in place of i to derive $\Sigma_{i-1}^{1,b}$-IND in $\Sigma_{i-1}^{1,b}$-DC $+\Sigma_{i-2}^{1,b}$-IND and hence also $\Sigma_i^{1,b}$-IND in $\Sigma_i^{1,b}$-DC $+\Sigma_{i-2}^{1,b}$-IND. Iterating this process i-times gets the wanted result.   Q.E.D.

We shall now state precisely the relation between first order subsystems of S_2 and second order systems like U_1^i, V_1^i implicit in the proof of Lemma 5.5.4.

Definition 5.5.12. *The theory R_2^i is a first order theory in the language L axiomatized by BASIC and the axiom scheme*

$$\phi(0) \wedge \forall x \left(\phi \left(\left\lfloor \frac{x}{2} \right\rfloor \right) \rightarrow \phi(x) \right) \rightarrow \forall x \phi(|x|)$$

for all Σ_i^b-formulas ϕ.
    R_2 is the theory $\bigcup_i R_2^i$.

Analogously with Lemma 5.2.8 $S_2^{i-1} \subseteq R_2^i \subseteq S_2^i$, hence $R_2 = S_2$, holds. The relation of T_2^{i-1} to R_2^i is open.

The reason for defining a theory like this is the following result.

Theorem 5.5.13 (Takeuti 1993, Razborov 1993). *There exists a translation of second order bounded formulas into first order bounded formulas*

$$A \in \Sigma_\infty^{1,b} \mapsto A^1 \in \Sigma_\infty^b$$

and a translation of first order bounded formulas into second order bounded formulas

$$B \in \Sigma^b_\infty \mapsto B^2 \in \Sigma^{1,b}_\infty$$

having the following properties:

1. *if* $S^i_2 \vdash B$ *then* $V^i_1 \vdash B^2$
2. *if* $V^1_1 \vdash A$ *then* $S^1_2 \vdash A^1$

and

3. $S^1_2 \vdash B \equiv (B^2)^1$
4. $V^1_1 \vdash A \equiv (A^1)^2$

The same properties hold for R^i_2 *and* U^i_1 *in place of* S^i_2 *and* V^i_1, *respectively.*

We shall omit the proof as it is a direct generalization of the argument in the proof of Lemma 5.5.4. Proof-theoretic arguments can be found in Takeuti (1993) and Razborov (1993). The pair of translations is called *RSUV-isomorphism.*

We shall use this theorem in Chapter 8 in characterizing functions and functionals definable in second order systems.

Note that the second order *choice principle* AC translates in the RSUV-isomorphism into the sharply bounded collection scheme BB; hence, for example, a result like Lemma 5.2.12 is essentially equivalent to Lemma 5.5.10. We shall see more examples of such translations of statements.

A very important closure property is the closure under *counting functions* (cf. Section 2.2).

Let a $\Sigma^{1,b}_0$-formula Enum(f, u, α) denote "f is an increasing bijection from u onto α," that is, the conjunction of conditions

1. $\forall x < y < u, f(x) < f(y)$
2. $\forall x < u, f(x) \in \alpha$
3. $\forall z \in \alpha \exists x < u, f(x) = z$

Lemma 5.5.14. U^1_1 *proves that every set can be enumerated in increasing order*

$$\forall \alpha \exists u \exists f, \ Enum(f, u, \alpha)$$

Proof. Let a bound all elements of α and consider the formula

$$A(t) := \forall x, y \le a \exists \beta \exists f, u, \ y \le x + 2^t$$
$$\to (\beta = \{z \in \alpha | x \le z \le y\} \wedge \ Enum(f, u, \beta))$$

This is a $\Sigma^{1,b}_1$-formula clearly satisfying induction assumptions $A(0)$ and $\forall t$, $A(t) \to A(t+1)$, and hence by $\Sigma^{1,b}_1$-LIND $A(|a|)$ holds: That is, f, u witnessing $A(|a|)$ for $x = 0$ and $y = a$ increasingly enumerate α.   Q.E.D.

The lemma implies that in U^1_1 one can $\Delta^{1,b}_1$-define the cardinality of a set

$$|\alpha| := u \quad \text{iff} \quad \exists f, \ Enum(f, u, \alpha) \ \text{iff} \ \forall f, v, \ Enum(f, v, \alpha) \to u = v$$

and hence also all combinatorial principles provable by rudimentary counting are provable in U_1^1. We shall return to this in Chapters 13 and 15.

We shall conclude this section by characterizing a limited use of exponentiation in first order proofs using second order systems. This is of interest as many combinatorial or number-theoretic arguments using exponentiation (or equivalently power set operation) use it only once. For example, the usual proof of the infinitude of primes requires one to compute the number $n! + 1$, or a combinatorial argument requires one to consider the set of all graphs on a given vertex set of size n and so forth. In the following discussion we will consider only a particular case of the theory S_2^1, but the arguments apply to other subsystems as well.

Definition 5.5.15. *The set $S_2^1 + 1-Exp$ consists of all Σ_∞^b- formulas $\phi(a)$ such that there is a term $t(a)$ for which S_2^1 proves the implication*

$$t(a) < |c| \to \phi(a)$$

where c is a free variable not occurring in $t(a)$ or ϕ.

Theorem 5.5.16. *Let $\phi(a)$ be a first order bounded formula. Then*

$$\phi(a) \in S_2^1 + 1-Exp \quad iff \quad V_2^1 \vdash \phi(a)$$

Proof. Assume first that $V_2^1 \nvdash \phi(a)$. That is, there is a model of V_2^1 such that

$$(K, \mathcal{X}) \models \neg\phi(m)$$

for some $m \in K$.

The construction form the proof of Lemma 5.5.4 gives a model $M \models S_2^1$ with $\text{Log}(M) = K$. Assume

$$S_2^1 \vdash t(a) < |c| \to \phi(a)$$

As $(K, \mathcal{X}) \models V_2^1$, $t(m) \in K$ so there is $2^{t(m)} \in M$, and hence $\phi(m)$ should hold in M. But by the claim in the proof of Lemma 5.1.3 the truth value of $\phi(m)$ has to be the same in K and M: a contradiction.

Assume now $\phi(a) \notin S_2^1 + 1-$ Exp. By compactness we can find a model M of S_2^1 with a cut $I \subseteq_e M$ such that
  (i) $I \models S_2^1$
  (ii) $\exists c \in M \setminus I \; \forall b \in I, M \models 2^b < c$
Consider a family $\mathcal{X}$ of subsets $\alpha \subseteq I$ coded by some $a \in M$, $M \models a \le c$.

We claim that $(I, \mathcal{X})$ is a model of V_2^1. This is because in the translation of second order bounded formulas over $(I, \mathcal{X})$ into first order ones over M, second order quantifiers $\exists\phi, \forall\phi$ can be replaced by $\exists a \le c$ and $\forall a \le c$, and bounded first order ones into sharply bounded quantifiers of the form $\exists x \le |c|, \forall x \le |c|$. Thus $\Sigma_1^{1,b}$-IND in $(I, \mathcal{X})$ follows from Σ_1^b-LIND in M. Q.E.D.

We refer the reader to Krajíček (1990) for more on this topic.

5.6. Bibliographical and other remarks

Definition 5.1.1 is from Parikh (1971) (he used Q in place of PA^-). The theory $I\Delta_0$ was studied by Paris and Wilkie in a series of papers (e.g., Paris and Wilkie 1981a,b; 1987b) where its extensions by axioms Ω_k and Exp were also defined.

Definitions and most lemmas in Sections 5.2 up to 5.2.12 are due to Buss (1986, 1990a). Lemma 5.2.13 is due to Ressayre (1986). Definition of PV in 5.3 is due to Cook (1975), and definition of PV_i's follows Krajíček, Pudlák, and Takeuti (1991).

The presentation of second order systems in 5.5 follows the model-theoretic construction of Krajíček (1990), developed proof-theoretically in Takeuti (1993) and Razborov (1993). Originally the systems were defined in Buss (1986), who observed Lemma 5.5.6. Lemmas 5.5.7–5.5.10 are from Krajíček and Takeuti (1992).

Theories R_2^i were defined in Takeuti (1993). Theorem 5.5.13, Definition 5.5.15, and Theorem 5.5.16 are from Krajíček (1990).

6

Definability of computations

This chapter presents important definability results for fragments of bounded arithmetic.

A *Turing machine* M will be given by its set of states Q, the alphabet Σ, the number of working tapes, the transition function, and its clocks, that is, an *explicit time bound*. Most results of the form "Given machine M the theory T can prove ..." could be actually proved in a bit stronger form: "For any k the theory T can prove that for any M running in time $\leq n^k$ ...." A natural formulation for such results is in terms of models of T and computations within such models, but in this chapter we shall omit these formulations.

An *instantaneous description* of a computation of machine M on input x consists of the current state, the positions of the heads, the content of all tapes, and the current time: That is, it is a sequence of symbols whose length is proportional to the time bound for $n := |x|$.

A computation will be *coded* by the sequence of the consecutive instantaneous descriptions.

Now we shall consider several bounded formulas defining these elementary concepts. They are all Δ_1^b in the language L^+ and thus also (by Lemma 5.4.1) in L.

Subsequently oracles will be represented by formulas, but to make the presentation uniform we shall augment the language L by new unary predicate symbols $\alpha(x)$; the new language will be denoted $L(\alpha)$. Definitions of classes $\Sigma_i^b(\alpha)$, $\Pi_i^b(\alpha)$, $\Delta_i^b(\alpha)$ and theories $S_2^i(\alpha), T_2^i(\alpha)$ are straightforward generalizations of the original definitions for the language L. The characterizations from Section 3.2 generalize to $\Sigma_i^b(\alpha)$ too: For example, subsets of N defined by $\Sigma_1^b(\alpha)$-formulas are exactly those in NP^α, NP with oracle $\{n \mid \alpha(n)\}$.

Let M be an oracle Turing machine with the explicit time bound $t(n)$ coded in the description of M. The property that u is an *instantaneous description* of a computation of M is defined by a simple Δ_1^b-formula

$$\text{Instan}_M(u)$$

saying that u is a tuple of sequences, one coding the current state, several coding contents of tapes with positions of heads, one coding the content of the query tape, and one coding the current time.

Another Δ_1^b-formula

$$\text{Init}_M(x, u)$$

expresses that u is an initial instantaneous description of M with the input x (i.e., in the initial state and positions of heads, with x on the input tape, with empty working tapes and empty query tape, and time 0).

A *halting configuration* is defined by the Δ_1^b-formula

$$\text{Halt}_M(u)$$

The property that u and v are two *consecutive configurations* of a computation of M with oracle α is expressed by the formula

$$\text{Consec}_M(u, v, \alpha)$$

which is $\Delta_1^b(\alpha)$; it says that both u and v are instantaneous descriptions, the time in v is 1 plus the time in u, v was obtained from u by the transition function of M, and, in particular, if u was in a query state then v answers the query according to α.

Finally, the formula

$$\text{Comp}_M(x, w, \alpha) := \text{Init}_M(x, (w)_1) \wedge$$
$$\forall i \leq t(|x|), \ \text{Consec}_M((w)_i, (w)_{i+1}, \alpha) \wedge \exists j \leq t(|x|), \ \text{Halt}_M((w)_j)$$

expresses that w is a *computation* of M on the input x with the oracle α, and the function

$$\text{Output}_M(x, \alpha)$$

is the content of the first working tape in the halting configuration of a computation of M on x with the oracle α.

It will follow from results in Section 6.1 that all these notions are, in fact, $\Delta_1^b(\alpha)$ in $\text{PV}_1(\alpha)$.

6.1. Polynomial time with oracles

In this section we shall define polynomial time computations with oracles from the polynomial time hierarchy PH in fragments of S_2.

Lemma 6.1.1. *For every polynomial time Turing machine M the theory PV_1 proves*

$$\forall x \exists! w, \; \text{Comp}_M(x, w, \emptyset)$$

Proof. Let the time bound (which is an explicit part of the definition of M) be $\leq n^k$.

Then the function

$$f(x, y) := \text{ the sequence of first } |y| \text{ instantaneous}$$
$$\text{descriptions of the computation of } M \text{ on } x$$

is defined by the limited recursion on notation

$$f(x, 0) := u \text{ iff Init}_M(u)$$

and

$$f(x, y) := \begin{cases} f(x, \lfloor \frac{y}{2} \rfloor) \frown u & \text{where Consec}_M(f(x, \lfloor \frac{y}{2} \rfloor)_{|y|-1}, u) \\ f(x, \lfloor \frac{y}{2} \rfloor) & \text{if } |y| \geq |x|^k \end{cases}$$

with the implicit bound

$$|f(x, y)| \leq |x|^{O(k)}$$

that is,

$$f(x, y) \leq x \# \ldots \# x$$

The uniqueness of a computation is obviously provable by LIND for open formulas of PV_1, available in PV_1 by its definition. Q.E.D.

Theorem 6.1.2. *Let $i \geq 0$, let $\phi(a)$ be a Σ_i^b-formula and assume $T = PV_1$ if $i = 0$ and $T = T_2^i$ if $i > 0$. Assume that M is an oracle polynomial time Turing machine. Then theory T proves*

$$\forall x \exists! w, \; \text{Comp}_M(x, w, \phi(a))$$

Proof. Consider a formula $A(x, v)$ defined as

$$A(x, v) := \exists u, w \leq v; v = \langle u, w \rangle \wedge \text{Comp}_M(x, w, u) \wedge \forall i \leq |w|$$
$$\forall b \leq w, (\text{ "if the } i\text{th oracle query was } [\phi(b)?]\text{" } \wedge \text{ bit}(u, i) = 1)$$
$$\rightarrow \phi(b)$$

Formula A is Σ_i^b for $i > 0$ and Δ_1^b for $i = 0$ (as u, w are uniquely determined by v and b is uniquely determined by i, w). We think of u as coding the oracle $\{i \mid i \in u\}$.

By Lemma 6.1.1, PV_1 proves that there is w_0 such that $\text{Comp}_M(x, w_0, \emptyset)$ holds. Put $v_0 := \langle 0, w_0 \rangle$.

By Lemma 5.2.6 the theory T proves that there exists maximal $v_1 \leq 2^{n^k}$ satisfying $A(x, v)$; note that such v_1 is then bounded by a term in x. Let $v_1 :=$ $\langle u_1, w_1 \rangle$.

Claim. *The theory T proves that for any $i \leq |w_1|, b \leq w_1$, if $[\phi(b)?]$ was the ith oracle query then*

$$[\text{bit}(u_1, i) = 1] \equiv \phi(b)$$

The claim holds as all the positive answers of u_1 are correct by the definition of the formula A, and if some negative answer would be wrong, then we could change it into a positive one and all later answers into negative ones. In this way we would construct $u_2 > u_1$ such that for w_2 for which $\text{Comp}_M(x, w_2, u_2)$ holds the number $v_2 = \langle u_2, w_2 \rangle$ is bigger than v_1, contradicting the choice of v_1.

It follows from the *claim* that $\text{Comp}_M(x, w_1, \phi(a))$ holds. The uniqueness of w_1 is simple.   Q.E.D.

Corollary 6.1.3. *For all $i \geq 1$, the theory PV_{i+1} is fully conservative over T_2^i.*

Proof. The idea is to augment T_2^i by function symbols for all Σ_{i+1}^b-definable functions. By Theorem 6.1.2 all $\square_{i+1}^p$-functions are Σ_{i+1}^b-definable in T_2^i, hence every function of PV_{i+1} will have a counterpart in this conservative extension of T_2^i. Moreover, the defining equations for the PV_{i+1}-functions determine a polynomial time oracle Turing machine computing that function using the defining equation, and hence open induction on the computation proves in T_2^i the defining equation. Finally, open induction in PV_{i+1} translates into open induction in the extension of T_2^i, which in turn translates into Δ_{i+1}^b-IND, available in T_2^i by Lemma 5.2.2.
                                                                         Q.E.D.

Corollary 6.1.4. *For every $i \geq 0$ there is a Σ_{i+1}^b-formula $UNIV_{i+1}(x, y, z)$ such that for any Σ_{i+1}^b-formula $\phi(a)$ there are a term $t(a)$ and a natural number e such that S_2^1 proves the equivalence*

$$\phi(a) \equiv UNIV_{i+1}(a, t(a), \underline{e})$$

($\underline{e}$ is the numeral of e).

Proof. The idea of the proof is to use the universal Turing machine. The problem is that this machine has to be itself in polynomial time, but then it cannot simulate machines running in a greater polynomial time. This is repaired by an extra input, produced by term $t(a)$: if the time of the universal machine is $p(n)$ and time of the machine to be simulated is $q(n)$, then we need term $t(a)$ so that

$$q(|a|) \leq p(|t(a)|)$$

Number e codes the description of a machine computing ϕ.   Q.E.D.

6.2. Bounded number of queries

Bounded query classes are defined similarly to oracle computations except that the machine is equipped with an extra bound to the number of oracle queries it can ask (cf. Krentel 1986). We shall consider only one type of such classes.

Definition 6.2.1. *Class $P^{\Sigma_i^p}[O(\log n)]$ is the class of languages recognized by a polynomial-time Turing machine querying at most $O(\log n)$-times a Σ_i^p-oracle.*

Theorem 6.2.2. *Let $i \geq 1$, let $\phi(a)$ be a Σ_i^b-formula, and let M be a bounded query oracle polynomial time Turing machine allowing $O(\log n)$ queries. Then the theory S_2^i proves*

$$\forall x \exists! w, \ Comp_M(x, w, \phi(a))$$

Proof. The proof of this theorem is completely analogous to the proof of Theorem 6.1.2 except that we can use the *length-maximum principle* in the *Claim* instead of the original *maximum principle*, which is by Lemma 5.2.6 provable in S_2^i. This is because the length of u's is now bounded by the log of the $|x|$; that is, u's are sharply bounded.   Q.E.D.

Although the definitions of these classes seem a bit ad hoc, they, in fact, coincide with several other, perhaps more naturally defined classes.

Theorem 6.2.3. *The class of sets $P^{NP}[O(\log n)]$ is identical with the following three classes:*

(i) sets log-space Turing reducible to SAT: L^{NP}
(ii) sets truth-table reducible to SAT: $\leq_{tt} (NP)$
(iii) sets definable by $\Sigma_0^b(\Sigma_1^b)$-formulas.

The interest in considering these classes in connection with bounded arithmetic stems from Corollary 7.3.6.

6.3. Interactive computations

In this section we define the notion of *counterexample computations*, and then we formalize a slightly more general notion in 6.3.2.

The intended environment for the counterexample computations are *optimization problems*. An *optimization problem* is a binary relation $R(x, y)$ and a function $c(x)$ which are both polynomial time computable and where the relation $R(x, y)$ satisfies

$$R(x, y) \rightarrow |y| \leq |x|^{O(1)}$$

Call any y such that $R(x, y)$ holds a *solution* to x, and given two solutions y_1 and y_2 to x call y_1 *better* than y_2 if $c(y_1) > c(y_2)$. A solution is *optimal* if there is no better one.

There are two prominent examples: the optimization problem CLIQUE with the relation

$$R(x, y) := \text{``}y \text{ is a clique in the graph } x\text{''}$$

with the cost function

$$c(y) := \text{``the size of } y\text{''}$$

and the problem TSP with the relation

$$R(x, y) := \text{``}y \text{ is a tour in the graph } x \text{ with}$$
$$\text{the edges labeled by natural numbers''}$$

and the cost function

$$c(y) := \text{``the sum of the labels of the tour } y\text{''}$$

We refer the reader to Krentel (1986) and Papadimitriou and Yannakakis (1988) for alternative definitions and treatment of optimization problems.

A specific way of searching for an optimal solution requires *interaction* between two players: *the teacher* and *the student*. The student will be a polynomial time machine, whereas the teacher has unlimited ability: It is an arbitrary function.

On input x the student computes a solution $y_1 : R(x, y_1)$, the candidate for an optimal solution. If y_1 is optimal then the teacher says so and the computation is finished with the output y_1. Otherwise the teacher presents the student with a better solution $y_2 : c(y_2) > c(y_1)$. Now knowing x and y_2 the student produces the solution y_3 and learns from the teacher that it is optimal or gets a better solution y_4.

In this way the interaction proceeds until the student finds an optimal solution. The total computational time of the student must be polynomial in the length of x.

In the example CLIQUE the student has a trivial strategy; first he produces a trivial clique consisting of one vertex and then he just repeats the teacher's counterexamples ($y_3 := y_2$, $y_5 := y_4$, ...). As there are only $\leq |x|$ possible cost values he always outputs a clique of maximal size. Note that a similar strategy fails in the second example TSP as there are, in general, exponentially many cost values.

There is a natural reducibility notion between optimization problems in which TSP is complete and in which CLIQUE is complete among those optimization problems whose cost function is bounded by a polynomial: $c(y) = |x|^{O(1)}$. One can even prove, assuming that NP does not have polynomial size circuits, a hierarchy theorem for computations allowing at most $f(|x|) \leq |x|^{1-\epsilon}$ counterexamples in a single computation. We refer the reader to Krajíček, Pudlák, and Sgall (1990) for details and proofs.

We shall now generalize the counterexample computations to a slightly more general context. Assume $\forall x \exists y \forall z, A(x, y, z)$ holds, with A a polynomial time predicate and with a polynomial bound $\leq t(x)$ to y and z implicit in A. Given a

the student should compute b such that $\forall z, A(a, b, z)$. A counterexample is any $c \leq t(a)$ for which $\neg A(a, b, c)$. Thus the student produces $b_1 \leq t(a)$ and either learns that it satisfies $\forall z, A(a, b_1, z)$ or receives from the teacher $c_1 \leq t(a)$ such that $\neg A(a, b_1, c_1)$. Having a, c_1 the student computes $b_2 \leq t(a)$ and again either learns that it is a solution or gets a counterexample c_2, and so on.

Note that the original situation with optimization problems is contained in this scheme as the property "b is an optimal solution to a" can be expressed as $\forall z, A(a, b, z)$: That is, it is a coNP-property.

A counterexample to the claim that $\forall z, A(a, b, z)$ holds is, in fact, a witness to the existential quantifier in $\exists z, \neg A(a, b, z)$. Hence the following notion of computation (cf. Krajíček 1993, Buss et al. 1993) generalizes the counterexample computations.

Definition 6.3.1. *A polynomial time Turing machine M with a* witness-oracle *$Q(x) = \exists y \leq t(x) R(x, y)$ is a polynomial time machine with a query tape for queries to Q that answers a query a as follows:*

1. *if $\exists y \leq t(a) R(a, y)$ holds, then it returns YES and some $b \leq t(a)$ such that $R(a, b)$: That is, it returns a witness to the affirmative answer.*
2. *if $\forall y \leq t(a) \neg R(a, y)$ holds, then it returns NO.*

Note that since there may be multiple witnesses to affirmative oracle answers, the computation is not uniquely determined and witness-oracle Turing machines thus generally compute only *multivalued* functions rather than functions. A multivalued k-ary function $f(\overline{x})$ is just a $(k + 1)$-ary relation $F(\overline{x}, y)$. We call any y such that $F(\overline{x}, y)$ a *value* of $f(\overline{x})$.

Definition 6.3.2. *Let $i \geq 1$ and let $q(n)$ be a function.*
$FP^{\Sigma_i^p}[wit, q(n)]$ is the class of multivalued functions f computable by a polynomial time witness-oracle Turing machine such that

1. *on an input of length n the machine M makes at most $q(n)$ oracle queries*
2. *the witness-oracle has the form*

$$\exists y \leq t(a), R(a, y)$$

with $R \in \Pi_{i-1}^b$ if $i > 1$ and in Δ_1^b if $i = 1$
3. *on an input x the machine M outputs some y such that $f(x) = y$*

Note that we do not require that all possible values of $f(x)$ appear as outputs of some computations. Also observe that without changing the strength of such machines we may allow the machine unlimited access to a Σ_{i-1}^p-oracle. This is because a question whether there is a computation with correct answers of a Σ_{i-1}^p-oracle is a Σ_i^p-query that always has a unique witness. Hence such a computation may be found by a single query to a Σ_i^p-witness-oracle.

Let

$$\text{WitComp}_M(x, w, Q)$$

formalizes that "w *is a computation of a witness-oracle machine M on an input x with an oracle Q.*"

Theorem 6.3.3. *Let* $i \geq 1$. *Every function* f *from* $\text{FP}^{\Sigma_i^p}[\text{wit}, c \cdot \log n]$ *is* Σ_{i+1}^b-*definable in* S_2^i.

That is, for each such function f *there are a polynomial time witness-oracle machine M and* Σ_i^p-*witness-oracle Q such that* S_2^i *proves*

$$\forall x \exists w, \ \text{WitComp}_M(x, w, Q)$$

Proof. Let $Q(a) = \exists y \leq t(a)$, $R(a, y)$ and assume R is a Δ_i^b-formula w.r.t. S_2^i. Let $p(n)$ be the time bound of the machine M computing f with witness-oracle Q. Define a formula $A(a, h, w)$ to be the conjunction of the following four conditions:

1. w satisfies all conditions posed on a computation of M on a with possibly incorrect oracle answers
2. $h = \langle (i_1, j_1), \ldots, (i_r, j_r) \rangle$ for $r \leq c \cdot ||a||$ such that $i_1 < i_2 < \ldots i_r \leq |w|$ and $j_1, \ldots, j_r = 0, 1$
3. oracle query in steps i_s, $s = 1, \ldots, r$, is answered YES if $j_s = 1$ and NO if $j_s = 0$
4. whenever $[Q(u_s)?]$ is the query in step i_s and $j_s = 1$ then w_s is a witness to it and it is a part of the instantaneous description

The first three conditions are Δ_1^b, the last is Δ_i^b, so the formula A is Δ_i^b.

Claim. *The theory* S_2^1 *proves that there is a lexicographically maximal e of the form*

$$e = \langle j_1, \ldots, j_r \rangle$$

such that

$$\exists h, w, \bar{i}, \ h = \langle (i_1, j_1), \ldots, (i_r, j_r) \rangle \wedge A(a, h, w)$$

The formula in the claim is Σ_i^b, the lexicographically maximal e is just the maximal e, and any e of such form satisfies

$$e \leq 2^r \leq 2^{c \cdot ||a||} \leq |a|^c$$

Hence the existence of such e follows by the Σ_i^b–LENGTH–MAX principle that is available in S_2^i by Lemma 5.2.7.

To conclude the proof, argue as in the proof of Theorem 6.1.2: For h and w witnessing the formula for the maximal e, all affirmative oracle answers are correct as they are witnessed, and all negative answers are also correct as otherwise a 0 in e could be changed to 1, all later 1's to 0, in this way creating $e' > e$ satisfying the claim. Hence w is a computation of M on input a.   Q.E.D.

Lemma 6.3.4. *For any $i \geq 1$, any $f \in FP^{\Sigma^p_i}[wit, q(n)]$ is computable by a machine M with a Σ^p_i-witness-oracle allowing $q(n) + 1$ oracle queries but requiring the oracle to return a witness* only *in the last query. Moreover, this last witness is an ordered pair whose first component is the output of the computation.*

Proof. Let M be a polynomial time machine M with witness-oracle Q. Consider machine M', which by a binary search constructs the maximal 0–1 sequence $e = (j_1, \ldots, j_r)$ satisfying the formula $B(a, e)$ from the claim from the previous proof. This requires $\leq q(n)$ queries to a Σ^p_i-oracle

$$\exists f, B(a, e \frown f$$

but it does not need the witnesses to the affirmative answers.

Having found such maximal e the machine M' states the query

$$[\exists \langle y, (h, w, \bar{i}) \rangle; h = \langle (i_1, j_1), \ldots, (i_2, j_r) \rangle$$
$$\wedge \ A(a, h, w) \wedge \text{``y is the output of w''?}]$$

The answer must be YES and a witness is an output of a computation of M on a, that is, a correct output. Q.E.D.

Corollary 6.3.5. *Predicates from the class*

$$\bigcup_c FP^{\Sigma^p_i}[wit, c \cdot \log n]$$

that is, functions with unique values from $\{0, 1\}$, are exactly the predicates from the class $P^{\Sigma^p_i}[O(\log n)]$ (cf. Definition 6.2.1).

Proof. That $P^{\Sigma^p_i}[O(\log n)]$ is included among predicates in $FP^{\Sigma^p_i}[wit, c \cdot \log n]$ is trivial. The opposite inclusion follows similarly to Lemma 6.3.4 (in the last query add condition $y = 1$ and do not require a witness to the oracle answer). Q.E.D.

6.4. Bibliographical and other remarks

Theorem 6.1.2 is due to Buss (1986), Definition 6.2.1 is from Krentel (1986), Theorem 6.2.2 is from Krajíček (1993), and Theorem 6.2.3 is due to Buss and Hay (1988) and Wagner (1990).

Counterexample computations (student–teacher) were defined explicitly in Krajíček et al. (1990) and used earlier in Krajíček et al. (1991). The witness-oracle computations were used in Krajíček (1993) and the classes based on them (6.3.2) were explicitly defined in Buss, Krajíček, and Takeuti (1993). Theorem 6.3.3 is from Krajíček (1993).

7

Witnessing theorems

This chapter considers various *witnessing theorems*, which are theorems characterizing functions definable in various systems of arithmetic in terms of their computational complexity. A prototype of such a theorem (and its proof) is the characterization of primitive recursive functions as provably total recursive functions in fragment $I\Sigma_1^0$ of PA (cf. Parsons 1970, Takeuti 1975, and Mints 1976).

There are other approaches to proving witnessing theorems, for example, skolemizing the given theory by Skolem functions from a particular class and then applying Herbrand's theorem. Or there are intrigued model-theoretic constructions. I shall mention these methods too, but my opinion is that one really has to know in advance which class of functions one targets before formulating an argument while the methods based on cut-elimination (Section 7.1) and generalizing Theorem 7.2.3 help to discover the right class. This certainly was the case for all witnessing theorems discussed in this chapter.

7.1. Cut-elimination for bounded arithmetic

We first extend the sequent predicate calculus by rules allowing the introduction of bounded quantifiers and by the induction rules and then we prove the cut-elimination for such a system.

The predicate calculus LK extends the propositional LK from Section 4.3 by four rules for introducing quantifiers to a sequent as in Definition 4.6.2:

1. $\forall$: **introduction**

$$\text{left } \frac{A(t), \Gamma \longrightarrow \Delta}{\forall x\, A(x), \Gamma \longrightarrow \Delta} \quad and \quad \text{right } \frac{\Gamma \longrightarrow \Delta, A(a)}{\Gamma \longrightarrow \Delta, \forall x\, A(x)}$$

2. $\exists$: **introduction**

$$\text{left } \frac{A(a), \Gamma \longrightarrow \Delta}{\exists x\, A(x), \Gamma \longrightarrow \Delta} \quad and \quad \text{right } \frac{\Gamma \longrightarrow \Delta, A(t)}{\Gamma \longrightarrow \Delta, \exists x\, A(x)}$$

where t is any term and variable a in $\forall$: **right**, and $\exists$: **left** must not occur in the lower sequent

3. and by instances of equality axioms

$$t_1 = s_1, \ldots, t_k = s_k \longrightarrow f(t_1, \ldots) = f(s_1, \ldots)$$

and

$$t_1 = s_1, \ldots, R(\bar{t}) \longrightarrow R(\bar{s})$$

for all function symbols f and predicate symbols R of the language.

For technical reasons (an easier formulation of Theorem 7.2.3)we stipulate that formula A in any initial sequent

$$A \longrightarrow A$$

must be atomic.

Definition 7.1.1. *LKB is a proof system for the predicate logic extending LK by the following four rules allowing introduction of bounded quantifiers:*

1. $\forall \leq$: **introduction**

$$\text{left} \quad \frac{A(t), \Gamma \longrightarrow \Delta}{t \leq s, \forall x \leq s\, A(x), \Gamma \longrightarrow \Delta} \quad \text{and} \quad \text{right} \quad \frac{a \leq t, \Gamma \longrightarrow \Delta, A(a)}{\Gamma \longrightarrow \Delta, \forall x \leq t\, A(x)}$$

2. $\exists \leq$: **introduction**

$$\text{left} \quad \frac{a \leq t, A(a), \Gamma \longrightarrow \Delta}{\exists x \leq t\, A(x), \Gamma \longrightarrow \Delta} \quad \text{and} \quad \text{right} \quad \frac{\Gamma \longrightarrow \Delta, A(t)}{t \leq s, \Gamma \longrightarrow \Delta, \exists x \leq s\, A(x)}$$

with the requirement that the variable a in $\forall \leq$: **right** *and* $\exists \leq$: **left** *does not occur in the lower sequent.*

The system LKB is sound and complete and, for that matter, any valid sequent formed from bounded formulas can be proved in LKB without the use of unbounded quantifiers. We shall see this later.

Definition 7.1.2.

(a) The IND-rule is the following inference rule

$$\frac{\Gamma, A(a) \longrightarrow A(a+1), \Delta}{\Gamma, A(0) \longrightarrow A(t), \Delta}$$

(b) The PIND-rule is the following inference rule

$$\frac{\Gamma, A(\lfloor a/2 \rfloor) \longrightarrow A(a), \Delta}{\Gamma, A(0) \longrightarrow A(t), \Delta}$$

(c) The LIND-rule is the following inference rule

$$\frac{\Gamma, A(a) \longrightarrow A(a+1), \Delta}{\Gamma, A(0) \longrightarrow A(|t|), \Delta}$$

where t is any term and where the variable a does not occur in the lower sequent.

The Σ_i^b-*IND rule denotes the rule restricted to* Σ_i^b-*formulas A, and similarly for other classes of formulas and the other two rules.*

Lemma 7.1.3. *Denote by BASIC^{LK} the set of sequents of the form*

$$\longrightarrow A$$

with A an axiom from BASIC.

Let $i \geq 1$ *and let B be any formula. Then*

$$T_2^i \vdash B$$

if and only if sequent

$$\longrightarrow B$$

is derivable in LKB $+\Sigma_i^b$-*IND from the initial sequents BASIC^{LK}.*

The same holds for PIND- and LIND-rules: That is, Σ_i^b–*PIND- (resp.* Σ_i^b– *LIND-) rules are* equivalent to Σ_i^b–*PIND- (resp.* Σ_i^b–*LIND-) axioms.*

The "if part" of the lemma is obvious and for the "only if part" it is sufficient to infer any instance of an Σ_i^b–IND-axiom in LKB using the Σ_i^b–IND-rule only; this is left to the reader (note that the restriction to Σ_i^b applies only to the induction formulas of the IND-*rule*, other formulas in a derivation can be arbitrary).

The next theorem, *the cut-elimination theorem*, is the crucial technical property of the sequent calculus. The cut-elimination for LK was proved by Gentzen and it is called *Gentzen's Hauptsatz* (see Gentzen 1969).

Recall from Section 4.3 (Definition 4.3.2) the notion of an *ancestor* and *successor* of a formula in a proof. A formula ϕ in an LKB derivation with induction rules is called *free* if ϕ is in an initial sequent or no successor of ϕ is identical to ϕ or to one of the two principal formulas of an induction rule.

Theorem 7.1.4. *Assume* $i \geq 1$ *and assume that the sequent*

$$\Gamma \longrightarrow \Delta$$

is provable in T_2^i. *Then*

$$\Gamma \longrightarrow \Delta$$

has a proof in LKB with the Σ_i^b–*IND-rule in which no cut-formula is free. The same is true for the* S_2^i *and the* Σ_i^b–*PIND-rule.*

The theorem is proved by double induction on the complexity of a proof in T_2^i and on the complexity of the cut-formulas in it. We refer the reader to Takeuti (1975) for a clear treatment of the cut-elimination that applies to LKB with the induction rules. We shall state explicitly one immediate corollary of cut-elimination: the *subformula property*.

Corollary 7.1.5. *Let $i \geq 1$ and assume that the sequent*

$$\Gamma \longrightarrow \Delta$$

is provable in T_2^i (resp. in S_2^i).
  Then there is a proof of

$$\Gamma \longrightarrow \Delta$$

in the same theory in which every formula is either a subformula of a Σ_i^b- or a Π_i^b-formula or a subformula of a formula from

$$\Gamma \longrightarrow \Delta$$

  In particular, if all formulas in Γ, Δ are Σ_i^b-formulas, then all formulas in the proof are Σ_i^b or Π_i^b.

Proof. Let ϕ be a formula in a proof of $\Gamma \longrightarrow \Delta$ without free cut-formulas; such proof exists by the previous theorem. Assume ϕ is not a subformula of a Σ_i^b-formula. It follows that ϕ is free, as are all its successors. If ϕ would not occur as a subformula in $\Gamma \longrightarrow \Delta$ then it would have to disappear before the end sequent by a cut, in which case the cut-formula would be free, a contradiction.   Q.E.D.

7.2. Σ_i^b-definability in S_2^i and oracle polynomial time

The idea of the *witness function method* is the following. Assume that

$$\Gamma \longrightarrow \Delta$$

is a valid sequent consisting of formulas starting with existential quantifiers, say

$$\exists x \psi(a, x) \longrightarrow \exists y \theta(a, y)$$

Assume that a are free variables that do not have to all appear in all formulas of the sequent. That means that for any x satisfying $\psi(a, x)$ there is y satisfying $\theta(a, y)$ so that the function

$$f(a, x) := \begin{cases} \text{some } y \text{ such that } \theta(a, y) & \text{if } \psi(a, x) \\ \text{any } y & \text{if } \neg\psi(a, x) \end{cases}$$

is well defined and *witnesses* the validity of the sequent.
  The witnessing functions for initial sequents are trivial so one hopes that progressing along a proof of the sequent allows an explicit description of the witnessing functions for each sequent in the proof and, in particular, an estimate of their computational complexity. Obviously this cannot be done with an arbitrary proof in which any formula may appear and so one has to appeal to the cut-elimination theorem to assure the existence of proofs for which there is information about the formulas in it.

The next definition introduces a method to deal with sequents whose formulas do not start with an existential quantifier. We use sequence coding from Section 5.4.

Definition 7.2.1 (Buss 1986). *Let $i \geq 1$ and A be a Σ_i^b-formula with free variables among $\overline{a}$. By induction on the logical complexity of A define the formula*

$$Witness_A^{i,\overline{a}}(w, \overline{a})$$

by:

1. *If $A \in \Sigma_{i-1}^b \cup \Pi_{i-1}^b$ then*

$$Witness_A^{i,\overline{a}}(w, \overline{a}) \equiv A(\overline{a})$$

2. *If $A = B \wedge C$ then*

$$Witness_A^{i,\overline{a}}(w, \overline{a}) \equiv (Witness_B^{i,\overline{a}}((w)_1, \overline{a}) \wedge Witness_C^{i,\overline{a}}((w)_2, \overline{a}))$$

3. *If $A = B \vee C$ then*

$$Witness_A^{i,\overline{a}}(w, \overline{a}) \equiv (Witness_B^{i,\overline{a}}((w)_1, \overline{a}) \vee Witness_C^{i,\overline{a}}((w)_2, \overline{a}))$$

4. *If $A(\overline{a}) = \forall x \leq |s(\overline{a})| B(\overline{a}, x)$ and $A \notin \Sigma_{i-1}^b \cup \Pi_{i-1}^b$ then*

$$Witness_A^{i,\overline{a}}(w, \overline{a}) \equiv (Seq(w) \wedge \forall x \leq |s(\overline{a})| Witness_B^{i,\overline{a},b}((w)_{x+1}, \overline{a}, x))$$

5. *If $A(\overline{a}) = \exists x \leq s(\overline{a}) B(\overline{a}, x)$ and $A \notin \Sigma_{i-1}^b \cup \Pi_{i-1}^b$ then*

$$Witness_A^{i,\overline{a}}(w, \overline{a}) \equiv (Seq(w) \wedge (w)_2 \leq s(\overline{a}) \wedge Witness_B^{i,\overline{a},b}((w)_1, \overline{a}, (w)_2))$$

6. *If $A = \neg B$ and $A \notin \Sigma_{i-1}^b \cup \Pi_{i-1}^b$ then use the prenex operations and de Morgan rules to push the negations into the Σ_{i-1}^b-subformulas and then apply one of clauses 1–4.*

Lemma 7.2.2. *Let $i \geq 1$ and let A be a Σ_i^b-formula. Then:*
(a) *The formula $Witness_A^{i,\overline{a}}(w, \overline{a})$ is Δ_i^b in S_2^i.*
(b)

$$PV_1 \vdash Witness_A^{i,\overline{a}}(w, \overline{a}) \to A(\overline{a})$$

(c)

$$S_2^1 + BB\Sigma_i^b \vdash A(\overline{a}) \to \exists w, \ Witness_A^{i,\overline{a}}(w, \overline{a})$$

Proof. Property (a) is obvious from the definition of the witness formula. Property (b) is proved by induction on the logical complexity of A, which is straightforward.

To prove property (c) proceed also by induction on the logical complexity of A. The only nontrivial cases of Definition 7.2.1 are those treating the quantifiers. Let

$$A(\overline{a}) = \exists x \leq t(\overline{a}) B(\overline{a}, x)$$

Argue in S_2^1: If $A(\overline{a})$ holds, then $B(\overline{a}, b)$ holds for some $b \leq t(\overline{a})$ and by the induction assumption

$$\text{Witness}_B^{i,\overline{a}}(w, \overline{a}, b)$$

holds for some w. Setting $w' := \langle w, b \rangle$, clearly

$$\text{Witness}_A^{i,\overline{a}}(w', \overline{a})$$

Now let

$$A(\overline{a}) = \forall x \leq |s(\overline{a})| B(\overline{a}, x)$$

By the induction assumption we already know

$$S_2^1 + BB\Sigma_i^b \vdash B(\overline{a}, b) \rightarrow \exists v_b \leq t(\overline{a}), \text{Witness}_B^{i,\overline{a},b}(v_b, \overline{a}, b)$$

So

$$S_2^1 + BB\Sigma_i^b \vdash A(\overline{a}) \rightarrow \forall x \leq |s(\overline{a})| \exists v_x \leq t(\overline{a}), \text{Witness}_B^{i,\overline{a},b}(v_x, \overline{a}, x)$$

Applying $BB\Sigma_i^b$ we get

$$S_2^1 + BB\Sigma_i^b \vdash A(\overline{a}) \rightarrow \exists w \forall x \leq |s(\overline{a})|,$$
$$\left(\text{Witness}_B^{i,\overline{a},b}((w)_{x+1}, \overline{a}, x) \wedge (w)_{x+1} \leq t(\overline{a}) \right)$$

and thus also

$$S_2^1 + BB\Sigma_i^b \vdash A(\overline{a}) \rightarrow \exists w, \ \text{Witness}_A^{i,\overline{a}}(w, \overline{a})$$

Q.E.D.

We are arriving at Buss's *witnessing theorem*; our formulation is a combination of the original version with a later improvement.

Theorem 7.2.3 (Buss 1986, 1990a). *Let $i \geq 1$ and assume that*

$$S_2^i \vdash A(\overline{a}) \longrightarrow B(\overline{a})$$

where A and B are Σ_i^b and $\overline{a}$ are all free variables in the sequent.
   Then there is a PV_i-function symbol $f(w, \overline{a})$ such that
   1. *the function f is $\square_i^p$ and it is Σ_i^b-definable in S_2^i*
   2.

$$PV_i \vdash \textit{Witness}_A^{i,\overline{a}}(w, \overline{a}) \longrightarrow \textit{Witness}_B^{i,\overline{a}}(f(w, \overline{a}), \overline{a})$$

Proof. From the assumption and Corollary 7.1.5 there is an LKB proof π of the sequent with the Σ_i^b–PIND rule in which all formulas are Σ_i^b or Π_i^b. A general sequent in π has the form (with possibly permuted formulas)

$$E_1, \ldots, E_k, U_1, \ldots, U_r \longrightarrow F_1, \ldots, F_\ell, V_1, \ldots, V_s$$

with E_i's and F_i's Σ_i^b and U_j's and V_j's Π_i^b.

By induction on the number of inferences in π above the sequent we show that there is a function $g(w, \overline{b})$ such that:

$$PV_i \vdash \text{Witness}_G^{i,\overline{b}}(w, \overline{b}) \longrightarrow \text{Witness}_H^{i,\overline{b}}(g(w, \overline{b}), \overline{b})$$

where

$$G := \bigwedge_i E_i \wedge \bigwedge_j (\neg V_j) \quad \text{and} \quad H := \bigvee_i F_i \vee \bigvee_j (\neg U_j)$$

and $\overline{b}$ are all free variables of the sequent.

In the induction step we shall distinguish cases according to the type of the last inference applied in π to derive the sequent. To simplify the notation we shall picture sequents in the form

$$E, \Gamma \longrightarrow \Delta \quad \text{or} \quad \Gamma \longrightarrow \Delta, F$$

thinking that all formulas are Σ_i^b, and we shall write, for example, $\text{Witness}_{E,\Gamma}^{i,\overline{b}}(w, \overline{b})$ instead of the conjunction G in the index. We shall also describe only the inductive definition of the function g leaving the provability of its properties in PV_i to the reader, and treat only the right rules of LKB as the left ones are treated dually.

Initial sequents. An initial sequent has either the form $C \longrightarrow C$ for C atomic or $\longrightarrow C$ for C an axiom from BASIC; in both cases C is open and by definition

$$C(\overline{b}) \equiv \text{Witness}_C^{i,\overline{b}}(w, \overline{b})$$

Structural inferences. Structural inferences are easy to handle, utilizing the permutation of the components of the witness (for *the exchange rule*), defining by cases distinguished by a Δ_i^b-property (for *the contraction rule*) or adding a dummy component (for *the weakening rule*).

Propositional inferences. Assume that for a $\bigwedge$: **right** inference

$$\frac{\Gamma \longrightarrow \Delta, C \quad \Gamma \longrightarrow \Delta, D}{\Gamma \longrightarrow \Delta, C \wedge D}$$

functions g_1 and g_2 witness the upper sequents where g_1 computes from a witness w a tuple $(\overline{w}_1, w_C)$ (a witness for the succedent of the left upper sequent), and g_2 computes on w a tuple $(\overline{w}_2, w_D)$. Then define the function g by

$$g(w) := \begin{cases} (\overline{w}_1, 0) & \text{if Witness}_\Delta^{i,\overline{b}}(\overline{w}_1, \overline{b}) \\ (\overline{w}_2, \langle w_C, w_D \rangle) & \text{otherwise} \end{cases}$$

Assume that w witnesses Γ. Then $(\overline{w}_1, w_C)$ witnesses Δ, C, so either w_1 witnesses Δ, in which case $g(w) = (\overline{w}_1, 0)$ witnesses $\Delta, C \wedge D$, or w_C witnesses C, in which case $g(w) = (\overline{w}_2, \langle w_C, w_D \rangle)$ witnesses $\Delta, C \wedge D$, as well, since either $\overline{w}_2$ witnesses Δ or pair $\langle w_C, w_D \rangle$ witnesses the conjunction $C \wedge D$.

Assume that for a $\bigvee$: **right** inference

$$\frac{\Gamma \longrightarrow \Delta, C}{\Gamma \longrightarrow \Delta, C \vee D}$$

function g_1 is witnessing the upper sequent, computing on w a tuple $(\overline{w}_1, w_C)$. Define function g by

$$g(w) := (\overline{w}_1, \langle w_C, 0 \rangle)$$

It trivially witnesses the lower sequent.

Finally assume that g_1 is a witnessing function for the upper sequent of a ¬ : **right** inference

$$\frac{C, \Gamma \longrightarrow \Delta}{\Gamma \longrightarrow \Delta, \neg C}$$

Since we assume that all formulas in π are Σ_i^b, C is in fact in $\Sigma_{i-1}^b \cup \Pi_{i-1}^b$ and hence $\text{Witness}_C^{i,\overline{b}}(w, \overline{b})$ is by definition formula C itself.

Define the function g by

$$g(w) := \langle g_1(\langle 0, w \rangle), 0 \rangle$$

Quantifier inferences. Let g_1 witness the upper sequent of the $\forall \leq$: **right** inference

$$\frac{c \leq t, \Gamma \longrightarrow \Delta, C(c)}{\Gamma \longrightarrow \Delta, \forall x \leq t\, C(x)}$$

We must consider two cases, when t is of the form $|s|$, and when it is not of this form.

In the latter case the formula $\forall x \leq t C(x)$ must be Π_{i-1}^b and hence its witness formula is identical with it and the definition of g poses no problem.

In the former case assume that g_1 computes on input $\langle (u, w), c \rangle$ the output (w^c, w_2). Define the function g by

$$g(w) := \begin{cases} w^c & \text{if } w^c \text{ witnesses } \Delta \text{ for some } c \leq |s| \\ \langle \overline{0}, (w_0, \ldots, w_{|s|}) \rangle & \text{otherwise} \end{cases}$$

If w witnesses Γ then $g(w)$ witnesses the lower sequent because each $g_1((u, w, c)), c = 0, \ldots, |s|$, witnesses either Δ or $C(c)$.

Note that the sequence $(w_0, \ldots, w_{|s|})$ exists by $BB\Sigma_i^b$ available in S_2^i.

Now assume that $g_1(w) = (w_1, v)$ witnesses the upper sequent of a $\exists \leq :$ **right** inference

$$\frac{\Gamma \longrightarrow \Delta, C(t)}{t \leq s, \Gamma \longrightarrow \Delta, \exists x \leq s\, C(x)}$$

Then define the function g by

$$g(u, w) := \begin{cases} 0 & \text{if } t > s \\ (w_1, (\mathrm{val}(t), v)) & \text{if } t \leq s \end{cases}$$

where $\mathrm{val}(t)$ is the value of the term t. If w witnesses Γ and $t \leq s$ then (u, w) witnesses $t \leq s, \Gamma$, in which case either w_1 witnesses Δ or v witnesses $C(t)$. That is, $(\mathrm{val}\,(t), v)$ witnesses $\exists x \leq c, C(x)$.

Cut-rule. Let $g_1(w) = (w_1, u)$ and $g_2(v, w) = w_2$, respectively, witness the left and the right upper sequents of a cut-rule inference

$$\frac{\Gamma \longrightarrow \Delta, C \qquad C, \Gamma \longrightarrow \Delta}{\Gamma \longrightarrow \Delta}$$

Define the function g by

$$g(w) := \begin{cases} w_1 & \text{if Witness}_\Delta^{i,\overline{b}}(w_1, \overline{b}) \\ g_2(u, w_2) & \text{otherwise} \end{cases}$$

The function $g(w)$ is either w_1, in which case it witnesses Δ, or otherwise u witnesses C and hence $g(u, w)$ witnesses Δ.

Σ_i^b–**PIND rule** Let $g_1(u, w, c) = (w_1, v)$ witness the upper sequent of a Σ_i^b–PIND inference

$$\frac{\Gamma, C(\lfloor c/2 \rfloor) \longrightarrow C(c), \Delta}{\Gamma, C(0) \longrightarrow C(t), \Delta}$$

We shall consider only the case when $C \notin \Sigma_{i-1}^b \cup \Pi_{i-1}^b$ as otherwise the construction of g is trivial.

The idea of the construction of the function $g(u, w)$ witnessing the lower sequent is to iterate the function g_1 for values $c = 0, \ldots, t$. We shall proceed by the limited recursion on notation

$$f(u, w, 1) := \begin{cases} 0 & \text{if } \neg\, \text{Witness}_{C(0)}^{i,\overline{b}}(u, \overline{b}) \\ g_1(u, w, 1) & \text{otherwise} \end{cases}$$

$$f(u, w, y) := g_1\left(\left(f\left(u, w, \left\lfloor \frac{y}{2} \right\rfloor\right)\right)_1, w, y\right)$$

and then we put

$$g(u, w) := f(u, w, t)$$

The definition of f is correct as f is a priori bounded by 2^ℓ, where ℓ is the maximal sum of the lengths of witnesses for $C(c)$ plus $|w|$ plus $|t|$.

To see that g witnesses the lower sequent verify by induction on the length of y that $f(u, w, y)$ witnesses the sequent

$$\Gamma, C(0) \longrightarrow C(y), \Delta$$

This completes the proof of the theorem. $\hspace{2cm}$ Q.E.D.

The witnessing Theorem 7.2.3 has several important corollaries. We shall mention four of them now.

Corollary 7.2.4. *The theory S^1_2 is Σ^b_1-conservative over the theory PV_1 and for $i \geq 1$ the theory S^{i+1}_2 is Σ^b_{i+1}-conservative over the theory PV_{i+1} and also over the theory T^i_2.*

Proof. Assume $A(a) \in \Sigma^b_{i+1}$ and

$$S^{i+1}_2 \vdash A(a)$$

By Theorem 7.2.3

$$PV_{i+1} \vdash \text{Witness}^{i,a}_A(f(a), a)$$

for some PV_{i+1}-function symbol f. But by Lemma 7.2.2 even

$$PV_1 \vdash \text{Witness}^{i,a}_A(f(a), a) \to A(a)$$

That is, together

$$PV_{i+1} \vdash A(a)$$

This proves a part of the corollary. The part about the conservativeness over T^i_2 follows from Theorem 5.3.5. $\hspace{2cm}$ Q.E.D.

Note that the theory T^i_2 has a $\forall \Sigma^b_{i+1}$-axiomatization; hence the preceding statement is the best possible unless in fact $T^i_2 = S^{i+1}_2$.

Corollary 7.2.5. *Let $i \geq 1$ and let $\phi(a)$ be a Σ^b_i-formula and $\psi(a)$ be a Π^b_i-formula. Assume that*

$$S^i_2 \vdash \phi(a) \equiv \psi(a)$$

Then the formula $\phi(a)$ defines a Δ^p_i-predicate in $\mathbf{N}$.

In particular, any property that is provably in S^1_2 in the class $NP \cap coNP$ is actually in the class P.

Proof. From the hypothesis it follows that

$$S^i_2 \vdash \phi(a) \vee \neg\psi(a)$$

where this disjunction is a Σ_i^b-formula. By Theorem 7.2.3 there is a $\square_i^p$-function f so that

$$\mathrm{PV}_i \vdash \mathrm{Witness}_{\phi \vee \neg \psi}^{i,a}(f(a), a)$$

That is, either $(f(a))_1$ witnesses $\phi(a)$ or $(f(a))_2$ witnesses $\neg\psi(a)$, and both cases cannot occur at the same time as

$$S_2^i \vdash \neg(\phi(a) \wedge \neg\psi(a))$$

Hence the characteristic function of ϕ

$$\Xi_\phi(a) := \begin{cases} 1 & \text{if } \mathrm{Witness}_\phi^{i,a}((f(a))_1) \\ 0 & \text{otherwise} \end{cases}$$

is a $\square_i^p$-function: That is, $\phi(a)$ defines a predicate in the Δ_i^p level of the polynomial time hierarchy PH.

For $i = 1$, $\Sigma_1^p = \mathrm{NP}$, $\Pi_1^p = \mathrm{coNP}$ and $\Delta_1^p = \mathrm{P}$ (cf. Section 2.2 and Theorem 3.2.12).   Q.E.D.

Corollary 7.2.6. *Let $i \geq 1$ and let $\phi(a, b)$ be a Σ_i^b-formula. Assume that*

$$S_2^i \vdash \forall x \exists y \phi(x, y)$$

Then there is a Σ_i^b-formula $\phi^*(a, b)$ such that S_2^i also proves
1. *$\phi^*(a, b) \rightarrow \phi(a, b)$*
2. *$\forall x \exists! y \phi^*(x, y)$*

Proof. By Theorem 7.2.3 the hypothesis implies

$$S_2^i \vdash \forall x, \ \mathrm{Witness}_\phi^{i,a}(f(x), x)$$

for some $\square_i^p$-function f Σ_i^b-definable in S_2^i. Define the formula ϕ^* by

$$\phi^*(a, b) := (b = (f(a))_1)$$

It clearly satisfies the requirement.   Q.E.D.

Corollary 7.2.7. *For $i \geq 1$ let $\phi \in \Sigma_{i+1}^b$ and $\psi \in \Pi_{i+1}^b$ be arbitrary, and assume*

$$T_2^i \vdash \phi(a) \equiv \psi(a)$$

Then

$$T_2^i \vdash (\phi(0) \wedge \forall x(\phi(x) \rightarrow \phi(x+1))) \rightarrow \forall x, \phi(x)$$

The same holds for $i = 0$ and PV_1 in place of T_2^0.
This scheme is called the Δ_{i+1}^b–IND scheme.

Proof. By Lemma 5.2.9 S^{i+1}_2 admits Δ^b_{i+1}–PIND. Using $\forall x, \phi(x) \equiv \psi(x)$ we can write the IND-axiom for ϕ as

$$\neg\psi(0) \vee \exists x \le a(\phi(x) \wedge \neg\psi(x+1)) \vee \phi(a)$$

which is a Σ^b_{i+1}-formula. By Corollary 7.2.4 T^i_2 proves all Σ^b_{i+1}-consequences of S^{i+1}_2. Q.E.D.

7.3. Σ^b_{i+2}- and Σ^b_{i+1}-definability in S^i_2 and bounded queries

We give two witnessing theorems for Σ^b_{i+1}-consequences of S^i_2. The first one uses a simple idea of introducing a symbol for a Herbrand function to reduce the quantifier complexity of the formula and then reinterpreting the Herbrand function.

Consider a formula of the form $\psi(a) = \exists x \forall y \phi(a, x, y)$. Let $f(a, x)$ be a new function symbol and define the formula $\psi_H(a)$ by

$$\psi_H(a) := \exists x \phi(a, x, y/f(a, x))$$

That is, ψ_H arises from ψ by reducing the quantifier complexity with the help of *one* Herbrand function for the first universal quantifier (a *herbrandization* of a formula would require introduction of Herbrand functions for all universal quantifiers and the new formula would be existential). The following lemma is trivial.

Lemma 7.3.1. *Let $\psi(a) = \exists x \forall y \phi(a, x, y)$. Then the implication*

$$\psi(a) \rightarrow \psi_H(a)$$

is logically valid.

We shall define a particular interpretation of the function $f(a, x)$

$$f^*(a, x) := \begin{cases} y & y \text{ is the minimal } y \text{ s.t. } \neg\phi(a, x, y) \text{ holds} \\ 0 & \text{there is no such } y \end{cases}$$

Lemma 7.3.2. *For every n the sentence $\exists x \phi(n, x, f^*(n, x))$ is valid if and only if $\exists x \forall y \phi(n, x, y)$ is valid.*

Proof. Assume $\exists x \phi(n, x, f^*(n, x))$ and let $x = a$ witness the existential quantifier, while $\forall x \exists y \neg\phi(n, x, y)$ and, in particular $\exists y \neg\phi(n, a, y)$. But the definition of $f^*(n, a)$ also implies that $\neg\phi(n, a, f^*(n, a))$, which is a contradiction.

Assume now $\exists x \forall y \phi(n, x, y)$ with $x = a$ witnessing the existential quantifier: $\forall y \phi(n, a, y)$. If $\forall x \neg\phi(n, x, f^*(n, x))$ would hold, then also $\neg\phi(n, a, f^*(n, a))$, which is again a contradiction. Q.E.D.

Now we use the Herbrand function to derive from Theorem 7.2.3 a witnessing theorem for the Σ^b_{i+2}-consequences of S^i_2.

Theorem 7.3.3. *Let $\phi(a, b, c)$ be a Σ_i^b-formula and assume that*

$$S_2^i \vdash \exists x \forall y \leq a\, \phi(a, x, y)$$

Then there is a function $g(a)$ such that

$$N \models \forall n \forall y \leq n, \phi(n, g(n), y)$$

and g is computable by a counterexample computation with the student *being a polynomial time Turing machine with a Σ_{i-1}^p-oracle and* the teacher *producing counterexamples to $\forall y \leq a\phi(a, b, y)$, that is, some $c \leq a$ for which $\neg\phi(a, b, c)$, if such c exists.*

In particular, $g \in FP^{\Sigma_{i+1}^p}[wit, poly]$. (See Section 6.3 for the definitions of the computational models involved.)

Proof. Consider the formula

$$(\exists x \forall y \leq a\phi(a, x, y))_H := \exists x, f(a, x) \leq a \rightarrow \phi(a, x, f(a, x))$$

The hypothesis of the theorem implies that

$$S_2^i(f) \vdash \exists x, f(a, x) \leq a \rightarrow \phi(a, x, f(a, x))$$

where $S_2^i(f)$ is a theory defined like S_2^i but in the language $L \cup \{f\}$ and with axiom $f(a, x) \leq a$. For such a theory we have a statement straightforwardly analogous to the witnessing Theorem 7.3.2 with the witnessing functions now being from the class $\Box_i^p(f)$. That is: The witnessing function is computable by a polynomial time machine querying a Σ_{i-1}^p-oracle and querying also for values of the function f. Such a witnessing *functional* F then satisfies

$$f(a, F(a, f)) \leq a \rightarrow \phi(a, F(a, f), f(a, F(a, f)))$$

Now substitute into the algorithm computing $F(a, f)$ for oracle f a particular function f^*

$$f^*(a, b) := \begin{cases} \min c \leq a \text{ s.t.} \neg\phi(a, b, c) & \text{if it exists} \\ a + 1 & \text{if } \forall y \leq a\phi(a, b, y) \end{cases}$$

Then

$$f^*(a, b) \leq a \rightarrow \phi(a, b, f^*(a, b))$$

implies

$$\forall y \leq a, \phi(a, b, y)$$

and hence the preceding implication yields

$$\forall y \leq a, \phi(a, F(a, f^*), y)$$

Note that the algorithm computing $F(a, f^*)$ specifies a counterexample computation with the required properties.   Q.E.D.

In principle we may apply the herbrandization to bounded formulas of arbitrary complexity and obtain a $\Sigma^b_i(f_1, \ldots, f_k)$-formula. The Herbrand functions, however, do not appear to have a nice interpretation (cf. Krajíček 1992 and Lemma 11.1.2).

Corollary 7.3.4. *Let $i \geq 1$ and assume that a function f is Σ^b_{i+1}-definable in S^i_2. Let the graph of f be defined by a formula of the form $\exists x \forall y \leq a A(a, b, x, y)$, where $\forall y \leq a A(a, b, c, y)$ is Π^b_i.*

Then there is a function g that can be computed by a counterexample computation where the student is a polynomial time machine with a Σ^b_{i-1}-oracle and receives from the teacher counterexamples to the formula

$$\forall y \leq a, A(a, b, c, y)$$

and such that the function $f(a)$ is a projection of $g(a)$. That is, for any a the value $g(a)$ has the form $g(a) = (f(a), x)$, for some x.

Proof. Let f be defined by a Σ^b_{i+1}-formula

$$\exists x \forall y \leq a A(a, b, x, y)$$

The statement then follows from Theorem 7.3.3.   Q.E.D.

If a function f is Σ^b_{i+1}-definable in S^i_2 and thus also in S^{i+1}_2, then it is a $\Box^p_{i+1}$-function. Corollary 7.3.4 implies that such a function is, in particular, in $FP^{\Sigma^b_i}[\text{wit}, \text{poly}]$. The class $FP^{\Sigma^b_i}[\text{wit}, \text{poly}]$ contains, however, the class $\Box^p_{i+1}$ trivially. Thus the only new information the corollary gives is the form of the formula to which the teacher offers counterexamples. That appears to be rather poor information to yield some estimate of the computational complexity of the functions Σ^b_{i+1}-definable in S^i_2 and we need another approach to it. We shall use witness-oracle computations to obtain a witnessing theorem for Σ^b_{i+1}-consequences of S^i_2.

Theorem 7.3.5 (Krajíček 1993). *Let $i \geq 1$ and let $A(a, b)$ be a Σ^b_{i+1}-formula such that*

$$S^i_2 \vdash \forall x \exists y A(x, y)$$

Then there is a multivalued function $f(x) \in FP^{\Sigma^b_i}[\text{wit}, O(\log n)]$ that is Σ^b_{i+1}-definable in S^i_2 such that

$$S^i_2 \vdash \forall x, A(x, f(x))$$

This means that any value $y = f(x)$ satisfies $A(x, y)$.

Note that the converse of this theorem, Theorem 6.3.3, is also true.

Proof. We shall show that there is a function $f \in \mathrm{FP}^{\Sigma_i^b}[\mathrm{wit}, O(\log n)]$; its Σ_{i+1}^b-definability in S_2^i follows from Theorem 6.3.3.

Assume for simplicity that $A \in \Pi_i^b$. By Corollary 7.1.5 the hypothesis of the theorem then implies that there is a derivation π of the sequent

$$\longrightarrow \exists y A(a, y)$$

in which every sequent has the form

$$\Gamma_1, \Delta_1 \longrightarrow \Gamma_2, \Delta_2$$

where

(i) $\Gamma_1 \cup \Gamma_2 \subseteq \Sigma_i^b \cup \Pi_i^b$
(ii) Δ_1 has the form

$$\exists y_1 \alpha_1(b, y_1), \ldots, \exists y_r \alpha_r(b, y_r)$$

(iii) Δ_2 has the form:

$$\exists z_1 \beta_1(b, z_1), \ldots, \exists z_s \beta_s(b, z_s)$$

where α_j's and β_j's are Π_i^b-formulas.

We will assume for simplicity that bounds to y_j's and z_j's are parts of the formulas α_j and β_j, respectively.

Now we modify the notion of a *witness* from Section 7.2. We say that u is a witness to Γ_1, Δ_1 (for parameters b) if u has the form

$$u := (b, y_1, \ldots, y_r)$$

and the conjunction

$$\bigwedge \Gamma_1(b) \wedge \bigwedge_j \alpha_j(b, y_j)$$

is true. We say that v of the form

$$v := (b, z_1, \ldots, z_s)$$

is a witness to Γ_2, Δ_2 if the disjunction

$$\bigvee \Gamma_2(b) \vee \bigvee_j \beta_j(b, z_j)$$

is true.

By induction on the number of inferences above the sequent in π we prove that there is a multivalued function $g(u) \in \mathrm{FP}^{\Sigma_i^b}[\mathrm{wit}, O(\log n)]$ having the property that if u witnesses Γ_1, Δ_1, then any value of $g(u)$ witnesses Γ_2, Δ_2.

This is proved analogously with the proof of Theorem 7.2.3 with an additional case in the $\exists \leq$: introduction-rules and a bit different treatment of the PIND-rule.

The case of $\exists \leq$: **right** only involves evaluating a term; this is done by a polynomial time machine. In the case of $\exists \leq$: **left** there are two different situations: The inference introduces one of the $\exists y_j$'s or it introduces an existential quantifier into a Σ_i^b-formula.

In the former case the witnessing algorithm remains the same, except that a variable treated as a parameter now moves into the nonparametrical part of the witness.

In the latter case assume that the formula introduced by the rule was

$$\exists t \gamma(b, t)$$

and that it was inferred from $\gamma(b, c)$, c a parameter not occurring among $\overline{b}$ (by the proviso of the $\exists \leq$: **left** rule). Assume that $M(b, c, \overline{y})$ computes the witness function for the upper sequent.

Consider a new algorithm M': on the input $\langle b, \overline{y} \rangle$ it first asks the witness-oracle a Σ_i^b-query $[\exists t \gamma(b, t) ?]$. If the answer is negative, M' outputs 0 and halts. This is correct as $\langle b, \overline{y} \rangle$ was not a witness to the antecedent of the lower sequent.

If the answer is affirmative then M' is also provided with a witness c to it: $\gamma(b, c)$. M' then forms a tuple $u = (\overline{b}, c, \overline{y})$ and runs as M on u. Clearly if (b, y) witnessed Γ_1, Δ_1, then the output of this computation must witness Γ_2, Δ_2.

The principal formula of a $\forall \leq$: **right** inference is Π_i^b. This case is treated dually to the $\exists \leq$: **left** inferences.

Now consider a PIND-inference

$$\frac{\gamma(\lfloor c/2 \rfloor) \to \gamma(c)}{\gamma(0) \to \gamma(t)}$$

with $\gamma \in \Sigma_i^b$ (we omit the side formulas). Let M be the witness-oracle machine witnessing the upper sequent. Define machine M' as follows.

On input $u' = \langle b', \ldots \rangle$ it computes the value v of the term t and asks the query $[\gamma(v)?]$. If the answer is affirmative it outputs 0 (any tuple witnesses true $\gamma(t)$). If it is negative M' asks query $[\gamma(0)?]$. If the answer to this query is negative it answers 0 and again stops (no tuple can witness false $\gamma(0)$).

In the remaining case (answers to $\gamma(v)$ and $\gamma(0)$ negative and affirmative, respectively) it finds by binary search some c such that $\gamma(\lfloor c/2 \rfloor)$ is true while $\gamma(c)$ is false. Note that this takes $O(\log n)$ Σ_i^b-queries as we deal with PIND (and where $n = |t|$).

Then M' forms $u = (\overline{b}, c, \ldots)$ and computes as M on u. Any output v of this computation witnesses the succedent of the upper sequent, but as $\gamma(c)$ is false it must witness the disjunction of the remaining formulas in the succedent. These formulas are also in the lower sequent, so v witnesses the succedent of the lower sequent too.   Q.E.D.

Corollary 7.3.6. *For $i \geq 1$, a predicate is Σ^b_{i+1}-definable in S^i_2 if and only if it is a predicate from the class* $P^{\Sigma^p_i}[O(\log n)]$.

Proof. By Corollary 6.3.5 the class $P^{\Sigma^p_i}[O(\log n)]$ equals the class of predicates in $FP^{\Sigma^p_i}[\text{wit}, O(\log n)]$. Theorem 7.3.5 then implies the corollary.   Q.E.D.

In the rest of this section we shall consider a witnessing theorem for a theory extending S^1_2 by a particular combinatorial principle. The witnessing functions will be computed by probabilistic algorithms.

A *pigeonhole principle* is the obvious fact that m pigeons cannot sit in n holes if $m > n$, each hole accommodating at most one pigeon. There are several ways to formalize this principle in arithmetic (see Dimitracopoulos and Paris 1986), and we will now consider one of them. Later (in Chapters 11, 12, and 15) we shall consider some other ways.

The *pigeonhole principle* $PHP(\Sigma^b_\infty)$ is the scheme

$$\forall x \leq n \exists y < n, A(x, y) \;\rightarrow\; \exists x_1 < x_2 \leq n, y < n, A(x_1, y) \wedge A(x_2, y)$$

for bounded formulas A. Observe that $PHP(\Sigma^b_\infty)$ proves (over BASIC) all induction axioms of T_2. It is unknown whether T_2 proves all instances of $PHP(\Sigma^b_\infty)$.

A *weak pigeonhole principle* for function f, $WPHP(f)$, is the formula

$$\exists y \leq 2a \forall x \leq a, f(x) \neq y$$

stating that no map from $[0, a]$ can be *onto* $[0, 2a]$. The word *weak* indicates that a stronger version is also valid with $a + 1$ in place of $2a$.

Let $WPHP(PV_1)$ denote the set of axioms $WPHP(f)$ for every PV_1-function symbol $f(x)$ ($f(x)$ may have other arguments besides x and they are treated as parameters in the axioms). Recall from Section 5.3 that $S^1_2(PV_1)$ is the theory S^1_2 in the language of PV_1; it is a conservative extension of S^1_2 by Theorem 5.3.4 and Corollary 7.2.4. Later (Theorem 11.2.4) we shall see that $T^4_2(PV_1)$ proves the scheme $WPHP(PV_1)$. It is open whether $S^1_2(PV_1)$ proves it too.

Theorem 7.3.7. *Let $\exists z \leq t \, A(a, z)$ be a $\Sigma^b_1(PV_1)$-formula and assume that*

$$S^1_2(PV_1) + WPHP(PV_1) \vdash \exists! z \leq t, A(a, z)$$

Then there is a multifunction g, computable by a probabilistic polynomial time algorithm with bounded error, that witnesses the formula $\forall x \exists z \leq t, A(x, z)$

$$\forall x, g(x) \leq t \wedge A(x, g(x))$$

In particular, predicates Σ^b_1-definable in $S^1_2(PV_1) + WPHP(PV_1)$ are witnessed by a function from R (random polynomial time; cf. Balcazár, Diáz and Gabarró 1988).

We should remark on what we mean by a *bounded error probabilistic* computation of a function g. This means that there is a probabilistic polynomial time algorithm that on input x outputs some $g(x)$ with probability $\geq (3/4)$.

Proof. Assume that there is a $S^1_2(PV_1)$-proof of

$$\exists! z \leq t, A(a, z)$$

using the WPHP(PV_1) axioms for functions $f_1, \ldots, f_k$. For simplicity assume $k = 1$; case $k > 1$ is similar.

It follows that there is a $S^1_2(PV_1)$-proof of

$$(\exists b \forall y \leq 2b \exists x \leq b, f(x) = y) \vee (\exists! z \leq t, A(a, z))$$

Introduce a Herbrand function h to get rid of the universal quantifier:

$$S^1_2(PV_1, h) \vdash (\exists b \exists x \leq b, h(b) \leq 2b \rightarrow f(x) = h(b)) \vee (\exists! z \leq t, A(a, z))$$

By the relativization of Theorem 7.2.3 there is a polynomial time machine M querying an oracle for values of the function h and computing from a a witness

$$M(a, h) := ((b, x), z)$$

to the disjunction, where either (b, x) are witnesses to the existential quantifiers in the first disjunct or z witnesses the second disjunct.

For a particular function $h = h^*$ computing counterexamples

$$h^*(b) := \text{some } y \leq 2b \text{ s.t. } \forall x \leq b, f(x) \neq y$$

it holds (for reasons analogous to those in the proof of Theorem 7.3.3) that the algorithm M outputs $((b, x), z)$ such that z is always a witness to the second disjunct: That is,

$$\forall x, A(x, (M(x, h^*))_2)$$

Assume that $p(n)$ is the time bound of the algorithm M. It is therefore sufficient to describe a probabilistic algorithm for h^* with the probability of error at most $\leq (1/4p(n))$: The probability that algorithm answers at least one query of M about h^* wrongly is then $\leq (1/4)$. Such algorithm is, however, trivial: Pick randomly with a uniform distribution ℓ numbers $\leq 2b$. Each of them has probability $\leq 1/2$ to be in the range of f; hence the probability that all of them are in the range is $\leq (1/2^\ell)$. So it is sufficient to pick $\ell > 2 + \log(p(n))$ such numbers.

Assume now that $B(x)$ is a predicate whose characteristic function is Σ^b_1-definable in $S^1_2(PV_1) + WPHP(PV_1)$. Then the algorithm can check whether $g(a)$ is actually a witness for $B(a)$: That is, it can make an error only if a does not satisfy $B(x)$. This shows $B \in R$.   Q.E.D.

7.4. Σ_{i+2}^b-definability in T_2^i and counterexamples

In this section we prove that Theorem 7.3.3 can be substantially improved when the theory S_2^i is replaced by PV_i, that is, by T_2^{i-1} (if $i > 1$).

Theorem 7.4.1 (Krajíček, Pudlák, and Takeuti 1991). *Let $i \geq 1$ and assume that $\phi(a, x, y)$ is an $\exists\Pi_i^b$-formula. Suppose*

$$T_2^i \vdash \exists x \forall y, \phi(a, x, y)$$

Then there are $\square_{i+1}^p$-functions $f_1(a), f_2(a, b_1), \ldots, f_k(a, b_1, \ldots, b_{k-1})$ with all free variables shown such that T_2^i proves

$$\phi(a, f_1(a), b_1) \vee \phi(a, f_2(a, b_1), b_2) \vee \ldots \vee \phi(a, f_k(a, b_1, \ldots, b_{k-1}), b_k)$$

This is also true for PV_{i+1} in place of T_2^i and for PV_1 if $i = 0$.

Proof. Take PV_{i+1}, a universal conservative extension of T_2^i by Theorem 5.3.5. Assume that

$$\phi(a, x, y) = \exists z \psi(a, x, y, z)$$

with $\psi \in \Pi_i^b$. By the definition of PV_{i+1}, the formula ψ is in PV_{i+1} equivalent to an open formula, say to a formula $\eta(a, x, y, z)$.

Since PV_{i+1} is a universal theory we may apply the Herbrand theorem to obtain from a PV_{i+1}-proof of $\exists z \, \eta(a, x, y, z)$ the terms t_u, s_{uv} such that the disjunction

$$[\eta(a, t_1(a), b_1, s_{1,1}) \vee \ldots \vee \eta(a, t_1(a), b_1, s_{1,n})]$$
$$\vee \ldots \vee [\eta(a, t_k(a, b_1, \ldots, b_{k-1}), b_k, s_{k,1})$$
$$\vee \ldots \vee \eta(a, t_k(a, b_1, \ldots, b_{k-1}), b_k, s_{k,n})]$$

is PV_{i+1}-provable. Terms $t_1, \ldots, t_k$ depend only on the variables shown; terms $s_{u,v}$ may depend on all variables $a, b_1, \ldots, b_k$.

From the disjunction follows by a repeated introduction of $\exists$-quantifiers instead of terms $s_{u,v}$ (and contracting occurrences of identical formulas) the disjunction

$$\exists z \, \eta(a, t_1(a), b_1, z) \vee \ldots \vee \exists z \, \eta(a, t_k(a, b_1, \ldots, b_{k-1}), b_k, z)$$

The functions

$$f_j := t_j$$

are then the desired functions.   Q.E.D.

Corollary 7.4.2. *For $i \geq 1$, a function Σ_{i+2}^b-definable in PV_i (or in T_2^{i-1} for $i > 1$) is computable by a counterexample computation where* the student *is a polynomial time machine with a Σ_{i-1}^p-oracle and* the teacher *answers at most ℓ queries for counterexamples to a Π_{i+1}^b-property, where ℓ is a constant.*

Proof. Let $f(a) = b$ be defined by

$$\exists x \le t \forall y \le t, \phi(a, b, x, y)$$

with $\phi \in \Sigma_i^b$, that is,

$$\mathrm{PV}_i \vdash \exists b \exists x \forall y, \phi(a, b, x, y)$$

(we leave out the bounds $\le t$ for simplicity of notation). Write this as

$$\mathrm{PV}_i \vdash \exists c \forall y, \psi(a, c, y)$$

where

$$\psi(a, c, y) := \phi(a, (c)_1, (c)_2, y)$$

It is obviously sufficient to describe a computation of c.

By the previous theorem there are $\square_i^p$-functions $f_1, \ldots, f_k$ such that

$$\psi(a, f_1(a), y_1) \vee \ldots \vee \psi(a, f_k(a, y_1, \ldots, y_{k-1}), y_k)$$

The counterexample computation, with $\ell = k - 1$, will look like this: The student computes $c_1 := f_1(a)$ and submits to the teacher that $\forall y \psi(a, c_1, y)$. If he fails he also receives a counterexample: y_1 s.t. $\neg \psi(a, c_1, y_1)$. Then the student computes $c_2 := f_2(a, y_1)$ and submits it to the teacher, and so on. As the preceding disjunction is valid, one of $c_1, \ldots, c_k$ computed with the help of at most ℓ counterexamples must work.   Q.E.D.

This corollary will play a crucial role in Section 10.2.

7.5. Σ_1^b-definability in T_2^1 and polynomial local search

In this section we shall characterize the Σ_1^b-consequences of T_2^1 in terms of *polynomial local search* problems. These problems occur naturally in polynomial time version only, but their obvious generalizations to $\square_i^p$-functions would, via a statement analogous to Theorem 7.5.3, characterize Σ_i^b-consequences of T_2^i.

The following definition is from Johnson, Papadimitriou, and Yannakakis (1988).

Definition 7.5.1. *A polynomial local search problem (PLS-problem) P is an optimization problem satisfying the following conditions:*

1. *Instances of the problem P are $x \in \{0, 1\}^*$, and for any x there is a set of solutions $F_P(x)$ satisfying*

   (i) the binary predicate $s \in F_P(x)$ is polynomial time

   (ii) $\forall x \forall s \in F_P(x), |s| \le p(|x|)$, some polynomial p

   (iii) $\forall x, 0 \in F_P(x)$.

2. A cost function *is a polynomial time function*

$$c_P(s, x) \; : \; \{0, 1\}^* \times \{0, 1\}^* \to N$$

3. A neighborhood function $N_P(s, x)$ *is polynomial-time and it satisfies*

$$\forall x, s, \; N_P(s, x) \in F_P(x)$$

4. *The cost and the neighborhood functions satisfy*

$$\forall x, s, \; N_P(s, x) \neq s \to c_P(s, x) < c_P(N_P(s, x), x)$$

5. *The* task *is, given* x *find a* locally optimal solution $s \in F_P(x)$, *that is, a solution* $s \in F_P(x)$ *for which*

$$N_P(s, x) = s$$

It follows from the definition that there is a polynomial time computable function $M_P(x)$ such that $M_P(x) > c_P(s, x)$ for all $s \in F_P(x)$.

A PLS-problem P can be expressed by a Π_1^b-sentence: the conjunctions of the first four conditions. If this is provable in T_2^1 then we say P is *definable* in T_2^1. The formula $\mathrm{Opt}_P(x, s)$ is the Δ_1^b-formula formalizing $N_P(s, x) = s$.

Lemma 7.5.2. *Let P be a PLS-problem definable in T_2^1. Then*

$$T_2^1 \vdash \forall x \exists y, \; \mathrm{Opt}_P(x, y).$$

Proof. By Lemma 5.2.7 T_2^1 proves the Σ_i^b–MAX axioms. Hence in T_2^1, for all x, there is a maximum value $c_0 < M_P(x)$ satisfying

$$\exists s \in F_P(x), c_P(s, x) = c_0$$

Taking s to be a witness for this last formula, s is even globally optimal and hence, specifically, satisfies $\mathrm{Opt}_P(x, s)$.   Q.E.D.

We shall use the Witness formula from Section 7.2.

Theorem 7.5.3 (Buss and Krajíček 1994). *Let $R(a)$ be a Σ_1^b-formula such that $T_2^1 \vdash \forall x \, R(x)$. Then there is a PLS-problem P definable in T_2^1 such that T_2^1 proves*

$$\forall x \forall s, \; \mathrm{Opt}_P(x, s) \to \; \mathrm{Witness}_R^{1,a}(s, x).$$

Proof. Assume that

$$T_2^1 \vdash R(a)$$

and that R is a strict Σ_1^b-formula. By Theorem 7.2.3 there is a T_2^1-proof π in the system LKB of the sequent $\longrightarrow R(a)$ such that every sequent in π has the form

$$A_1(\overline{b}), \ldots, A_k(\overline{b}) \longrightarrow B_1(\overline{b}), \ldots, B_\ell(\overline{b})$$

where $\bar{b}$ are all r (including a) free variables and where all the formulas A_i and B_i are strict Σ_1^b-formulas.

We shall prove by induction on the number of inferences in π above the sequent that the sequent corresponds computationally to a PLS-problem. Namely, there is a PLS-problem P' such that

1. inputs to P' are $k + r$-tuples $\langle m_1, \ldots, m_r, v_1, \ldots, v_k \rangle$ where $m_1, \ldots, m_r$ are values for the variables $b_1, \ldots, b_r$
2. for an input tuple $\langle \overline{m}, \overline{v} \rangle$, the locally optimal solutions are the $k + r + 1$-tuples of the form $\langle \overline{m}, \overline{v}, w \rangle$ such that if each v_i witnesses $A_i(\overline{m})$ then w is a witness for one of the formulas $B_1(\overline{m}), \ldots, B_\ell(\overline{m})$.

From such a problem P' we get problem P satisfying the requirement of the theorem by adding to each P'-solution $\langle \overline{v}, w \rangle$ a new neighbor w with a higher cost, provided w is a witness to R.

The existence of the PLS-problem is obvious for the initial sequents. The cases of the propositional inferences and the structural inferences are obvious, requiring only minor changes to the PLS-problem.

The case of an $\exists \le :$ **right** inference

$$\frac{\Gamma \longrightarrow \Delta, A(t)}{t \le s, \Gamma \longrightarrow \Delta, \exists x \le s A(x)}$$

is simple too: By the induction hypothesis there is a PLS problem P_0 that applies to the upper sequent. We modify P_0 to get a PLS problem P' that works for the lower sequent. First, let

$$c_{P'}(s, x) = c_{P_0}(s, x) + 1$$

for $s \in F_{P_0}(x)$.

Inputs $\langle \overline{m}, v_0, \overline{v} \rangle$ to P' that provide witnesses to Γ are assigned cost 0 and have as neighbor the input $\langle \overline{m}, \overline{v} \rangle$ to P_0. An output $\langle \overline{m}, \overline{v}, w \rangle$ of P_0 has as its P'-neighbor a tuple $\langle \overline{m}, v_0, \overline{v}, w' \rangle$ with the cost $M_{P_0}(\langle \overline{m}, \overline{v} \rangle) + 1$, where $w' = w$ or $w' = \langle t(\overline{m}), w \rangle$, depending on which one of them provides a witness to a formula in the succedent

$$\Delta, \exists x \le s A$$

Clearly P' has the desired properties.

The cases of $\exists \le :$ **left** and $\forall \le :$ **left** are handled by simple modifications to the PLS-problem too. The case where the final inference is a $\forall \le :$ **right** is analogous to the case of the induction rule IND treated later.

Consider the case where the sequent was inferred using the cut-inference

$$\frac{\Gamma \longrightarrow A \qquad A \longrightarrow \Delta}{\Gamma \longrightarrow \Delta}$$

By the induction hypothesis, two PLS-problems P_1 and P_2 for the upper sequents satisfy the requirements.

A PLS problem for the lower sequent is formed as a "composition" of PLS problems. To simplify this case, we assume w.l.o.g. that the cut formula A is the only formula in the succedent (antecedent) of the left (resp. right) upper sequent. W.l.o.g. assume that domains F_{P_1} and F_{P_2} are disjoint. The local optima of the problem P_1 will have as neighbors the instances of P_2. By adding M_{P_1} to the cost function of P_2, the cost of any P_2-solution is greater than the cost of any P_1-solution. This allows us to arrange that any local optimum of the combined problem can be found by applying P_2 to a local optimum of P_1. The details are left to the reader.

Finally consider the case when the sequent was inferred by the induction inference Σ_1^b–IND

$$A(b_0, \overline{b}) \longrightarrow A(b_0 + 1, \overline{b})$$
$$\overline{A(0, \overline{b}) \longrightarrow A(t(b), \overline{b})}$$

Assume w.l.o.g. that there are no side formulas. Given a PLS-problem P_0 for the upper sequent, define problem P' for the lower sequent by an exponentially long iteration of instances of P_0.

First, the set $F_{P'}(\langle \overline{m}, v \rangle)$ is the set of tuples $\langle m_0, z, s \rangle$ where $m_0 < t(\overline{m})$ and $s \in F_{P_0}(\langle m_0, \overline{m}, z \rangle)$; thus $F_{P'}$ is a disjoint union of the solution sets of the instances of P_0. Then define

$$c_{P'}(\langle m_0, z, s \rangle, \langle \overline{m}, v \rangle) = m_0 \cdot M(m, \overline{z}) + c_P(s, \langle m_0, \overline{m}, z \rangle)$$

where M is large enough to dominate $M_{P_0}(s)$ whenever $m_0 < t(\overline{m})$ and $s \in F_{P_0}(\langle m_0, m, z \rangle)$. The neighborhood function is defined such that

$$N_{P'}(\langle m_0, z, s \rangle, \langle \overline{m}, v \rangle) = \langle m_0, z, N_{P_0}(s, \langle m_0, \overline{m}, z \rangle) \rangle$$

with the exception of the case when $s = N_{P_0}(s, \langle m_0, \overline{m}, z \rangle)$, in which case we set

$$N_{P'}(\langle m_0, z, s \rangle, \langle \overline{m}, v \rangle) = \langle m_0 + 1, z', \langle m_0 + 1, \overline{m}, z' \rangle \rangle$$

for $m_0 < t(\overline{m}) - 1$, where z' is the last component of s, that is, the witness for $A(m_0 + 1, \overline{m})$. If $m_0 = t(\overline{m}) - 1$, then

$$N_{P'}(\langle m_0, z, s \rangle, \langle \overline{m}, v \rangle) = \langle \overline{m}, v, z' \rangle$$

This last case gives a local optimum for P'. It is easy to verify that P' gives a PLS-problem that satisfies the requirement of the theorem for the sequent.

Q.E.D.

Corollary 7.5.4. *A multivalued function f is Σ_1^b-definable in T_2^1 iff f can be expressed as a PLS-problem composed with a projection function.*

Proof. Any PLS-problem (and hence also its projection) is by Lemma 7.5.2 Σ_1^b-definable in T_2^1. On the other hand, by Theorem 7.5.3, witnesses to a Σ_1^b-definable function can be computed by a PLS-problem (i.e., its local optima are all witnesses) and a witness to the existential quantifier of the Σ_1^b-formula is obtainable by a projection from the witness.   Q.E.D.

It is a bit unnatural to speak about definable multivalued functions but in this situation it appears inevitable. It is an open question whether any PLS-problem P can be reduced to a PLS-problem P' that has a *unique* optimal solution in every instance. Reduction refers to an obvious notion: By polynomial time functions translate P-instances into P'-instances and then P'-local optima back to P-local optima (cf. Johnson et al. 1988 for details).

On the side of bounded arithmetic the related problem (by Lemma 7.5.2 and Theorem 7.5.3) is whether for every Σ_1^b-formula A for which

$$T_2^1 \vdash \forall x \exists y, A(x, y)$$

one can find another Σ_1^b-formula A^* such that
   (i) $T_2^1 \vdash A^*(x, y) \rightarrow A(x, y)$
   (ii) $T_2^1 \vdash \forall x \exists! y, A^*(x, y)$
(an analogous statement is open for T_2^i too). Note that by Corollary 7.2.6 this implication holds for S_2^1.

Several other *search classes* were introduced by Papadimitriou and Yannakakis (1988) and Papadimitriou (1990, 1994), in particular the classes PPA, PPAD, and PPP. A search problem in PPA is defined as a PLS-problem except that there is no cost function and the neighborhood function satisfies
   1. $\forall x, s,\ N_P(s, x) \subseteq F_P(x) \setminus \{s\} \wedge |N_P(s, x)| \leq 2$
   2. $\forall x, s, s',\ s' \in N_P(s, x) \equiv s \in N_P(s', x)$
   3. $\forall x \exists!\ s,\ s \in N_P(0, x)$.
The task is to find $s \in F_P(x) \setminus \{0\}$ such that

$$\exists!\ s',\ s' \in N_P(s, x)$$

Thinking about the condition $s' \in N_P(s, x)$ as defining the edge (s, s') in a graph $G_P(x)$ with the set of vertices $F_P(x)$, the conditions imply that $G_P(x)$ is a symmetric graph of degree ≤ 2 in which vertex 0 has degree 1. Hence the task is to find another degree 1 vertex.

The problems in the class PPAD are defined similarly but the graph $G_P(x)$ is now directed and satisfies the conditions
   1. $\text{indeg}(0) = 0 \wedge \text{outdeg}(0) = 1$
   2. $\text{indeg}(v) \leq 1 \wedge \text{outdeg}(v) \leq 1$ for any vertex v
and the task is to find a vertex $v \neq 0$ such that

$$\text{indeg}(v) + \text{outdeg}(v) \leq 1$$

A problem in class PPP is specified by a polynomial time function

$$f_P(s, x) : F_P(x) \to F_P(x) \setminus \{0\}$$

and the task is to find two $s, s' \in F_P(x)$, $s \neq s'$, such that

$$f_P(s, x) = f_P(s', x)$$

In other words: the task is to find two witnesses to the fact that $f_P(s, x)$ does not violate the pigeonhole principle.

Denote by $\mathrm{PPA}(PV_1)$ the set of $\forall \Sigma_1^b(PV_1)$-sentences expressing that no PV_1-function $N_P(s, x)$ can define a symmetric graph of degree ≤ 2 with $\deg(0) = 1$ and *no* other degree 1 node. Similarly denote by $\mathrm{PPAD}(PV_1)$ and $\mathrm{PPP}(PV_1)$ the corresponding principles underlying the classes PPAD and PPP.

The following theorem is proved similarly to Theorem 7.3.7 (interpreting the Herbrand function h as a function h^* witnessing the principle).

Theorem 7.5.5. *A multivalued function f is Σ_1^b-definable in $S_2^1(PV_1) + PPA(PV_1)$ iff it can be witnessed by PPA problem.*
Similarly for PPAD and PPP.

A reason for studying these classes is that numerous other seemingly different search problems are reducible to one of them (cf. Papadimitriou 1990).

7.6. Model-theoretic constructions

In this section we shall give a few model-theoretic constructions for some results proved earlier in the chapter by proof-theoretic methods and obtain a new proposition.

We begin with a general discussion of complexity classes in models of PV. This is elaborated on more in Sections 15.2–3.

In model M of PV the class P is the class of subsets of M definable by an atomic PV-formula with parameters from M (in S_2^1 this would be provably Δ_1^b-formulas with parameters), or equivalently, recognizable by a *standard* DTM with an extra input (the parameter) that may be nonstandard, or equivalently, recognizable by a DTM possibly with a nonstandard description but whose time is bounded by a standard degree polynomial from $M[n]$.

The class P/poly is defined in the same way except that the parameter of the standard DTM may vary with the length of the inputs, the class $\mathrm{P}^{\mathrm{stan}}$ is defined as the class P but allowing no parameters, and the classes NP, NP/poly, $\mathrm{NP}^{\mathrm{stan}}$ are defined analogously using NDTMs.

The first theorem is a strengthening of Corollary 7.2.5.

Theorem 7.6.1. *Let M be a countable model of PV. Then there exists a model M^* of PV, a Σ_1^b-elementary extension of M, such that*

$$M^* \models \text{``} P = NP \cap coNP \text{''}$$

Proof. Let $\alpha(x) \in \Sigma_1^b$ and $\beta(x) \in \Pi_1^b$ be two formulas with parameters from M. There are two cases:

(a)

$$PV + \text{Th}_{\forall \Pi_1^b}(M) \vdash \forall x (\alpha(x) \equiv \beta(x))$$

in which case there is by the Herbrand theorem a PV-symbol $f(x, y)$ and $b \in M$ such that

$$PV + \text{Th}_{\forall \Pi_1^b}(M) \vdash \forall x (\alpha(x) \equiv (f(x, b) = 1))$$

so the set defined by $\alpha(x)$ is in the class P of M, or

(b)

$$PV + \text{Th}_{\forall \Pi_1^b}(M) \nvdash \forall x (\alpha(x) \equiv \beta(x))$$

in which case *either*

(1)

$$PV + \text{Th}_{\forall \Pi_1^b}(M) \nvdash \forall x (\neg \alpha(x) \vee \beta(x))$$

which means that

$$M \models \alpha(c) \wedge \neg \beta(c)$$

some $c \in M$, and this will remain true in every extension of M, *or*

(2)

$$PV + \text{Th}_{\forall \Pi_1^b}(M) \nvdash \forall x (\alpha(x) \vee \neg \beta(x))$$

In case (2) take a *new* constant c and form a model M' satisfying

$$PV + \text{Th}_{\forall \Pi_1^b}(M) + \neg \alpha(c) + \beta(c)$$

Any Σ_1^b-elementary extension of M' will satisfy this theory and hence also

$$\neg \forall x (\alpha(x) \equiv \beta(x))$$

Enumerate with infinite repetitions $(\alpha_0, \beta_0), (\alpha_1, \beta_1), \ldots$ all pairs of a Σ_1^b and a Π_1^b-formula in the language of PV augmented by a name for every element of M and by countably many new constants. Construct a countable chain of Σ_1^b-elementary extensions

$$M = M_0 \subseteq M_1 \subseteq \ldots$$

where M_{i+1} is obtained from M_i as M' from M for pair $(\alpha, \beta) = (\alpha_i, \beta_i)$, and such that every element of any M_i has a name among the constants of the language.

Then the union of the chain

$$M^* := \bigcup_i M_i$$

is the desired model.   Q.E.D.

The nonuniform analog of this theorem for P/poly, NP/poly, and coNP/poly also holds as one can treat every length from $\text{Log}(M)$ separately in the construction of the chain.

Next we give a model-theoretic proof for the counterexample witnessing Theorem 7.4.1.

Lemma 7.6.2. *Let M be a model of T_2^i, for $i \geq 1$, or of PV_1 for $i = 0$. Assume that $M^* \subseteq M$ is a subset of M closed under all standard $\square_{i+1}^p$-functions definable in M using parameters from M^*.*

Then it holds that

   1. M^* is a Σ_i^b-elementary substructure of M
   2. $M^* \models T_2^i$ for $i \geq 1$, or $M^* \models PV_1$ for $i = 0$

Proof. The Skolem functions for the Σ_i^b-formulas are Σ_{i+1}^b-definable in T_2^i, as T_2^i proves the Σ_i^b–MAX principle (by Lemma 5.2.7). This shows condition 1.

For condition 2 we want to show that IND holds for any Σ_i^b-formula ϕ

$$\neg\phi(0) \vee \phi(a) \vee (\exists x < a, \phi(x) \wedge \neg\phi(x + 1))$$

Since M^* is a Σ_i^b-elementary substructure of M, it suffices to find a $\square_{i+1}^p$-function in M that on the input a for which $\phi(0) \wedge \neg\phi(a)$ holds outputs $x < a$ for which

$$\phi(x) \wedge \neg\phi(x + 1)$$

Such a function is available in PV_{i+1} by definition (cf. Section 5.3): Take the function $(h(b, b))_1$.   Q.E.D.

With the help of this lemma we give an alternative proof of Theorem 7.4.1.

Let ϕ satisfy the hypothesis of that theorem and assume for the sake of contradiction that for no k and $f_1, \ldots, f_k \in \square_{i+1}^p$, T_2^i proves the disjunction required in the theorem.

Let $f_1, f_2, \ldots$ be an enumeration of all $\square_{i+1}^p$-functions with the following properties:

   (i) f_j is $\leq j$-ary,
   (ii) each $\square_{i+1}^p$-function occurs in the list infinitely many times.

Let $c, d_1, d_2 \ldots$ be new constants. By compactness then the theory

$$T_2^i + \neg\phi(c, f_1(c), d_1) \vee \neg\phi(c, f_2(c, d_1), d_2) \vee \ldots$$

is consistent. Take M as a model of this theory.

Let $M^* \subseteq M$ be the set

$$M^* := \{f_1(c), f_2(c, d_1), \ldots\}$$

It holds that

1. $\{c, d_1, \ldots\} \subseteq M^*$
2. M^* is closed under the $\Box_{i+1}^p$-functions.

This is because the projections are $\Box_{i+1}^p$-definable and because each $\Box_{i+1}^p$-function occurs infinitely many times.

Lemma 7.6.2 then implies that $M^* \models T_2^i$ and it is Σ_i^b-elementary. That is,

$$M^* \models T_2^i + \forall x \exists y, \neg\phi(c, x, y)$$

(for $x = f_j(c, d_1, \ldots, d_{j-1})$ take $y = d_j$). This contradicts the hypothesis, and hence Theorem 7.4.1 is proved.

The last theorem in this section is a model-theoretic statement implying the witnessing Theorem 7.2.3. In proving that theorem, one would like to reason as follows. Assume that for $A \in \Sigma_i^b$, S_2^i does not prove $A(a, f(a))$ for any $f \in \Box_i^p$. By compactness then there is a model M and $a \in M$ such that

$$M \models S_2^i + \neg A(a, f(a))$$

for all $f \in \Box_i^p$. Take $M^* \subseteq M$ to be the subset of M generated from a by all $\Box_i^p$-functions. M^* is a substructure of M and

$$M^* \models \exists x \forall y \neg A(x, y)$$

Unfortunately the argument fails at this point as there is no apparent reason why M^* should satisfy S_2^i.

However, a chain construction similar to the one underlying Theorem 7.6.1 and Lemma 7.6.2 works.

Theorem 7.6.3. *Any countable model M of PV_i has a Σ_i^b-elementary extension M' such that*

1.

$$M' \models S_2^i(PV_i)$$

2. *for any open PV_i-formula $\phi(x, y)$ with parameters from M' there is a PV_i-term $f(x)$ with parameters from M' for which*

$$M' \models \forall x \exists y \phi(x, y) \rightarrow \forall x \phi(x, f(x))$$

Proof. Let M be a countable model of PV_i and let ϕ be an open PV_i-formula with parameters from M. Let $\mathrm{Th}_{\Pi_1^b}(M)$ denote the Π_1^b-theory of M. If

$$PV_i + \mathrm{Th}_{\Pi_1^b}(M) \vdash \forall x \exists y \leq t(x)\phi(x, y)$$

then by the Herbrand theorem (analogously to the proof of Theorem 7.6.1) there
is a PV_i-term $f(x)$ such that

$$\forall x, \; f(x) \leq t(x) \wedge \phi(x, y)$$

will hold in every Σ_1^b-elementary extension of M. Otherwise there is either $a \in M$
such that

$$M \models \forall y \leq t(a) \neg \phi(a, y)$$

and that will remain valid in all Σ_1^b-elementary extensions of M or, if there is
no such element a, take a *new* constant c and form a model M_1 of the theory
$PV_i + Th_{\Pi_1^b}(M) + \forall y \leq t(c) \neg \phi(c, y)$. Hence $\forall x \exists y \leq t(x) \phi(x, y)$ will fail in all
extensions of M_1.

Iterating this construction countably many times as in the proof of Theorem
7.6.1 yields a model M' satisfying condition 2. However, that already implies that
M' is also a model of $S_2^i(PV_i)$: if ψ is an open PV_i-formula and $b \in M'$ such that

$$M' \models \psi(0, u) \wedge \forall x < |b| \forall v \exists w, \; \psi(x, v) \rightarrow \psi(x + 1, w)$$

holds (with the bounds to v, w implicit in ψ) then by condition 2 also

$$M' \models \forall x < |b| \forall v \psi(x, v) \rightarrow \psi(x + 1, f(x, v))$$

for some PV_i-term f. As M' is a model of PV_i there is $w \in M'$ of the form

$$w = \langle u_0, \ldots, u_{|b|} \rangle$$

where

$$u_0 := u \quad \text{and} \quad u_{j+1} := f(j, u_j)$$

But then

$$M' \models \forall j < |b|, \; \psi(j, u_j) \rightarrow \psi(j + 1, u_{j+1})$$

and by induction for the formula $\psi(x, u_x)$ available in PV_i $\psi(|b|, u_{|b|})$ follows:
That is, $\exists v \psi(|b|, v)$ holds in M'.   Q.E.D.

Following Zambella (1994) we derive Corollary 7.2.4 from Theorem 7.6.3.
Assume that $S_2^i(PV_i)$ proves $\forall x \exists y \leq t(x) \phi(x, y)$, where ϕ is an open PV_i-
formula (every Σ_i^b-formula is equivalent to a $\Sigma_1^b(PV_i)$-formula). Assume for the
sake of contradiction that for no PV_i-term $f(x)$ does PV_i prove $\forall x, \; f(x) \leq t(x)$
$\wedge \phi(x, f(x))$. By compactness we may take a countable model M of PV_i in which

$$M \models \neg(f(a) \leq t(a) \wedge \phi(a, f(a)))$$

holds for some $a \in M$ and all PV_i-terms f. Take M' to be the extension of M guaranteed by Theorem 7.6.3. Clearly $\forall x \exists y \leq t(x)\phi(x, y)$ fails in M', contradicting the assumption that the formula is provable in S_2^i.

7.7. Bibliographical and other remarks

The proof-theoretic method for proving the witnessing theorems for systems S_2^i and T_2^i was developed by Buss (1986) and Theorem 7.1.4 is stated there. The *witness* formula (7.2.1) is also defined there. Theorem 7.2.3 follows from Buss (1986, 1990a) from which Corollaries 7.2.4–7.2.7 were also obtained.

The use of Herbrand functions for witnessing Theorem 7.3.3 comes from Krajíček (1992); it was obtained by using a direct method in Pudlák (1992b). Theorem 7.3.5 and Corollary 7.3.6 are from Krajíček (1993).

Theorem 7.3.7 was proved by Wilkie (unpublished). Theorem 7.4.1 is from Krajíček et al. (1991). Section 7.5 is based on Buss and Krajíček (1994). Topics mentioned at the end of Section 7.5 are considered in Chiari and Krajíček (1994).

Theorem 7.6.1 is new. I do not know whether it is valid for P^{stan}, NP^{stan} and $coNP^{stan}$ in place of P, NP, and coNP. Lemma 7.6.2 and the model-theoretic proof of Theorem 7.4.1 follow Krajíček et al. (1991).

The first model-theoretic proof of Corollary 7.2.4 was obtained by Wilkie (unpublished). The proof here follows Zambella (1994).

8

Definability and witnessing in second order theories

This chapter is devoted primarily to proving several definability and witnessing theorems for the second order system U_2^i and V_2^i, analogous to those in Chapters 6 and 7.

Our tool is the RSUV isomorphism (Theorem 5.5.13), or rather the definition of V_1^i (Definition 5.5.3), together with the model-theoretic construction of Lemma 5.5.4.

The first section discusses and defines the second order computations. In the second section are proved some definability and witnessing theorems for the second order systems and further conservation results for first order theories (Corollaries 8.2.5–8.2.7). The proofs are sketched and the details of the RSUV isomorphism arguments are left to the reader.

8.1. Second order computations

Let $A(a, \beta^{t(b)})$ be a second order bounded formula and $(K, \mathcal{X})$ a model of V_1^1. By Definition 5.5.3 we may think of K as of $K = \text{Log}(M)$ for some $M \models S_2^1$, with $\mathcal{X}$ being the subsets of K coded in M. Pick some $a, b \in K$ of length n and some $\beta^{t(b)}$. Then

$$(K, \mathcal{X}) \models A(a, \beta^{t(b)})$$

if and only if (see Theorem 5.5.13 for the notation)

$$M \models A^1(a, u)$$

where u codes $\beta^{t(b)}$. The length of u is thus $2^{|t(b)|} = 2^{O(n)}$.

Moreover, if $A \in \Sigma_i^{1,b}$ then $A^1 \in \Sigma_i^b$; hence a second order query to a $\Sigma_i^{1,b}$ oracle with the first order inputs of length n and the second order inputs (i.e., the characteristic function of $\beta^{t(b)}$) of length $2^{O(n)}$ is just a query to a Σ_i^b-oracle of length $2^{O(n)}$.

Hence the second order computations in a model of V_1^1 are analogous to the first order computations in a model M of S_2^1 stipulating that some inputs and outputs are from $\text{Log}(M)$. With such a notion we shall be able to use the RSUV-isomorphism 5.5.13 to characterize the functions definable in the second order systems.

One more remark concerns the time bound. If $(K, \mathcal{X}) \models V_2^1$ then $M \models S_3^1$, where S_3^1 is defined as S_2^1 is but in the extended language $L(\#_3) := L \cup \{\#_3\}$, where

$$x \#_3 y = 2^{|x||\#|y|}$$

Compare the definition of the function $\omega_2(x)$ in Theorem 5.1.5. This is completely analogous to Lemma 5.5.4 as K is closed under # iff the lengths in M are closed under # iff M itself is closed under $\#_3$. Similarly R_3^i is defined as R_2^i (Definition 5.5.12) but in the language $L(\#_3)$. We use a statement analogous to Theorem 5.5.13 as a lemma, as we shall use it later.

Lemma 8.1.1. *The RSUV-isomorphism (Theorem 5.5.13) holds identically for pairs of theories S_3^i and V_2^i, and R_3^i and U_2^i.*

In addition definability and witnessing theorems of Chapters 6 and 7 straightforwardly translate to the theories in the language $L(\#_3)$. Call $\#_3$-*time* a time bound $t(n)$ where t is a term of language L. We have, for example, a statement analogous to Theorems 6.1.2 and 7.2.3: The theory $T_3^i \Sigma_{i+1}^b(\#_3)$-defines exactly the functions computable in the $\#_3$-time using a Σ_i^b-oracle.

In the following definition we identify a finite set ψ^y with its characteristic function on the interval $[0, y]$.

Definition 8.1.2. *Let $Q(x, \psi^y)$ be a second order oracle set, with x its first order input and ψ^y its second order input.*

A Turing machine M with the time bound $t(n)$, with first and second order inputs and with the oracle Q, is required to satisfy the following conditions:

1. *M has two read-only input tapes: a first order one with inputs of size n and a second order one with inputs of size $\leq 2^{t(n)}$.*

2. *M writes the queries on a pair of write-only tapes. On the first order tape M writes a string x of size $\leq t(n)$, on the second order tape a string ψ^y of size $\leq 2^{t(n)}$. The oracle answers YES/NO according to whether $Q(x, \psi^y)$ holds or not, and erases the query tapes.*

3. *The total computational time is bounded by $2^{t(n)}$.*

4. *M writes the outputs on two write-only tapes: one for a first order output, which can have size at most $\leq t(n)$, and one for a second order output, which may have size $\leq 2^{t(n)}$.*

$TIME(2^{t(n)})^Q$ denotes the class of the functionals computable by a Turing machine M satisfying the conditions stated. Set

$$EXP^Q := \bigcup_k TIME(2^{n^k})^Q$$

and

$$EXP^{\Sigma_i^{1,b}} := \bigcup_{Q \in \Sigma_i^{1,b}} EXP^Q$$

The space complexity of such a computation is the total space used on working tapes of M. $SPACE(s(n))^Q$ denotes the class of the functionals computable by such computations with the space complexity $\leq s(n)$ and

$$PSPACE^{\Sigma_i^{1,b}} := \bigcup_{Q \in \Sigma_i^{1,b}} \bigcup_k SPACE(n^k)^Q$$

We say that a machine from the definition computes a *function*, if it has no second order inputs or outputs (but it may ask second order queries).

Lemma 8.1.3. Let $i \geq 0$. The functionals from the class $EXP^{\Sigma_i^{1,b}}$ that are functions are precisely the functions from the class $EXP^{\Sigma_i^p}$; for $i = 0$ this is EXP.

Similarly the functionals from the class $PSPACE^{\Sigma_i^{1,b}}$ that are functions are precisely the functions from $PSPACE^{\Sigma_i^p}$, and for $i = 0$ this is just PSPACE.

Proof. The statement follows from the discussion before Lemma 8.1.1. The cases for $i = 0$ follow from the observation that any $\Sigma_0^{1,b}$-property can be decided in PSPACE.   Q.E.D.

8.2. Definable functionals

The first result is analogous to Theorem 7.2.3.

Theorem 8.2.1. Let $i \geq 1$. The functionals from the class $EXP^{\Sigma_{i-1}^{1,b}}$ are precisely those $\Sigma_i^{1,b}$-definable in V_2^i. In particular, the functions $\Sigma_1^{1,b}$-definable in V_2^1 are precisely the functions from the class EXP.

Proof. Let $F(a, \alpha) = (b, \beta)$ be a functional (for simplicity we suppress the superscripts in the second order variables). Let $(K, \mathcal{X})$ be a model of V_2^i, $a, b \in K$, $\alpha, \beta \in \mathcal{X}$ such that

$$(K, \mathcal{X}) \models F(a, \alpha) = (b, \beta)$$

Take $M \models S_3^i$ s.t. $K = \mathrm{Log}(M)$ and s.t. $\mathcal{X}$ are the subsets of K coded in M, and let $\alpha = \tilde{u}, \beta = \tilde{v}$ for $u, v \in M$.

Assume that a $\Sigma_i^{1,b}$-formula $A(a, \alpha, b, \beta)$ defines the graph of F in $(K, \mathcal{X})$. Then the Σ_i^b-formula $A^1(a, u, b, v)$ defines the graph in M (cf. Theorem 5.5.13 for the definition of A^1). As A defines F provably in V_2^i, by Theorem 7.2.3 A^1 defines a $\square_i^p(\#_3)$-function f provably in S_3^i. On the input a, u with $|a| = n$ and $|u| = 2^{n^{O(1)}}$ the running time is $2^{n^{O(1)}}$, and it queries a Σ_{i-1}^b-oracle.

The same machine then computes F when α, β are identified with their characteristic functions., and the oracle becomes, by the remarks before Lemma 8.1.1, a $\Sigma_{i-1}^{1,b}$-oracle. Thus F is in $\mathrm{EXP}^{\Sigma_{i-1}^{1,b}}$.

The extra statement for $i = 1$ follows then from Lemma 8.1.3. Q.E.D.

If we would augment the machine in Definition 8.1.2 by a witness-oracle rather than just an oracle (cf. Definition 6.3.1) then it would not change the class $\mathrm{EXP}^{\Sigma_i^{1,b}}$. This is because the number of queries is unlimited so M may ask consecutively for all $\leq 2^{t(n)}$ bits of a witness. However, when the number of queries is bounded, then this trivial simulation does not apply and we get an apparently distinct class.

Definition 8.2.2. Class $\mathrm{EXP}^{\Sigma_i^{1,b}}[wit, poly]$ is a class of multivalued functionals computable by a Turing machine M in time $2^{n^{O(1)}}$ equipped with a witness-oracle from $\Sigma_i^{1,b}$ satisfying all conditions of Definition 8.1.2 and in addition satisfying

1. If the oracle answer to the query (x, ψ) is YES then the oracle writes on a special write-only witness tape a witness for the $\Sigma_i^{1,b}$-formula $Q(x, \psi)$.
2. The total number of queries made is bounded by a polynomial $n^{O(1)}$.

Lemma 8.2.3. Let $i \geq 0$ and assume that values of $f \in \mathrm{EXP}^{\Sigma_i^{1,b}}[wit, poly]$ are only first order.
Then, in fact

$$f \in \mathrm{PSPACE}^{\Sigma_i^{1,b}}$$

Proof. Computing a particular bit of the value of f does not require witnesses to the oracle answers, analogously to Lemma 6.3.4 and Corollary 6.3.5.

For f with only first order values we may compute each bit of the output separately. By Definition 8.1.2 there are at most $n^{O(1)}$ bits in the output and in a computation of each the machine asks $n^{O(1)}$ queries. So f is computable by an $\mathrm{EXP}^{\Sigma_i^{1,b}}$-machine with only *polynomially* many queries to the oracle: Call the class of the functionals computable in this way $\mathrm{EXP}^{\Sigma_i^p}[poly]$. Completely analogously with Theorem 6.2.3 (condition (i) of that theorem) we get

$$\mathrm{PSPACE}^{\Sigma_i^{1,b}} = \mathrm{EXP}^{\Sigma_i^{1,b}}[poly]$$

Q.E.D.

Theorem 8.2.4. For $i \geq 1$, multivalued functionals $\Sigma_i^{1,b}$-definable in U_2^i are precisely those from the class $\mathrm{EXP}^{\Sigma_{i-1}^{1,b}}[wit, poly]$. In particular, the functions

$\Sigma_i^{1,b}$-definable in U_2^i are those from the class $PSPACE^{\Sigma_{i-1}^{1,b}}$, which is PSPACE for $i = 1$.

For $i \geq 2$ the same is true with V_2^{i-1} in place of U_2^i.

Proof. In the RSUV-isomorphism (Theorem 5.5.13) the theory U_2^i corresponds to R_3^i, V_2^{i-1} to S_3^{i-1} and $\Sigma_i^{1,b}$ to Σ_i^b.

By statements analogous to Theorems 6.3.3 and Corollary 7.3.6 for the $\#_3$-time, the multivalued functions Σ_i^b-definable in S_3^{i-1} are precisely those from the class $TIME(\#_3)^{\Sigma_{i-1}^b}[poly]$. Hence the RSUV-isomorphism implies the part of the theorem for V_2^{i-1}.

To prove the part for U_2^i we establish a statement analogous to Theorem 6.3.3 and Theorem 7.3.6 for the theory R_3^i in place of S_3^{i-1}. In the proof of the witnessing Theorem 7.3.5 the number of queries in the functions witnessing the sequents in a proof increases by a constant for all LKB rules except for the cut, by a factor of 2 for the cut rule, and by $O(\log n)$ queries in the case of the Σ_{i-1}^b–PIND rule. Now, if the rule

$$\frac{\phi(\lfloor a/2 \rfloor) \longrightarrow \phi(a)}{\phi(0) \longrightarrow \phi(|t|)}$$

of R_3^i from Definition 5.5.12 is used instead of the Σ_{i-1}^b–PIND, the witness contains a witness to the existential quantifiers of the Σ_i^b-formula ϕ, and hence the function g witnessing the lower sequent is obtained from the function g_1 witnessing the upper sequent by $\leq ||t||$ iterations. The value of $||t||$ is $(\log n)^{O(1)}$ for the language $L(\#_3)$. Hence if the number of queries asked in the computation of g_1 is $(\log n)^{O(1)}$, so is the number of queries asked in the computation of g.

Thus a witnessing argument identical to that in the proof of Theorem 7.3.5 shows that the multivalued functions Σ_i^b-definable in R_3^i are in the class $TIME(\#_3)^{\Sigma_{i-1}^b}[(\log n)^{O(1)}]$.

The proof of Theorem 6.3.3 needs no change at all to show that S_3^{i-1} can Σ_i^b-define all functions from $TIME(\#_3)^{\Sigma_{i-1}^b}[(\log n)^{O(1)}]$, and hence also R_3^i as $S_3^{i-1} \subseteq R_3^i$ (see the remark after Definition 5.5.12).

This gives the wanted characterization of the functions Σ_i^b-definable in R_3^i, and hence the theorem is proved.   Q.E.D.

Corollary 8.2.5. *For $i \geq 1$, the theory U_2^{i+1} is $\forall \Sigma_{i+1}^{1,b}$-conservative over V_2^i.*
Also the theory R_3^{i+1} is $\forall \Sigma_{i+1}^b$-conservative over S_3^i.

Proof. The corollary follows from the previous theorem similarly to the way Corollary 7.2.4 follows from Theorem 7.2.3.   Q.E.D.

From this statement and its proof we can deduce three more interesting corollaries. Recall that $\mathcal{B}(\Sigma_i^{1,b})$ denotes the class of Boolean combinations of $\Sigma_i^{1,b}$-formulas, AC is *the axiom of choice* scheme (cf. Lemma 5.5.8), and $BB\Sigma_i^b$ is the *sharply bounded collection scheme* (cf. Definition 5.2.11).

Corollary 8.2.6. *For $i \geq 1$, the theory U_2^{i+1} is $\forall B(\Sigma_{i+1}^{1,b})$-conservative over the theory $V_2^i + \Sigma_{i+1}^{1,b}$–AC.*
Also the theory R_3^{i+1} is $\forall B(\Sigma_{i+1}^b)$-conservative over the theory $S_3^i + BB\Sigma_{i+1}^b$.

Proof. Work with the first order systems. As any Boolean combination can be written in a conjunctive normal form, it suffices to prove the statement for a formula of the form

$$\pi \vee \sigma$$

with $\pi \in \Pi_{i+1}^{1,b}$ and $\sigma \in \Sigma_{i+1}^{1,b}$, that is, for sequents of the form

$$\neg\pi \longrightarrow \sigma$$

This is a sequent of $\Sigma_{i+1}^{1,b}$-formulas, and hence the witnessing theorem for R_3^{i+1} (outlined in the proof of Theorem 8.2.4) implies that R_3^{i+1} proves the sequent

$$\exists w \ \mathrm{Witness}_{\neg\pi}^{i+1,a}(w, a) \longrightarrow \exists w' \ \mathrm{Witness}_\sigma^{i+1,a}(w', a)$$

which is, by Corollary 8.2.5, provable in S_3^i too.

Hence it would be enough to show in S_3^i the equivalence

$$\exists w' \ \mathrm{Witness}_\sigma^{i+1,a}(w', a) \equiv \sigma(a)$$

By Lemma 7.2.2 (a statement analogous to it for the $\#_3$-time), however, this is provable in $S_3^1 + BB\Sigma_{i+1}^b$.

This proves the theorem for the first order case and by RSUV-isomorphism also for the second order case. Q.E.D.

The next corollary is related to Lemma 5.2.9.

Corollary 8.2.7. *For $i \geq 1$, both S_3^i and S_2^i prove the Δ_{i+1}^b–PIND scheme.*
Also V_2^i proves the $\Delta_{i+1}^{1,b}$–PIND scheme.

Proof. As with Lemma 5.2.9 one can show

$$R_3^i \vdash \Delta_{i+1}^b\text{–PIND}$$

An instance of PIND for a Δ_{i+1}^b-formula is $\forall\Sigma_{i+1}^b$; hence Corollary 8.2.5 implies that

$$S_3^i \vdash \Delta_{i+1}^b\text{–PIND}$$

too.

This proof does not apply to S_2^i but we may argue as follows. By Lemma 5.2.13 we know that $S_2^i + BB\Sigma_{i+1}^b$ is $\forall\Sigma_{i+1}^b$-conservative over S_2^i. It is thus sufficient to prove Δ_{i+1}^b–PIND in $S_2^i + BB\Sigma_{i+1}^b$. We have

$$S_2^i + BB\Sigma_{i+1}^b \vdash \forall x \exists y \forall t \leq |x|, \sigma(t) \equiv t \in y$$

for any Δ^b_{i+1}-formula σ, as follows from $BB\Sigma^b_{i+1}$ applied to the formula

$$\forall t \leq |x| \exists y, (y = 1 \wedge \sigma(t)) \vee (y = 0 \wedge \neg\sigma(t))$$

Hence LIND for σ follows from Δ^b_1–LIND for $t \in y$, which is provable in S^1_2.
$\Delta^{1,b}_{i+1}$–PIND is a special case of $\Sigma^{1,b}_{i+1}$–PIND, and hence provable in U^{i+1}_2, and
its instance is a $\forall\Sigma^{1,b}_{i+1}$-formula, hence by Corollary 8.2.5 also provable in V^i_2.

Q.E.D.

We conclude this section with a useful statement.

Corollary 8.2.8. *For $i \geq 1$*

$$V^i_2 \vdash \Delta^{1,b}_{i+1}\text{–}AC$$

Proof. By Lemma 5.2.8 U^{i+1}_2 proves $\Delta^{1,b}_{i+1}$–AC and by Corollary 8.2.5 it is provable in V^i_2 too.   Q.E.D.

8.3. Bibliographical and other remarks

The treatment of the second order computations follows Buss (1986) and Buss et al. (1993). Theorem 8.2.1 is from Buss (1986); Definition 8.2.2 and the statements 8.2.3–8.2.8 are from Buss et al. (1993).

We did not use the proof-theoretic treatment of the second order systems but we at least remark on the sequent calculus formalization. Namely, the $\Sigma^{1,b}_0$–CA scheme becomes provable from the introduction rules for the second order $\exists$

$$\frac{\Gamma \longrightarrow \Delta, A(\phi/B)}{\Gamma \longrightarrow \Delta, \exists\phi\, A(\phi)}$$

where A is any formula and B is $\Sigma^{1,b}_0$. The CA for B then follows by applying the rule to

$$\longrightarrow \forall x \leq a, B(x) \equiv B(x)$$

quantifying only one occurrence of B.

9

Translations of arithmetic formulas

We shall define in this chapter two translations of bounded arithmetic formulas into propositional formulas and, more importantly, we shall also define translations of proofs in various systems of bounded arithmetic into propositional proofs in particular proof systems.

In the first section we shall consider the case when the language of $I\Delta_0$ is augmented by new predicate or function symbols, and the case of the theories U_1^1 and V_1^1. In the second section we treat formulas in the language L and the theories S_2^i, T_2^i, and U_2^1.

In the third section we study the provability of the reflection principles for propositional proof systems in bounded arithmetic and the relation of these reflection principles to the polynomial simulations. In the fourth section we present some model-theoretic proofs for statements obtained earlier. The final section then suggests another relation of arithmetic proofs to Boolean logic, namely the relation between witnessing arguments and test (decision) trees.

9.1. Bounded formulas with a predicate

First we shall treat the theory $I\Delta_0(R)$ and then generalize the treatment to the theories U_1^1 and V_1^1. Instead of $I\Delta_0(R)$ we could consider the theory $I\Sigma_0^{1,b}$ but the presentation for the former is simpler. The language $L_{PA}(R)$ of $I\Delta_0(R)$ is the language L_{PA} augmented by a new binary predicate symbol $R(x, y)$.

Definition 9.1.1. *Let $\theta(a_1, \ldots, a_k)$ be a bounded formula in the language $L_{PA}(R)$ and let p_{ij} be propositional atoms, one for each pair $(i, j) \in N \times N$.*

For $n_1, \ldots, n_k \in N$ we define the propositional formula $\langle\theta\rangle_{(n_1,\ldots,n_k)}$ by induction on the logical depth of θ:

139

1. *if θ is the atomic formula $s(\bar{a}) = t(\bar{a})$ then*

$$\langle\theta\rangle_{(\bar{n})} := \begin{cases} 1 & \text{if } s(\bar{n}) = t(\bar{n}) \text{ is true} \\ 0 & \text{if } s(\bar{n}) = t(\bar{n}) \text{ is false} \end{cases}$$

2. *if θ is the atomic formula $s(\bar{a}) \leq t(\bar{a})$ then*

$$\langle\theta\rangle_{(\bar{n})} := \begin{cases} 1 & \text{if } s(\bar{n}) \leq t(\bar{n}) \text{ is true} \\ 0 & \text{if } s(\bar{n}) \leq t(\bar{n}) \text{ is false} \end{cases}$$

3. *if θ is the atomic formula $R(s(\bar{a}), t(\bar{a}))$, $s(\bar{n}) = i$ and $t(\bar{n}) = j$ then*

$$\langle\theta\rangle_{(\bar{n})} := p_{ij}$$

4. *if $\theta = \neg\xi$ then*

$$\langle\theta\rangle_{(\bar{n})} := \neg\langle\xi\rangle_{(\bar{n})}$$

5. *if $\theta = \nu \circ \xi$, $\circ = \vee, \wedge$ then*

$$\langle\theta\rangle_{(\bar{n})} := \langle\nu\rangle_{(\bar{n})} \circ \langle\xi\rangle_{(\bar{n})}$$

6. *if $\theta = \exists x \leq s(\bar{a})\, \nu(\bar{a}, x)$ and $s(\bar{n}) = u$ then*

$$\langle\theta\rangle_{(\bar{n})} := \bigvee_{m \leq u} \langle\nu\rangle_{(\bar{n},m)}$$

7. *if $\theta = \forall x \leq s(\bar{a})\, \nu(\bar{a}, x)$ and $s(\bar{n}) = u$ then*

$$\langle\theta\rangle_{(\bar{n})} := \bigwedge_{m \leq u} \langle\nu\rangle_{(\bar{n},m)}$$

In clause 6 (resp. 7) the disjunction (resp. the conjunction) is formed from the binary connectives with the brackets associated, for example, to the left.

The next lemma is proved by induction on the complexity of θ.

Lemma 9.1.2. *Let $\theta(\bar{a})$ be a bounded formula in the language $L_{PA}(R)$. Then there are d and ℓ such that for every $\bar{n}$*

1. $dp(\langle\theta\rangle_{(\bar{n})}) \leq d$
2. $|\langle\theta\rangle_{(\bar{n})}| \leq (\max(\bar{n}) + 2)^{\ell}$

Note that we used *the depth* of θ and not *the logical depth*, as the connectives in clauses 6 and 7 of the preceding definition are of unbounded arity.

The theory $I\Delta_0(R)$ is defined exactly as the theory $I\Delta_0$ is in Section 5.1 except that the axiom scheme of *induction* IND is accepted for all bounded formulas in $L_{PA}(R)$. The theories $T_2^i(R)$, $S_2^i(R)$, and $S_2(R)$ are defined analogously.

Theorem 9.1.3. *Let $\theta(a)$ be a bounded formula in $L_{PA}(R)$ and assume*

$$I\Delta_0(R) \vdash \forall x \theta(x)$$

Then there are d and ℓ such that every propositional formula $\langle\theta\rangle_{(n)}$ has a depth d F-proof of size at most n^ℓ.

Moreover, there is a polynomial time algorithm producing on input $1\ldots 1(n$-times) a depth d F-proof of $\langle\theta\rangle_{(n)}$.

Proof. We shall describe the construction of a depth d size n^ℓ LK-proof of the sequent

$$\longrightarrow \langle\theta\rangle_{(n)}$$

and it will be obvious that the required algorithm exists. This is equivalent to the required task by Lemma 4.4.15.

By cut-elimination Theorem 7.1.4 (straightforwardly modified for $I\Delta_0(R)$) there is an LKB-proof π using the $\Delta_0(R)$–IND rule of the sequent

$$\longrightarrow \theta(a)$$

A sequent in the proof π has the form

$$\phi_1(\overline{b}), \ldots \phi_r(\overline{b}) \longrightarrow \psi_1(\overline{b}), \ldots \psi_s(\overline{b})$$

where all ϕ_i, ψ_j are $\Delta_0(R)$ (by Corollary 7.1.5). By induction on the number of inferences above the sequent in π, prove that there are d and ℓ such that for any tuple $\overline{n}$ the sequent

$$\langle\phi_1\rangle_{(\overline{n})}, \ldots, \langle\phi_r\rangle_{(\overline{n})} \longrightarrow \langle\psi_1\rangle_{(\overline{n})}, \ldots, \langle\psi_s\rangle_{(\overline{n})}$$

has a depth d size $(\max(\overline{n}) + 2)^\ell$ LK-proof.

By Definition 7.1.1 all initial sequents have the form $A \longrightarrow A$, A atomic, or $\longrightarrow A$, A an axiom of PA^-. In the former case the propositional translation is either $0 \longrightarrow 0$, $1 \longrightarrow 1$, or $p_{ij} \longrightarrow p_{ij}$. In the latter case the translation is of the form $\longrightarrow \tau$, where τ is a true Boolean sentence (i.e., without atoms). Moreover, the depth of τ is constant ($=$ the maximal logical depth of a PA^--axiom). Any such sentence has a $dp(\tau)$ LK-proof of size $O(|\tau|)$.

The case when the sequent was obtained by structural or propositional rules or by the cut-rule is obvious: The same rules of propositional logic should be applied to the propositional translations of the upper sequents.

For the closed terms $t \leq s(\overline{n})$ is a formula of the form $\langle\eta(t)\rangle_{(\overline{n})}$ one of the disjuncts of $\langle\exists x \leq s, \eta(x)\rangle_{(\overline{n})}$; thus $\exists \leq :$ *right* rule is simulated by repeated (polynomially many times) $\bigvee :$ *right* rule of LK.

For the $\forall \leq :$ *right* inference

$$\frac{a \leq t, \Gamma \longrightarrow \Delta, A(a)}{\Gamma \longrightarrow \Delta, \forall x \leq t\ A(x)}$$

assume that for each $a = 0, 1, \ldots, \text{val}(t)$ there is an LK-proof of the translation of the upper sequent with the required properties. Then all $a \leq t$ translate to 1, and thus can be cut out with the initial sequent $\longrightarrow 1$, and the sequents obtained are joined by repeated applications of the $\bigwedge$: *right* rule for $a = 0, 1, \ldots,$ val(t). Hence the size of this translation is val$(t)^{O(1)} = \max(\bar{n})^{O(1)}$. The left quantifier rules are treated analogously.

Finally, the IND-rule

$$\frac{A(a), \Gamma \longrightarrow \Delta, A(a+1)}{A(0), \Gamma \longrightarrow \Delta, A(t)}$$

is simulated by applying the cut rule to the LK-proofs of the translations of the upper sequent for $a = 0, 1, \ldots,$ val$(t) - 1$. Q.E.D.

Corollary 9.1.4. *Assume that T is a theory defined as $I\Delta_0(R)$ except that the language of T might be richer than $L_{PA}(R)$ and that T contains finitely many bounded axioms besides the induction axioms. Let $f(n)$ be a non-decreasing function such that any term in the language of T can be majorized by some iteration of $f(n)$.*

Then if $\theta(a)$ is a bounded formula in the language of T and

$$T \vdash \theta(a)$$

there are d and ℓ such that every propositional formula $\langle\theta\rangle_{(n)}$ has depth d LK-proof of size $\leq f^{(\ell)}(n)$ (the ℓth iteration of f).

Proof. The argument is entirely the same as in the preceeding proof, estimating the values of the terms in the axioms, in the quantifier rules, and in the IND-rule by iterations of the function $f(n)$ rather then by the terms themselves. Q.E.D.

We shall extend the simulation from Theorem 9.1.3 to the second order theories U_1^1 and V_1^1.

First we have to amend Definition 9.1.1 as the language of U_1^1 and V_1^1 does not contain symbol $R(x, y)$ but has second order variables α^t and new atomic formulas $\alpha^t = \beta^s$ and $s \in \alpha^t$.

For each variable $\alpha^{t(a)}$ and $u \in \omega$ introduce new atoms $p_0, p_1, \ldots, p_v$ where $v = $ val$(t(u))$, and define

1. $\langle\alpha^{s(x)} = \beta^{t(y)}\rangle_{(n,m)} := \bigwedge_{i \leq s(n)}(p_i \equiv q_i) \wedge \bigwedge_{s(n) < i \leq t(m)} \neg q_i$ where $s(n) \leq t(m)$. *The case $s(n) > t(m)$ is defined analogously.*

2.

$$\langle s(x) \in \beta^{t(y)}\rangle_{(n,m)} := \begin{cases} q_u & \text{if } u = s(n) \leq t(m) \\ 0 & \text{otherwise} \end{cases}$$

where atoms p_i and q_j are associated with α^s and β^t.

A sequence of formulas $\theta_1, \ldots, \theta_k$ is called an EF-*sequence* iff it satisfies the conditions of Definition 4.5.2 to be an EF-proof with the condition that no extension atom appears in θ_k dropped.

Theorem 9.1.5. *Let $A(x)$ be a $\Sigma_0^{1,b}$-formula and assume that*

$$V_1^1 \vdash \forall x \, A(x).$$

Then the formulas $\langle A(x) \rangle_n$ have polynomial size EF-proofs.

Proof. We shall consider V_1^1 formalized in the sequent calculus with the $\Sigma_1^{1,b}$–IND-rule in place of the $\Sigma_1^{1,b}$–IND-axioms and with the introduction rules for the second order quantifiers replacing $\Sigma_0^{1,b}$–CA (cf. Lemma 5.5.4 and the discussion of $\Sigma_0^{1,b}$–CA in Section 8.3).

Assume that π is a V_1^1-proof of the sequent $\to A(a)$; w.l.o.g. we may assume that all formulas in π are *strict*$\Sigma_1^{1,b}$. These are $\Sigma_1^{1,b}$-formulas in which all second order quantifiers precede all first order quantifiers and all connectives (a notion analogous to strictΣ_1^b from Lemma 5.2.15).

By induction on the number of steps in π above a sequent show that if

$$\exists \psi_1 B_1(\overline{x}, \overline{\alpha}, \psi_1), \ldots, \exists \psi_u B_u(\overline{x}, \overline{\alpha}, \psi_u)$$
$$\to \exists \xi_1 C_1(\overline{x}, \overline{\alpha}, \xi_1), \ldots, \exists \xi_v C_v(\overline{x}, \overline{\alpha}, \xi_v)$$

is a sequent in π, then there is a constant k such that for all $\overline{m}$ there is an EF-sequence of size at most $(\max(\overline{m}) + 2)^k$ ending with the sequent

$$\langle B_1 \rangle_{\overline{m}}(\overline{p}^{\alpha}, \overline{p}^{\psi_1}), \ldots, \langle B_u \rangle_{\overline{m}}(\overline{p}^{\alpha}, \overline{p}^{\psi_u})$$
$$\longrightarrow \langle C_1 \rangle_{\overline{m}}(\overline{p}^{\alpha}, \overline{p}^{\xi_1}), \ldots, \langle C_v \rangle_{\overline{m}}(\overline{p}^{\alpha}, \overline{p}^{\xi_v})$$

and such that in this EF-sequence none of the atoms $p_t^{\alpha_i}$ or $p_t^{\psi_j}$ corresponding to a free second order variable α_i (resp. to a second order variable ψ_j) from an antecedent is an extension atom.

The construction follows the proof of Theorem 9.1.3 and we need to treat only two new rules: the introduction of the second order $\exists$ to the succedent and $\Sigma_1^{1,b}$–IND (the introduction of the second order $\exists$ to the antecedent does not change the translation).

Assume that in the former case the minor formula of the inference is

$$C\left(\overline{x}, \overline{\alpha}, \frac{t \in \xi}{E(\overline{x}, t, \overline{\alpha})}\right)$$

with both $C, E \in \Sigma_0^{1,b}$, and that the principal formula is $\exists \xi \, C(\overline{x}, \overline{\alpha}, \xi)$. Introduce a new atom $p_t^{\xi} \equiv \langle E \rangle_{\overline{m},t}(\overline{p}^{\alpha})$. Then the equivalence

$$\langle C \rangle_{\overline{m}}\left(\overline{p}^{\alpha}, \frac{p_t^{\xi}}{\langle E \rangle_{\overline{m},t}}\right) \equiv \langle C \rangle_{\overline{m}}(\overline{p}^{\alpha}, \overline{p}^{\xi})$$

can be derived from the new extension axioms by an F-derivation of size

$$O\left(\left(|\langle C \rangle_{\overline{m}}| + |\langle E \rangle_{\overline{m},t}|\right)^2\right) = \left(\max(\overline{m} + 2)^{O(1)}\right),$$

as t is implicitly bounded in E by a power of $\max(\overline{m})$. This concludes the first case.

Now consider the $\Sigma_1^{1,b}$–IND inference

$$\frac{\exists \xi_b C(b, \xi_b) \rightarrow \exists \xi_{b+1} C(b + 1, \xi_{b+1})}{\exists \xi_0 C(0, \xi_0) \rightarrow \exists \xi_n C(n, \xi_n)}$$

(the other free variables and the side formulas are omitted for simplicity). By the induction hypothesis we have polynomial size EF-sequences ending with the formulas

$$\langle C \rangle_{\overline{m},u} (\overline{p}^{\,\xi_u}) \rightarrow \langle C \rangle_{\overline{m},u+1} (\overline{p}^{\,\xi_{u+1}})$$

for $u = 0, 1, \ldots, n-1$. Joining these sequences by $n-1$ cuts gives an EF-sequence ending with the implication

$$\langle C \rangle_{\overline{m},0} (\overline{p}^{\,\xi_0}) \rightarrow \langle C \rangle_{\overline{m},n} (\overline{p}^{\,\xi_n})$$

of total size polynomial in $\max(\overline{m}, n)$.

As the formula A is $\Sigma_0^{1,b}$, atoms in $\langle A \rangle_{(n)}$ correspond to free second order variables in A and hence cannot be the extension atoms. Thus the final EF-sequence is, in fact, an EF-proof.   Q.E.D.

The proof can be modified to yield the following statement.

Theorem 9.1.6. *Let $A(x)$ be a $\Sigma_0^{1,b}$-formula and assume that*

$$U_1^1 \vdash \forall x \, A(x)$$

Then the formulas $\langle A(x) \rangle_n$ have F-proofs of size $n^{(\log n)^{O(1)}}$

Proof. By Theorem 9.1.5 there is an EF-proof of $\langle A(x) \rangle_n$ of size $n^{O(1)}$. Moreover, by its proof in these EF-proofs will be introduced only $(\log n)^{O(1)}$ extension atoms as the U_1^1-proof π uses $\Sigma_1^{1,b}$–PIND instead of $\Sigma_1^{1,b}$–IND. There will be $O((\log n)^k)$ extension atoms if there are k nested induction inferences.

Having such an EF-proof, replace in it the last introduced extension atom by its definition: This yields a new EF-proof of size $n^{O(1)} \cdot n^{O(1)}$ with one less application of the extension rule. Repeating this procedure until all extension axioms are eliminated produces an F-proof of size $n^{(\log n)^{O(1)}}$.   Q.E.D.

9.2. Translation into quantified propositional formulas

We define a translation of Σ_∞^b-formulas of the language L into the quantified propositional formulas.

Let $A(a_1, \ldots, a_k)$ be a Σ_∞^b-formula. As all terms in L are polynomial time functions there exists a polynomial $p_A(x)$ such that for $n_1, \ldots, n_k$ the truth value of $A(n_1, \ldots n_k)$ can be computed in the interval $[0, 2^{p_A(m)}]$, where $m = \max_i(|n_i|)$.

Some canonical $p_A(x)$ is easy to define by induction on the logical complexity of A and we shall assume that such polynomials are fixed. Any polynomial $q(x)$ majorizing $p_A(x)$ is called a *bounding polynomial of A*.

We shall also need to express the bits of numbers, and for better readability we shall write the function $\text{bit}(a, i)$ as $a(i)$. Hence

$$n = \sum_{i \le |n|} n(i) \cdot 2^i$$

Define $a(i) := 0$ for $i > |a|$.

For B a quantified propositional formula with atoms $\overline{p} = (p_0, p_1, \ldots)$, $B(n)$ denotes B with the bits $n(i)$ substituted for p_i.

Any term $t(a_1, \ldots, a_k)$ can be computed for $|a_1|, \ldots, |a_k| \le m$ by a Boolean circuit C_t (by Theorem 3.1.4) with $k \cdot (m + 1)$ bits of input and poly(m) bits of output (the bits of the value of $t(\overline{a})$). Moreover, the size of C_t is polynomial in m. If we introduce new atoms for every node of the circuit, the statement "C_t on inputs $\overline{p}_1, \ldots, \overline{p}_k$ outputs $\overline{q}$" can be expressed by a Σ_1^q-propositional formula (saying that there exists a computation of C_t), which we shall denote

$$B_t^m(\overline{p}_1, \ldots, \overline{p}_k, \overline{q})$$

Such a formula is defined precisely by induction on the complexity of the term t and we leave it to the reader. Note that

$$|B_t^m(\overline{p}_1, \ldots, \overline{p}_k, \overline{q})| = m^{O(1)}$$

Finally, recall Definition 4.6.2 of the systems G, G_i, and G_i^*.

Definition 9.2.1. *Let $A(a_1, \ldots, a_k)$ be a bounded formula in the language L of S_2. Let $q(x)$ be a bounding polynomial for the formula A.*

For every m we construct a quantified propositional formula

$$\|A\|_{q(m)}^m$$

with the atoms $\overline{p}_i$, $i = 1, \ldots, k$ where each $\overline{p}_i = (p_i^0, \ldots, p_i^{q(m)})$. We proceed by induction on the logical complexity of A:

(a) For A the atomic formula $t(\overline{a}) = s(\overline{a})$ define

$$\|A\|_{q(m)}^m := \exists x_0, \ldots, x_{q(m)}, y_0, \ldots, y_{q(m)}, B_t(\overline{p}_1, \ldots, \overline{p}_k, q_j/x_j)$$
$$\wedge B_s(\overline{p}_1, \ldots, \overline{p}_k, q_j/y_j) \wedge \bigwedge_{i \le q(m)} x_i \equiv y_i$$

where we define that tuples $\overline{p}_1, \ldots, \overline{p}_k, \overline{q}$ in B have $q(m) + 1$ bits, with p_i^j for $j > m$ not occurring and with not all q^j necessarily occurring.

(b) *For A the atomic formula $t(\overline{a}) \leq s(\overline{a})$ define*

$$\|A\|_{q(m)}^{m} := \exists x_0, \dots, x_{q(m)}, y_0, \dots, y_{q(m)}, B_t(\overline{p}_1, \dots, \overline{p}_k, q_j/x_j)$$

$$\wedge B_s(\overline{p}_1, \dots, \overline{p}_k, q_j/y_j)$$

$$\wedge \bigwedge_{0 \leq i \leq q(m)} \left(\left(\bigwedge_{i+1 \leq j \leq q(m)} x_j \equiv y_j \right) \wedge x_i \rightarrow y_i \right)$$

(the last conjunct defines the lexicographic order on $\overline{x}, \overline{y}$).

(c) *For $A = \neg A_1$ define*

$$\|A\|_{q(m)}^{m} := \neg \|A_1\|_{q(m)}^{m}$$

(d) *For $A = A_1 \circ A_2$, $\circ = \vee, \wedge$, define*

$$\|A\|_{q(m)}^{m} := \|A_1\|_{q(m)}^{m} \circ \|A_2\|_{q(m)}^{m}$$

(e) *For $A(a) = \exists x \leq |t| A_1(a, x)$ define*

$$\|A\|_{q(m)}^{m} := \bigvee_{\overline{\epsilon}} \|b \leq |t| \wedge A_1(a, b)\|_{q(m)}^{m}(\overline{q}/\overline{\epsilon})$$

and for $A(a) = \forall x \leq |t| A_1(a, x)$ define

$$\|A\|_{q(m)}^{m} := \bigwedge_{\overline{\epsilon}} \|b \leq |t| \rightarrow A_1(a, b)\|_{q(m)}^{m}(\overline{q}/\overline{\epsilon})$$

where $\overline{\epsilon}$ range over $(q(m) + 1)$-tuples of $0, 1$ s.t. $\epsilon_i = 0$ for $i > |q(m)|$.

(f) *For $A(a) = \exists x \leq t A_1(a, x)$ with t not of the form $|s|$ define*

$$\|A\|_{q(m)}^{m} := \exists x_0, \dots, x_{q(m)} \|b \leq t \wedge A_1(a, b)\|_{q(m)}^{m}(\overline{q}/\overline{x})$$

and for $A(a) = \forall x \leq t A_1(a, x)$ with t not of the form $|s|$ define

$$\|A\|_{q(m)}^{m} := \forall x_0, \dots, x_{q(m)} \|b \leq t \rightarrow A_1(a, b)\|_{q(m)}^{m}(\overline{q}/\overline{x})$$

where $\overline{q}$ is the tuple associated with b.

Lemma 9.2.2. *Let A be a bounded formula and $q(x)$ its bounding polynomial. Then it holds that*

1. *There is a polynomial $r(x)$ such that for all m*

$$|\,\|A\|_{q(m)}^{m}| \leq r(m)$$

2. *If $A \in \Sigma_0^b$ then $\|A\|_{q(m)}^{m}$ is Δ_1^q in G_1^*. That is: Provably in G_1^* the formula $\|A\|_{q(m)}^{m}$ is equivalent to a Σ_1^q-formula and to a Π_1^q-formula.*
3. *If $A \in \Sigma_i^b$ then $\|A\|_{q(m)}^{m}$ is provably in G_i^* a Σ_i^q-formula.*

Proof. Condition 1 is obvious from the definition of $\|A\|_{q(m)}^m$. For condition 2 it is sufficient to verify the claim for atomic formulas, as translations of the Σ_0^b-formula are formed from translation of atomic formulas by Boolean connectives only.

For A of the form $t = s$, $\|A\|_{q(m)}^m$ can be expressed as

$$\forall x_0, \ldots, x_{q(m)}, y_0, \ldots, y_{q(m)}, B_t(\overline{p}_1, \ldots, \overline{p}_k, q_j/x_j)$$
$$\wedge B_s(\overline{p}_1, \ldots, \overline{p}_k, q_j/y_j) \to \bigwedge_{i \leq q(m)} x_i \equiv y_i$$

and for $A = t \leq s$, $\|A\|_{q(m)}^m$ can be expressed as

$$\forall x_0, \ldots, x_{q(m)}, y_0, \ldots, y_{q(m)}, B_t(\overline{p}_1, \ldots, \overline{p}_k, q_j/x_j)$$
$$\wedge B_s(\overline{p}_1, \ldots, \overline{p}_k, q_j/y_j) \to \bigwedge_{i \leq q(m)} \left(\left(\bigwedge_{i+1 \leq j \leq q(m)} x_i \equiv y_i \right) \wedge x_i \to y_i \right)$$

That these formulas are equivalent to the original ones depends on the fact that a circuit has a unique computation on an input, and the provability of the equivalence rests on the provability of the formulas

$$\forall x_0, \ldots, x_{q(m)}, y_0, \ldots, y_{q(m)}, B_t(\overline{p}_1, \ldots, \overline{p}_k, q_j/x_j)$$
$$\wedge B_t(\overline{p}_1, \ldots, \overline{p}_k, q_j/y_j) \to \bigwedge_{i \leq q(m)} x_i \equiv y_i$$

(and similarly for B_s).

To see that this formula is provable in G_1^* in size poly(m) assume that B_t has the form

$$B_t := \exists w_1, \ldots, w_{r(m)} D_t(\overline{p}_1, \ldots, \overline{p}_k, \overline{w}, \overline{x})$$

where D_t is quantifierfree, the bits $w_1, \ldots$ correspond to the nodes of the circuit C_t, and D_t is a conjunction of $\leq r(m)$ conditions saying that the value at the node w_i is computed from the values at the incoming nodes according to the definition of C_t, and that $\overline{p}_i$'s are the inputs and $\overline{x}$'s are the outputs. Now assume w.l.o.g. that C_t computes w_i's in the order $w_1, w_2, \ldots$. Then by induction on ℓ construct G_0-proofs of the implications

$$D_t(\overline{p}_1, \ldots, \overline{p}_k, \overline{w}, \overline{x}) \wedge D_t(\overline{p}_1, \ldots, \overline{p}_k, \overline{v}, \overline{y}) \to \bigwedge_{i \leq \ell} w_i \equiv v_i$$

This needs $O(\ell)$ steps, each of which has size poly(m): That is, the total size is poly(m). This proves condition 2.

Condition 3 follows from 2 by induction on the logical complexity of A.

Q.E.D.

Lemma 9.2.3. *Let $A \in \Sigma_0^b$, t be a term and $q(x)$ a bounding polynomial for $A(t)$. Then for every m there are size $m^{O(1)}$ G_1^*-proofs of*

$$\|t = a \rightarrow A(a)\|_{q(m)}^m \rightarrow \|A(t)\|_{q(m)}^m$$

Moreover, these proofs are constructible from m by a polynomial time function definable in S_2^1.

Proof. The lemma is proved by induction on the logical complexity of the formula A and the term t, and we leave it to the reader.

The definability of the proofs in S_2^1 follows immediately.   Q.E.D.

Lemma 9.2.4. *Let A be an axiom of BASIC and $q(x)$ a bounding polynomial of A. Then for all m there are size $m^{O(1)}$ G_1^*-proofs of the formula $\|A\|_{q(m)}^m$. Moreover, these proofs are constructible by a polynomial time function definable in S_2^1.*

Proof. This lemma is a special case of a theorem of Cook (1975) . That theorem says that the translations of all axioms of PV have polynomial size extended resolution proofs and hence by Lemmas 4.5.8 and 4.6.3 also G_1^*-proofs.

Such proofs are constructed in two steps: First find a PV-derivation of A (A expressed as an equation) consisting of the equations $E_1, \ldots, E_\ell$, and then by induction on i construct size poly(m) proofs of the formulas $\|E_i\|_{r(m)}^m$, where $r(m)$ is a bounding polynomial for all $E_1, \ldots, E_\ell$.

The first step is straightforward as the axioms of BASIC mostly state just the recursive properties of functions that are used in their PV-definition by the limited recursion on notation.

The second step is also simple. The only nontrivial part is a simulation of rule $R5$ in Definition 5.3.2, which is handled in the same way as the proofs of the last implication in the proof of Lemma 9.2.2. That is: The equality $\|f_1 = f_2\|_{q(m)}^m$ of the new functions introduced in the rule is proved by induction on the length of y. Q.E.D.

Theorem 9.2.5. *Let $i \geq 1$ and $A(a)$ be a Σ_i^b-formula. Assume that*

$$T_2^i \vdash A(a)$$

Then there is a bounding polynomial $q(x)$ for A such that for all m formulas $\|A\|_{q(m)}^m$ have size $m^{O(1)}$ G_i-proofs.

Moreover, these G_i-proofs are definable in S_2^1.

Proof. For this proof it is convenient to extend the system G by two derived rules for introduction of implication:

Impl : left

$$\frac{\Gamma \longrightarrow \Delta, B \qquad C, \Gamma \longrightarrow \Delta}{B \rightarrow C, \Gamma \longrightarrow \Delta}$$

Impl : right

$$\frac{B, \Gamma \longrightarrow \Delta, C}{\Gamma \longrightarrow \Delta, B \to C}$$

Recall that $B \to C$ is an abbreviation of $\neg B \vee C$ and the preceding rules just abbreviate these two derivations, **left** : from $\Gamma \longrightarrow \Delta, B$ derive $\neg B, \Gamma \longrightarrow \Delta$ and then $B \to C, \Gamma \longrightarrow \Delta$, and **right** : from $B, \Gamma \longrightarrow \Delta, C$ derive $\Gamma \longrightarrow \Delta, \neg B, C$ and by two $\bigvee$: right rules and a contraction derive $\Gamma \longrightarrow \Delta, B \to C$.

Assume that

$$T_2^i \vdash A(a)$$

By Corollary 7.1.5 there is an LKB-proof π with the Σ_i^b–IND rule of the sequent

$$\longrightarrow A(a)$$

in which all formulas are in $\Sigma_i^b \cup \Pi_i^b$.

Choose $q(x)$ to be a bounding polynomial of all formulas occurring in the proof π. The idea of the simulation of π by a G_i-proof is to translate every formula B in π into $\| B \|_{q(m)}^m$, and possibly fill in some derivations to obtain a valid G_i-proof.

As usual we proceed by induction on the number of inferences above a sequent in π of the form

$$\Gamma \longrightarrow \Delta$$

to show that the sequent

$$\| \Gamma \| \longrightarrow \| \Delta \|$$

has size $m^{O(1)}$ G_i-proof, where for $\Gamma = (A_1, \ldots, A_k)$ the symbol $\| \Gamma \|$ abbreviates the cedent $\| A_1 \|_{q(m)}^m, \ldots, \| A_k \|_{q(m)}^m$.

The case of the initial sequents, that is, the axioms of BASIC, the equality axioms or the logical axioms of the form $B \longrightarrow B$, B atomic, are obvious with the exception of the axioms of BASIC, for which the statement follows from Lemma 9.2.4.

The case of the structural or of the propositional rules or the cut-rule is handled by the same rules of G_i.

Consider now the $\forall \leq$: **right** rule

$$\frac{a \leq s, \Gamma \longrightarrow \Delta, B(a)}{\Gamma \longrightarrow \Delta, \forall x \leq s B(x)}$$

There are two possibilities: s is or is not of the form $|t|$. For the former let the set X be

$$X := \left\{ \bar{\epsilon} \in \{0, 1\}^* \mid \epsilon_i = 0 \quad \text{for } i > |q(m)| \right\}$$

and first derive the sequent

$$\bigvee_{\overline{\epsilon} \in X} \|a = \overline{\epsilon} \wedge \overline{\epsilon} \le |t| \| \longrightarrow \|a \le |t| \| \tag{S_1}$$

and the sequent

$$\|B(a)\| \longrightarrow \|B(a)\| \tag{S_2}$$

and from these two derive by the introduction of the implication to the left and to the right

$$\|a \le |t| \to B(a)\| \longrightarrow \| \bigvee_{\overline{\epsilon} \in X} a = \overline{\epsilon} \wedge \overline{\epsilon} \le |t| \to B(a)\| \tag{S_3}$$

Then derive the sequent

$$\bigvee_{\overline{\epsilon} \in X} \|a = \overline{\epsilon} \wedge \overline{\epsilon} \le |t| \| \to \|B(a)\| \longrightarrow \bigwedge_{\overline{\epsilon} \in X} \|a = \overline{\epsilon} \wedge \overline{\epsilon} \le |t| \to B(a)\| \tag{S_4}$$

and by the cut rule applied to (S_3) and (S_4) derive

$$\|a \le |t| \to B(a)\| \longrightarrow \bigwedge_{\overline{\epsilon} \in X} \|a = \overline{\epsilon} \wedge \overline{\epsilon} \le |t| \to B(a)\| \tag{S_5}$$

Separately derive

$$\bigwedge_{\overline{\epsilon} \in X} \|a = \overline{\epsilon} \wedge \overline{\epsilon} \le |t| \to B(a)\| \longrightarrow \bigwedge_{\overline{\epsilon} \in X} \|a \le |t| \to B(a)\|(\overline{p}/\overline{\epsilon}) \tag{S_6}$$

and by the cut-rule applied to (S_5) and (S_6) also

$$\|a \le |t| \to B(a)\|(\overline{p}) \longrightarrow \bigwedge_{\overline{\epsilon} \in X} \|a \le |t| \to B(a)\|(\overline{p}/\overline{\epsilon}) \tag{S_7}$$

From the upper sequent of the simulated inference infer by **Impl : right**

$$\|\Gamma\| \longrightarrow \|\Delta\|, \|a \le s \longrightarrow B(a)\| \bigwedge_{\overline{\epsilon} \in X} \|a \le |t| \to B(a)\|(\overline{p}/\overline{\epsilon})$$

and by the cut with (S_7) infer the translation of the lower sequent of the simulated inference.

This shows the simulation if $s = |t|$. Assume now that s does not have the form $|t|$. From the upper sequent by **Impl : right** derive

$$\|\Gamma \longrightarrow \|\Delta\|, \|a \le s \to B(a)\|$$

and by $q(m) + 1$ inferences $\forall : right$ applied to $\overline{p}$ associated with a derive the wanted sequent

$$\|\Gamma\| \longrightarrow \|\Delta\|, \|\forall x \le s B(x)\|$$

Now consider the rule $\forall \leq$: **left**

$$\frac{B(t), \Gamma \longrightarrow \Delta}{t \leq s, \forall x \leq s\, B(x), \Gamma \longrightarrow \Delta}$$

Assume first that $t = |r|$ for some term r. In both cases we shall first derive the sequent

$$\|t \leq s\|, \|\forall x \leq s\, B(x)\| \longrightarrow \|B(t)\| \qquad (S_0)$$

and by the cut-rule with the upper sequent obtain the wanted sequent

$$\|t \leq s\|, \|\forall x \leq s\, B(x)\|, \|\Gamma\| \longrightarrow \|\Delta\|$$

The cases differ as to the derivation of (S_0). If $t = |r|$ derive

$$\|t \leq |r|\,\| \longrightarrow \bigvee_{\overline{\epsilon} \in X} \|t = a \wedge a \leq |r|\,\|(\overline{p}/\overline{\epsilon}) \qquad (S_1)$$

and then

$$\begin{aligned} \|\forall x \leq |r|B(x)\|, \bigvee_{\overline{\epsilon} \in X} \|t = a \wedge a \leq |r|\,\|(\overline{p}/\overline{\epsilon}) \\ \longrightarrow \bigvee_{\overline{\epsilon} \in X} \|t = a \wedge B(a)\|(\overline{p}/\overline{\epsilon}) \end{aligned} \qquad (S_2)$$

By the cut-rule applied to (S_1) and (S_2) derive

$$\|t \leq |r|\,\|, \|\forall x \leq |r|B(x)\| \longrightarrow \bigvee_{\overline{\epsilon} \in X} \|t = a \wedge B(a)\|(\overline{p}/\overline{\epsilon}) \qquad (S_3)$$

By Lemma 9.2.3 applied to the formula $B(a)$ derive the sequent

$$\bigvee_{\overline{\epsilon} \in X} \|t = a \wedge B(a)\|(\overline{p}/\overline{\epsilon}) \longrightarrow \|B(t)\| \qquad (S_4)$$

and by the cut derive (S_0) from (S_3) and (S_4).

Assume now that t is not of the form $|r|$. First derive

$$\|t \leq s\| \longrightarrow \exists \overline{x}\, \|a \leq s \wedge a = t\|(\overline{p}/\overline{x}) \qquad (S_1)$$

and

$$\|\forall x \leq s\, B(x)\|, \exists \overline{x}\, \|a \leq s \wedge a = t\|(\overline{p}/\overline{x}) \longrightarrow \exists \overline{x}\, \|a = t \wedge B(a)\|(\overline{p}/\overline{x}) \quad (S_2)$$

From (S_1) and (S_2) the cut-rule yields the sequent

$$\|t \leq s\|, \|\forall x \leq s\, B(x)\| \longrightarrow \exists \overline{x}\, \|a = t \wedge B(a)\|(\overline{p}/\overline{x}) \qquad (S_3)$$

Using Lemma 9.2.3 also derive

$$\exists \overline{x}\, \|a = t \wedge B(a)\|(\overline{p}/\overline{x}) \longrightarrow \|B(t)\| \qquad (S_4)$$

and the sequent (S_0) can be obtained from (S_3) and (S_4) by the cut-rule.

This concludes the treatment of the $\forall \leq$: introduction rules. The cases of the $\exists \leq$: introduction rules are dual and we leave them to the reader.

It remains to treat the Σ_i^b–IND rule

$$\frac{B(a), \Gamma \longrightarrow \Delta, B(a+1)}{B(0), \Gamma \longrightarrow \Delta, B(t)}$$

Assume that the atoms $\overline{p}$ are associated with the variable a and the atoms $\overline{q}$ with the term t. For simpler readability we shall omit the side formulas Γ, Δ.

The first idea would be to simulate the IND-rule by repeated cuts applied to the upper sequent for $a = 0, 1, \ldots, t-1$. This would, however, have an exponential size. We shall instead shorten the simulation by the substitution rule, which is by the proof of Lemma 4.6.3 available in G.

Assume that we have the sequent

$$\|B(a)\| \longrightarrow \|B(a+1)\| \tag{S}$$

Derive the sequents

$$\|B(a)\| \longrightarrow \|B(a+2^i)\| \tag{U_i}$$

for $i = 0, \ldots, q(m)$. The sequent (U_0) is just (S), and (U_{i+1}) is derived from (U_i) as follows: Let $\overline{r}$ be a $(q(m)+1)$-tuple of new atoms. By substituting r_j's for p in (U_i) derive

$$\|B(a)\|(\overline{p}/\overline{r}) \longrightarrow \|B(a+2^i)\|(\overline{p}/\overline{r}) \tag{V_1}$$

Using the equality axioms whose translations are shortly provable derive

$$\|a+2^i = b\|(\overline{p},\overline{r}), \|B(a+2^i)\|(\overline{p}) \longrightarrow \|B(a)\|(\overline{p}/\overline{r}) \tag{V_2}$$

where $\overline{r}$ is associated with the variable b. By the cut-rule applied to (U_i) and (V_2) get

$$\|a+2^i = b\|(\overline{p},\overline{r}), \|B(a)\|(\overline{p}) \longrightarrow \|B(a)\|(\overline{p}/\overline{r}) \tag{V_3}$$

By the cut obtain from (V_1) and (V_3) the sequent

$$\|a+2^i = b\|(\overline{p},\overline{r}), \|B(a)\|(\overline{p}) \longrightarrow \|B(a)\|(\overline{p}/\overline{r}) \tag{V_4}$$

Again using the equality axioms derive

$$\|a+2^i = b\|(\overline{p},\overline{r}, \|B(a+2^i)\|(\overline{p}/\overline{r}) \longrightarrow \|B(a+2^{i+1})\|(\overline{p}) \tag{V_5}$$

and by the cut from (V_4) and (V_5) get

$$\|a+2^i = b\|(\overline{p},\overline{r}), \|B(a)\|(\overline{p}) \longrightarrow \|B(a+2^{i+1})\|(\overline{p}) \tag{V_6}$$

Derive

$$\longrightarrow \exists \overline{x} \|a+2^i = b\|(\overline{p},\overline{x}) \tag{V_7}$$

and from (V_6) by $(q(m) + 1) \, \exists :$ **left** rules

$$\exists \overline{x} \, \|a + 2^i = b\|(\overline{p}, \overline{x}), \|B(a)\|(\overline{p}) \longrightarrow \|B(a + 2^{i+1})\|(\overline{p}) \qquad (V_8)$$

and by the cut-rule from (V_7) and (V_8) the sequent (U_{i+1}).

The next idea is to derive the sequent

$$\|2^i \geq b\|(\overline{r}), \|B(a)\|(\overline{p}) \longrightarrow \|B(a + b)\|(\overline{p}, \overline{r}) \qquad (S_i)$$

consecutively for $i = 0, \ldots, q(m)$. The sequent (S_0) follows from (U_0), using simple

$$\|2^0 \geq b\|(\overline{r}) \longrightarrow \|a = b \vee a + 1 = b\|(\overline{p}, \overline{r})$$

Assume that we have (S_i) and derive (S_{i+1}) as follows. Let c, d be new variables with the associated atoms $\overline{u}, \overline{v}$. By substituting $\overline{u}$ for $\overline{p}$ and $\overline{v}$ for $\overline{r}$ in (S_i) derive

$$\|2^i \geq c\|(\overline{u}), \|B(d)\|(\overline{v}) \longrightarrow \|B(d + c)\|(\overline{v}, \overline{u}) \qquad (Z_1)$$

From (U_i) derive

$$\|B(a)\|(\overline{p}), \|a + 2^i = d\|(\overline{p}, \overline{v}) \longrightarrow \|B(d)\|(\overline{v}) \qquad (Z_2)$$

and by the cut from (Z_1) and (Z_2) the sequent

$$\|2^i \geq c\|(\overline{u}), \|a + 2^i = d\|(\overline{p}, \overline{v}), \|B(a)\|(\overline{p}) \longrightarrow \|B(d + c)\|(\overline{v}, \overline{u}) \quad (Z_3)$$

From (Z_3) get

$$\|2^i \geq c\|(\overline{u}), \|b = 2^i + c\|(\overline{r}, \overline{u}), \|B(a)\|(\overline{p}) \longrightarrow \|B(a + b)\|(\overline{p}, \overline{r}) \quad (Z_4)$$

From the sequents (S_i) and (Z_4) infer by $\bigvee :$ **left**

$$\|2^i \geq b \vee (2^i \geq c \wedge b = 2^i + c)\|(\overline{r}, \overline{u}), \|B(a)\|(\overline{p}) \longrightarrow \|B(a+b)\|(\overline{p}, \overline{r}) \quad (Z_5)$$

By $(q(m) + 1)$-applications of $\exists :$ **left** to $\overline{u}$ get

$$\exists \overline{x} \, \|2^i \geq b \vee (2^i \geq c \wedge b = 2^i + c)\|(\overline{r}, \overline{x}), \|B(a)\|(\overline{p})$$
$$\longrightarrow \|B(a + b)\|(\overline{p}, \overline{r}) \qquad (Z_6)$$

Separately derive

$$\|2^{i+1} \geq b\|(\overline{r}) \longrightarrow \exists \overline{x} \, \|2^i \geq b \vee (2^i \geq c \wedge b = 2^i + c)\|(\overline{r}, \overline{x}) \qquad (Z_7)$$

and by the cut-rule applied to (Z_6) and (Z_7) derive the sequent (S_{i+1}).

Having $(S_{q(m)})$ substitute $\overline{0}$ for $\overline{p}$ and $\overline{q}$ for $\overline{r}$ to obtain

$$\|2^{q(m)} \geq t\|(\overline{q}), \|B(0)\| \longrightarrow \|B(t)\|(\overline{q})$$

Since the sequent

$$\longrightarrow \ \|2^{q(m)} \geq t\|(\overline{q})$$

has a simple proof, the cut-rule applied to these two sequents yields the wanted sequent

$$\|B(0)\| \ \longrightarrow \ \|B(t)\|$$

This concludes the proof of the theorem.   Q.E.D.

Theorem 9.2.6. *Let* $i \geq 1$ *and let* $A(a)$ *be a* Σ_i^b*-formula. Assume that*

$$S_2^i \vdash A(a)$$

Then there is a bounding polynomial $q(x)$ *for* A *such that for all* m *the formulas* $\|A\|_{q(m)}^m$ *have size* $m^{O(1)}$ G_i^**-proofs.*
Moreover, these proofs are Σ_1^b*-definable in* S_2^1.

Proof. The simulations constructed in the proof of Theorem 9.2.5 are all treelike except the simulation of the Σ_i^b–IND rule where the sequents (U_i) and (S_i) are used repeatedly. But in S_2^i the Σ_i^b–IND is replaced by the Σ_i^b–PIND rule

$$\frac{B(\lfloor a/2 \rfloor), \Gamma \ \longrightarrow \ \Delta, B(a)}{B(0), \Gamma \ \longrightarrow \ \Delta, B(t)}$$

which can be simply simulated by $\leq (q(m) + 1)$ cuts. Let π be a G_i^*-proof of

$$\left\| B\left(\left\lfloor \frac{a}{2} \right\rfloor\right) \right\| \ \longrightarrow \ \|B(a)\|$$

Substituting in the *whole* proof for atoms $\overline{p}$ associated with a consecutively the bits of t, $\lfloor t/2 \rfloor, \ldots, 1, 0$ get $\leq (q(m) + 1)$ proofs $\pi_t, \pi_{\lfloor \frac{t}{2} \rfloor}, \ldots, \pi_1, \pi_0$ each of size $m^{O(1)}$. Joining these proofs by the cut-rule entails the wanted sequent

$$\|B(0)\| \ \longrightarrow \ \|B(t)\|$$

Q.E.D.

Corollary 9.2.7. *Assume that the equation* $t = s$ *in the language of PV is provable in PV. Then there is a bounding polynomial* $q(x)$ *for* $t = s$ *such that for all* m *formulas* $\|t = s\|_{q(m)}^m$ *have size* $m^{O(1)}$ *EF-proof.*

Proof. From PV $\vdash t = s$ follows $S_2^1(\text{PV}) \vdash t = s$ and by the preceding theorem (trivially modified to the language of $S_2^1(\text{PV})$) the formula $\|t = s\|_{q(m)}^m$ has size $m^{O(1)}$ G_1^*-proofs. The corollary then follows from Lemma 4.6.3.   Q.E.D.

The rest of this section is devoted to the extension of the previous simulation theorems to the theory U_2^1 and to the full system G. Such a simulation is not unexpected as U_2^1 relates to PSPACE by Theorem 8.2.4 and the set of tautological quantified propositional formulas is PSPACE-complete (cf. Balcazár et al. 1988). In fact, Dowd (1979) proposed an equational theory PSA related to PSPACE analogous to the relation of PV to P and showed a simulation of PSA by G. His translation, however, was not the $\| \ldots \|$ translation. In his translation the quantifier complexity of a Σ_i^b-formula was not Σ_i^q but increased with the length of the input and with the space bound.

We shall prove the simulation using the $\| \ldots \|$ translation.

Theorem 9.2.8. *Let $i \geq 1$ and let $A(a)$ be a bounded first order Σ_i^b-formula. Assume that*

$$U_2^1 \vdash A(a)$$

Then there is a bounding polynomial $q(x)$ for A such that for all m the formula $\|A\|_{q(m)}^m$ has size $m^{O(1)}$ G-proof.

Moreover, these proofs are Σ_1^b-definable in S_2^1.

Proof. We shall prove, in fact, a stronger statement (the following claim), which is a propositional version of the witnessing Theorem 8.2.4 for the case of U_2^1. First we shall make some simplifications and conventions.

We shall assume that all $\Sigma_1^{1,b}$-formulas are, in fact, strict$\Sigma_1^{1,b}$, that is, a block of second order $\exists$-quantifiers followed by a $\Sigma_0^{1,b}$-formula. This is achieved by adding the pairing function $\langle x, y \rangle$ to the language and coding sequences of sets by

$$j \in (\alpha)_i := \langle i, j \rangle \in \alpha$$

and by enlarging BASIC by a few axioms implying $\Sigma_1^{1,b}$–AC (over $\Sigma_0^{1,b}$–PIND).

We shall also extend the translation $\| \ldots \|$ from Σ_∞^b- to $\Sigma_0^{1,b}$-formulas by stipulating that atomic formulas $a \in \alpha$ are translated as

$$\|\alpha\|_{q(m)}^m (p_1, \ldots, p_{q(m)})$$

where $\|\alpha\|$ is a new *metavariable* for quantified propositional formulas. These metavariables will actually never occur in a G-proof, but they allow convenient notation. We shall manipulate them freely; for instance, $\|A(a, \alpha/\beta)\|$ is the same as $\|A(a, \alpha)\|(\|\alpha\|/\|\beta\|)$.

Claim. *Let $\exists \phi^t A(a, \alpha^s, \phi^t)$ and $\exists \psi^r B(a, \alpha^s, \psi^r)$ be strict$\Sigma_1^{1,b}$-formulas such that*

$$U_2^1 \vdash \exists \phi^t A(a, \alpha^s, \phi^t) \longrightarrow \exists \psi^r B(a, \alpha^s, \psi^r)$$

Then for every m there is a Σ_∞^b-formula with the metavariables

$$W_m \left(\|\alpha^s\|, \|\phi^t\| \right)$$

and a bounding polynomial $q(x)$ for A, B such that for any two Σ_∞^q-formulas

C, D without metavariables the implication

$$\|A\|_{q(m)}^{m} \left(\overline{p}, \|\alpha^s\|/C, \|\phi^t\|/D \right)$$

$$\longrightarrow \|B\|_{q(m)}^{m} \left(\overline{p}, \|\alpha^s\|/C, \|\psi^r\|/W_m(\|\alpha^s\|/C, \|\phi^t\|/D) \right)$$

has a G-proof of size $poly(m, |C|, |D|)$.
The formulas C, D have $q(m) + 1$ atoms and may contain the atoms $\overline{p}$ too.

Under the hypothesis of the claim there is a U_2^1-proof π in the second order LKB with $\Sigma_1^{1,b}$–PIND in which all formulas are strict $\Sigma_1^{1,b}$, ending with the implication. Any sequent in π has the form

$$\ldots, \exists \phi_i A_i(a, \overline{b}, \alpha, \overline{\beta}, \phi_i), \ldots, \Gamma \longrightarrow \Delta, \ldots, \exists \psi_j B_j(a, \overline{b}, \alpha, \overline{\beta}, \psi_j), \ldots$$

where $\overline{b}, \overline{\beta}$ are the other free variables different from a, α, and Γ, Δ are the cedents of $\Sigma_0^{1,b}$-formulas. For simplicity of notation we shall omit $\overline{b}, \overline{\beta}$, and we shall write $\| \ldots \|$ instead of $\| \ldots \|_{q(m)}^{m}$.

A remark on the choice of polynomial $q(x)$: It is a bounding polynomial of all formulas in π; in the case of second order formulas with variables γ^s this means that $q(m)$ also bounds the length of all possible values of s needed in the evaluation of the formula for first order inputs of length $\leq m$.

We shall also write

$$\|A_i\|(C, D_i)$$

instead of

$$\|A_i\| (\|\alpha\|/C, \|\phi_i\|/D_i)$$

(and similarly for B_j), as there is no danger of confusion.

By induction on the number of inferences in π above such a sequent we construct Σ_∞^b-formulas without metavariables $C, \ldots, D_i, \ldots$ (with $(q(m) + 1)$ new atoms and with possible occurrences of other atoms free in the sequent) such that there are size $m^{O(1)}$ G-proofs of

$$\ldots, \|A_i\|(C, D_i), \ldots, \|\Gamma\| \longrightarrow \|\Delta\|, \ldots, \|B_j\|(C, W^j(C, \ldots, D_i, \ldots)), \ldots$$

Formulas W_k^j are called *witnessing formulas*.

We proceed now considering several cases according to the type of the inference giving the sequent.

The principal formulas of all propositional inferences must be $\Sigma_0^{1,b}$, and the only two nontrivial cases are $\vee$: **left** and $\wedge$: **right**.

Let a principal formula of an $\vee$: **left** inference be $E_1 \vee E_2$ with W_1^j (resp. W_2^j) the witnessing formulas for the upper sequent with E_1 (resp. with E_2). Define the witnessing formulas W^j for the lower sequent by the *definition by cases*

$$W^j := \left(\|E_1\| \wedge W_1^j \right) \vee \left(\|\neg E_1\| \wedge W_2^j \right)$$

If $\|E_1 \vee E_2\|$ is true, then W^j is correctly defined as a valid witness; otherwise W^j can be anything. Moreover, there are polynomial size G-proofs of the sequents

$$\|E_1\| \longrightarrow W^j \equiv W_1^j$$

and

$$\|\neg E_1\| \longrightarrow W^j \equiv W_2^j$$

Using these proofs one can construct from the corresponding proofs for the upper sequents polynomial size proofs of the lower sequent with the witness W^j.

The $\wedge$: **right** rule is treated dually.

The structural rules are simple; in particular, the *contraction : right* rule is treated by the definition by cases again: If two occurrences of formula $\exists \psi\, B(a, \alpha, \psi)$ in the succedent are contracted, and W_1^j, W_2^j are the witnessing formulas corresponding to these two occurrences, set

$$W^j := \left(\|B\|(\|\alpha\|, W_1^j) \wedge W_1^j \right) \vee \left(\neg \|B\|(\|\alpha\|, W_1^j) \wedge W_2^j \right)$$

Now let us consider the second order $\exists$: **right** inference (which in the sequent calculus represents $\Sigma_0^{1,b}$–CA [cf. Section 8.3])

$$\frac{\ldots \longrightarrow \ldots, B(a, \alpha, \psi/E(a, \alpha))}{\ldots \longrightarrow \ldots, \exists \psi\, B(a, \alpha, \psi)}$$

with $E \in \Sigma_0^{1,b}$. Then just define the new witness formula for new variable ψ by

$$W := \|E\|(\|\alpha\|)$$

The last nontrivial case to consider is $\Sigma_1^{1,b}$–PIND, where the induction formula $\exists \phi\, E(a, b, \alpha, \phi)$ is $\Sigma_1^{1,b}$ and not $\Sigma_0^{1,b}$ (the latter case is treated as in the proof of Theorem 9.2.5).

Assume that $\overline{W}(a, b, \|\alpha\|, \ldots, \|\phi_i\|, \ldots, \|\phi\|)$ is the witness formula associated with $\exists \phi$ in the succedent of the upper sequent of the induction inference

$$\frac{\ldots, \exists \phi\, E(a, \lfloor \frac{b}{2} \rfloor, \alpha, \phi) \longrightarrow \exists \phi\, E(a, b, \alpha, \phi), \ldots}{\ldots, \exists \phi\, E(a, 0, \alpha, \phi) \longrightarrow \exists \phi\, E(a, t, \alpha, \phi), \ldots}$$

Define terms $t_0, t_1, \ldots, t_k$ by $t_0 := t$ and $t_{i+1} := \lfloor t_i/2 \rfloor$, for $k = q(m)$. Note that $t_k \equiv 0$. Using these terms then define

$$V_1 := \overline{W}(a, b/t_{k-1}, \|\alpha\|, \ldots, \|\phi_i\|, \ldots, \|\phi\|)$$

and

$$V_{\ell+1} := \overline{W}(a, b/t_{k-\ell-1}, \|\alpha\|, \ldots, \|\phi_i\|, \ldots, \|\phi\|/V_\ell)$$

for $\ell = 1, \ldots, q(m) - 1$, and set

$$W := V_{q(m)}$$

It is clear, and easily G-proved, that V_ℓ witnesses the implication

$$\exists \phi E(a, 0, \alpha, \phi) \longrightarrow \exists \phi E(a, t_{k-\ell}, \alpha, \phi)$$

and hence W witnesses the lower sequent of the induction rule.

There is one more requirement that is not obviously satisfied: namely, W has to have a size of at most $m^{O(1)}$. Indeed, if $\|\phi\|$ has at least two occurrences in $\overline{W}$, then the size of V_ℓ grows exponentially with ℓ, whereas if $\|\phi\|$ occurs in $\overline{W}$ only once, the size of V_ℓ increases only proportionally. Hence a way to overcome this obstacle is first to put $\overline{W}$ into a G-equivalent form with only one occurrence of $\|\phi\|$. For example, replace $\overline{W}$ by

$$\forall x, x \equiv \|\phi\| \to \overline{W}(\|\phi\|/x)$$

This will increase the quantifier complexity of $\overline{W}$, but we do not care about that as the complexity of W is greater than that of $\overline{W}$ in any case.   Q.E.D.

9.3. Reflection principles and polynomial simulations

In this section we show that the provability of the reflection principles for propositional proof systems in bounded arithmetic implies polynomial simulation. As an illustration of this idea assume that we can verify in S_2^1 the soundness of a proof system P. Then the simulation of S_2^1 by EF (Corollary 9.2.7) allows to "prove" the soundness of P in EF, and then to use this proof to simulate P-proofs by EF-proofs. This idea is due to Cook (1975).

We shall apply the idea to systems considered in the first two sections of this chapter, and we shall also derive some corollaries about bounded arithmetic, not just about the propositional proof systems.

First we shall consider the language L_2 of V_1^1, and the translation of Section 9.1. In this language we have finite sets and all basic operations with sets are $\Sigma_0^{1,b}$-definable (in $I\Sigma_0^{1,b}$ using $\Sigma_0^{1,b}$–CA). In particular, recall that a sequence of sets can be coded by a set by

$$j \in (\alpha)_i \equiv \langle j, i \rangle \in \alpha$$

and again such coding applies CA to the definition of the sequence; this will always be $\Sigma_0^{1,b}$ or $\Delta_1^{1,b}$. Thus we can carry in $I\Sigma_0^{1,b}$ some usual set-theoretic coding of propositional formulas, say as finite binary trees with inner nodes labeled by the connectives and leaves labeled by atoms or constants. Proofs are then particular sequences of formulas, and for systems F or EF the definitions of F-proofs (resp. EF-proofs) are obviously also $\Sigma_0^{1,b}$. A truth evaluation of a formula will be coded a 0, 1-labeling of the nodes of the formula computed according to truth tables of the connectives. Moreover, these definitions allow us to prove in $I\Sigma_0^{1,b}$ such elementary syntactic properties as "A formula has unique immediate subformulas," and so on. We leave the reader to design her/his own definitions and to apply them to the following arguments. We just stipulate a certain notation.

Definition 9.3.1.

1. $Fla(\alpha)$ is a $\Sigma_0^{1,b}$-definition of "α is a propositional formula"
2. For $P = F, EF$, $Prf_P(\pi, \alpha)$ is a $\Sigma_0^{1,b}$-definition of "π is a P-proof of α"
3. $Assign(\eta, \alpha)$ is a $\Sigma_0^{1,b}$-definition of "η is a truth assignment to the atoms of the formula α," and $Assign(\eta, \alpha)$ implies in $I\Sigma_0^{1,b}$ $Fla(\alpha)$
4. $Eval(\eta, \alpha, \gamma)$ is a $\Sigma_0^{1,b}$-definition of "γ is the evaluation of the formula α over the truth assignment η to its atoms," and $Eval(\eta, \alpha, \gamma)$ implies in $I\Sigma_0^{1,b}$ the conjunction $Fla(\alpha) \wedge Assign(\eta, \alpha)$
5. $\eta \models \alpha$ is a $\Delta_1^{1,b}$-definition in U_1^1 of "η is a satisfying truth assignment to the atoms of the formula α," and it is in $I\Sigma_0^{1,b}$-defined by

$$\exists \gamma, \; Eval(\eta, \alpha, \gamma) \wedge \text{"}\gamma \text{ evaluates to } 1\text{"}$$

6. $TAUT(\alpha)$ is a $\Pi_1^{1,b}$-formula defined in $I\Sigma_0^{1,b}$ as

$$\forall \eta, \; Assign(\eta, \alpha) \rightarrow \eta \models \alpha$$

Formula $\eta \models \alpha$ is $\Delta_1^{1,b}$ in U_1^1 as even $I\Sigma_0^{1,b}$ can prove the implication

$$Eval(\eta, \alpha, \gamma_1) \wedge Eval(\eta, \alpha, \gamma_2) \rightarrow \gamma_1 = \gamma_2$$

(by induction on the size of γ_1, γ_2), and by the following lemma, which is *not* obvious.

Lemma 9.3.2. *The theory U_1^1 proves that every propositional formula can be evaluated over any truth assignment to its atoms*

$$\forall \eta, \; \alpha \exists \gamma, \; Assign(\eta, \alpha) \rightarrow Eval(\eta, \alpha, \gamma)$$

Proof. We would like to proceed by the induction on the logical depth of α, but this would require $\Sigma_1^{1,b}$–IND, whereas we have only $\Sigma_1^{1,b}$–PIND. We shall therefore make use of a proof of the Spira theorem 3.1.15, reducing the depth to a logarithmic one.

Let α be a formula and η an evaluation of its atoms. Assume that a is a bound to all nodes in α and for $b < a$ let α_b denote the subformula of α with the root b. For β a subformula of α and δ a set of some mutually incomparable nodes of α, let $\beta \downarrow \delta$ denote a formula consisting of those nodes of β not majorized by any node of δ, with $y \in \delta$ a leaf of $\alpha \downarrow \delta$ labeled by a new atom q_y. Thus $\beta \downarrow \delta$ has at most $|\delta|$ new atoms that do not occur in α.

Consider the formula

$$A(u) := [\forall x < a \forall \delta(|\delta| \geq |a| - u), Fla(\alpha_x \downarrow \delta) \wedge |\alpha_x \downarrow \delta| \leq (3/2)^u]$$
$$\rightarrow \forall \xi [Assign(\eta \cup \xi, \alpha_x \downarrow \delta) \rightarrow \exists \gamma, Eval(\eta \cup \xi, \alpha_x \downarrow \delta, \gamma)]$$

As the size of δ is bounded by $|a|$, ξ can be coded by a number bounded by a and hence $A(u) \in \Sigma_1^{1,b}$.

For $u = 0, |\alpha_x \downarrow \delta| = 1$ so $\alpha_x \downarrow \delta$ is just an atom or a constant and $A(0)$ clearly holds. The idea of the proof of implication $A(u) \rightarrow A(u+1)$ comes from Spira (1971), where the following claim was stated.

Claim. *Any formula α_1 has a subformula α_2 such that $|\alpha_2| \leq (2/3)|\alpha_1|$ and $|\alpha_1| - |\alpha_2| \leq (2/3)|\alpha_1|$.*

This claim is easy to formalize (we have a $\Delta_1^{1,b}$-definition of the cardinality of a set, by Lemma 5.5.14) and to prove by PIND on the size of the formula.

Assume $A(u)$ and let $\alpha_x \downarrow \delta$ fulfill the hypotheses of $A(u+1)$. From the claim it follows that there is a node y in $\alpha_x \downarrow \delta (= \psi_1)$ such that both $\alpha_y \downarrow \delta (= \psi_2)$ and $\alpha_x \downarrow (\delta \cup \{y\})$ have size at most $(3/2)^u$. By $A(u)$ then there is an evaluation γ' of $\alpha_y \downarrow \delta$ over any $\eta \cup \xi$ and an evaluation γ'' of $\alpha_x \downarrow (\delta \cup \{y\})$ over $\eta \cup \xi \cup \chi$, where χ assigns to q_y the value computed by γ'. It is then straightforward to $\Sigma_0^{1,b}$-define from γ', γ'' the evaluation γ of $\alpha_x \downarrow \delta$ over $\eta \cup \xi$. This entails $A(u+1)$.

By $\Sigma_1^{1,b}$–LIND, available in U_1^1, it follows then that $A(\log_{3/2}(a))$: That is, there is an evaluation γ of α over η.   Q.E.D.

Theorem 9.3.3. *The theory U_1^1 proves that F is a sound proof system*

$$\forall \alpha, \pi, \mathrm{Prf}_F(\pi, \alpha) \rightarrow TAUT(\alpha)$$

Proof. Argue in U_1^1. Let π, α satisfy $\mathrm{Prf}_F(\pi, \alpha)$, and let $(\pi)_1, \ldots, (\pi)_k = \alpha$ be the steps in π. Assume that η is a truth assignment to the atoms of α: Assign(η, α). We may assume that no other atoms occur in π (as otherwise they could be substituted for by 0, for example).

Consider the formula $A(u)$

$$A(u) := \eta \models (\pi)_u$$

This formula is by Definition 9.3.1 $\Delta_1^{1,b}$ in U_1^1. By Lemma 5.5.9 U_1^1 admits $\Delta_1^{1,b}$–IND; hence we have the induction for the formula $A(u)$.

Clearly $A(1)$ holds (as $(\pi)_1$ must be a logical axiom), and $A(u) \rightarrow A(u+1)$ (as all Frege rules are sound). Hence $A(k)$ holds: that is, $\eta \models \alpha$.

This applies to any η; hence α is a tautology.   Q.E.D.

Theorem 9.3.4. *The theory V_1^1 proves that EF is a sound proof system*

$$\forall \alpha, \pi, \mathrm{Prf}_{EF}(\pi, \alpha) \rightarrow TAUT(\alpha)$$

Proof. Argue in V_1^1. Let π be an EF-proof of α with steps $(\pi)_1, \ldots, (\pi)_k = \alpha$, where $(\pi)_1, \ldots, (\pi)_m$ are the extension atoms used in π, $m < k$.

Let $\text{Assign}(\eta, \alpha)$ hold and w.l.o.g. assume that in π only the atoms from α or the extension atoms occur. Take the formula $A(u)$

$$A(u) := \exists \xi \text{ "} \xi \text{ is a truth assignment to the extension atoms in } \pi \text{"}$$
$$\wedge \; \forall v \leq u, \; \eta \cup \xi \models (\pi)_v$$

This is a $\Sigma_1^{1,b}$-formula clearly satisfying $A(1)$ and $A(u) \rightarrow A(u + 1)$, giving to the extension atoms truth values computed from η by their definitions, for $u = 1, \ldots, m$. Hence $\Sigma_1^{1,b}$–IND implies $A(k)$, and $\eta \models \alpha$ follows. Q.E.D.

There is no $\Sigma_0^{1,b}$-definition of the relation $\eta \models \alpha$. This is because such a definition would allow us to express every Boolean formula by a constant depth circuit $\langle \eta \models \alpha \rangle_n$ of size polynomial in $n = |\alpha|$, which is impossible (e.g., the parity function $\oplus(x_1, \ldots, x_n)$ has a formula of size n^2 but it has no polynomial size constant depth circuits, cf. Theorem 3.1.10). Thus we cannot translate the previous arguments into $I\Sigma_0^{1,b}$. The theory $I\Sigma_0^{1,b}$ does not even prove that every formula can be evaluated

$$\forall \eta, \alpha, \text{Assign}(\eta, \alpha) \rightarrow \exists \gamma, \text{Eval}(\eta, \alpha, \gamma)$$

The unprovability of this implication is shown as follows: Assume that $I\Sigma_0^{1,b}$ proves the implication. Then the witnessing argument of Theorem 9.1.5 would give the witness formulas for γ in terms of the bits of η, α, and these witness formulas would be $\Sigma_0^{1,b}$. Hence again it would follow that formulas can be written as polynomially larger constant depth circuits: a contradiction with Theorem 3.1.10 (and with the existence of small formulas for the parity function; see the end of Section 3.1).

There is, however, a $\Sigma_0^{1,b}$-definition of $\eta \models \alpha$ *assuming* that the depth of α is bounded by a *standard* constant.

Lemma 9.3.5. *Let d be a constant and let $\text{Fla}_d(\alpha)$ be a $\Sigma_0^{1,b}$-definition of "α is a depth $\leq d$ formula." Then*

$$I\Sigma_0^{1,b} \vdash \forall \eta, \alpha, \text{Fla}_d(\alpha) \wedge \text{Assign}(\eta, \alpha) \rightarrow \exists \gamma, \text{Eval}(\eta, \alpha, \gamma)$$

Proof. Prove the statement by induction on d showing that the evaluation γ is actually (for fixed d) $\Sigma_0^{1,b}$-definable from η and α (the implication then follows by $\Sigma_0^{1,b}$–CA). This is because $I\Sigma_0^{1,b}$ can prove (by the remark before Definition 9.3.1) that a depth d formula with the outmost connective $\wedge$ is a conjunction of (arbitrarily bracketed) depth $d - 1$ formulas: That is, it is true iff "all these subformulas of depth $d - 1$ are true." Assuming that a truth definition for $d - 1$ formulas is already formed, this allows us to define the truth for depth d formulas with the help of a $\forall\leq$ quantifier; similarly when the outmost connective is $\vee$. If it is $\neg$ then first apply de Morgan rules to rewrite α so that all negations apply only to the atomic subformulas. Q.E.D.

Theorem 9.3.6. *Let $d > 0$ be a constant. Then the theory $I\Sigma_0^{1,b}$ proves that any Frege proof of depth $\leq d$ is sound*

$$\forall \pi, \alpha, \text{Prf}_F(\pi, \alpha) \wedge \text{``}dp(\pi) \leq d\text{''} \to TAUT(\alpha)$$

Proof. The proof goes by induction on the number of steps in π as in Theorem 9.3.3, using the $\Sigma_0^{1,b}$-definition of the satisfaction relation for depth $\leq d$ formulas provided by Lemma 9.3.5.   Q.E.D.

Before turning to the quantified propositional proof systems and the translation of Section 9.2 we formulate in model-theoretic terms a sufficient condition for a demonstration of superpolynomial lower bounds.

Lemma 9.3.7. *Assume that A is a $\Sigma_0^{1,b}$-formula. Then $I\Sigma_0^{1,b}$ proves the equivalence*

$$A(\alpha^x) \equiv (\tilde{\alpha} \models \langle A \rangle_x(\overline{p}))$$

where $\tilde{\alpha}$ is a truth assignment, $\Sigma_0^{1,b}$-definable in $I\Sigma_0^{1,b}$, assigning to p_i the value 1 iff $i \in \alpha^x$.

Proof. This is readily established by induction on the logical complexity of A.
                                                                          Q.E.D.

Recall that $n^\omega := \bigcup_{k<\omega} n^k$ for n, an element of a nonstandard model of arithmetic.

Theorem 9.3.8. *Let $A(a, \alpha)$ be a $\Sigma_0^{1,b}$-formula with a and α the only free variables. Let M be a nonstandard model of the true arithmetic $Th(\omega)$ and let $n \in M \setminus \omega$ be its nonstandard element.*

Assume that for every bounded set $\pi \subseteq n^\omega$ coded in M there is a family $\mathcal{X} \subseteq exp(n^\omega)$ of bounded subsets of n^ω and $\alpha \in \mathcal{X}$ such that

(i) $\pi \in \mathcal{X}$

(ii) $(n^\omega, \mathcal{X}) \models I\Sigma_0^{1,b}$

(iii) $(n^\omega, \mathcal{X}) \models \neg A(n, \alpha)$.

Then the formulas $\langle A(a) \rangle_m, m < \omega$, do not have polynomial size constant-depth F-proofs.

If $(n^\omega, \mathcal{X}) \models U_1^1$ then the formulas $\langle A(a) \rangle_m$ do not have polynomial size F-proofs, and if even $(n^\omega, \mathcal{X}) \models V_1^1$ then they do not have polynomial size EF-proofs.

Proof. Assume that the formulas $\langle A(a) \rangle_m, m < \omega$, do have polynomial size constant-depth F-proofs. As M satisfies the true arithmetic there is $k < \omega$ such that for every element $n \in M$, M codes a constant-depth F-proof of $\langle A(a) \rangle_n$ of size at most n^k. Let $\pi \subseteq n^k$ be such a proof.

Take $\mathcal{X}$ and $\alpha \in \mathcal{X}$ satisfying the conditions (i)–(iii). Then $(n^\omega, \mathcal{X})$ is a model of $I\Sigma_0^{1,b}$ in which the propositional formula $\langle A(a) \rangle_n$ has a depth d F-proof, some

$d \in \omega$. By Theorem 9.3.6 the formula $\langle A(a) \rangle_n$ must be a tautology. However, $\neg A(n, \alpha)$ is true, hence the assignment $\tilde{\alpha}$ defined in Lemma 9.3.7 does not satisfy that lemma. This is a contradiction.

The of U_1^1 and V_1^1 follow analogously, using Theorems 9.3.3 and 9.3.4 in place of Theorem 9.3.6.   Q.E.D.

Now we shall turn to the quantified propositional formulas and the translation of Section 9.2.

Lemma 9.3.9. *There are a Δ_1^b in S_2^1 definition $w \models_0 A$ of the satisfaction relation for Σ_0^q-formulas A, and a $\mathcal{B}(\Sigma_i^b)$ definition $w \models_i A$ of the satisfaction relation for $\Sigma_i^q \cup \Pi_i^q$-formulas.*

Moreover, S_2^1 proves Tarski's conditions for truth definitions ($i \geq 0$):

1. $(w \models_i A \vee B) \equiv ((w \models_i A) \vee (w \models_i B))$
2. $(w \models_i A \wedge B) \equiv ((w \models_i A) \wedge (w \models_i B))$
3. $(w \models_i \neg A) \equiv (\neg w \models_i A)$
4. $(w \models_i \exists x A(x)) \equiv ((w \models_i A(0)) \vee (w \models_i A(1)))$
5. $(w \models_i \forall x A(x)) \equiv ((w \models_i A(0)) \wedge (w \models_i A(1)))$

In fact, 1 and 4 are provable in a stronger form

6. $(w \models_i \bigvee_{\bar{\epsilon} \in X} A(\bar{\epsilon})) \equiv (\exists \bar{\epsilon} \in X, w \models_i A(\bar{\epsilon}))$, where $X \subseteq \{0, 1\}^n$
7.

$$(w \models_i \exists x_1, \ldots, x_n, A(x_1, \ldots, x_n))$$
$$\equiv (\exists w', \text{ "w' is a is a truth assignment to atoms}$$
$$p_1, \ldots, p_n \text{" } \wedge w \frown w' \models A(\bar{p}))$$

Moreover, S_2^1 proves ($i \geq 0$)

$$\forall A \in \Sigma_i^q \cup \Pi_i^q \, \forall w, (w \models_i A) \equiv (w \models_{i+1} A)$$

Definition 9.3.10. *For $i \geq 0$ define a $\forall \Pi_{i+1}^b$-formula*

$$Taut_i(A) := \forall w(|w| \leq |A|), w \models_i A$$

It defines the set of the tautological $\Sigma_i^q \cup \Pi_i^q$-formulas. The set is denoted $TAUT_i$.

The following definition extends Definitions 4.1.1 and 4.1.3 to systems for quantified propositional logic.

Definition 9.3.11.

(a) *A polynomial time computable binary relation $P(x, y)$ is a quantified propositional proof system iff*

$$(\exists \pi, P(\pi, A)) \rightarrow A \in \bigcup_{i \geq 0} TAUT_i$$

(b) *Let P and Q be two quantified propositional proof systems and $i \geq 0$. Then P i-polynomially simulates Q, $P \geq_i Q$ in symbols, iff there is a polynomial time function $f(x, y)$ such that*

$$\forall \pi, A, \, Q(\pi, A) \wedge A \in \Sigma_i^q \cup \Pi_i^q \rightarrow P(f(\pi, A), A)$$

(c) *Assume that a quantified propositional proof system P is Δ_1^b-defined in S_2^1. Then we define the formula*

$$i\text{-}RFN(P)$$

as

$$\forall \pi, A, \, P(\pi, A) \wedge A \in \Sigma_i^q \cup \Pi_i^q \rightarrow \, Taut_i(A)$$

Note that i–$RFN(P)$ is $\forall \Pi_1^b$ for $i = 0$ and $\forall \Sigma_i^b$ for $i \geq 1$.

(d) *Assume that a proof system P is Δ_1^b-defined in S_2^1. Then we define the formula*

$$\mathrm{Con}(P)$$

as

$$\forall \pi, \neg P(\pi, 0)$$

We do not require in the definition of a quantified propositional proof system that it is complete (with respect to all quantified tautologies). This is because we want the definition also to cover systems like F, EF, G_i, G_i^* that are not complete w.r.t. all tautological quantified propositional formulas.

Lemma 9.3.12.

(a) *Let $i \geq 0$ and assume that $\phi(a) \in \Sigma_i^b$ for $i \geq 1$ and that $\phi(a)$ is Δ_1^b in S_2^1 for $i = 0$. Let $\tilde{a}$ be a truth assignment assigning to the atom p_j the bit $a(j)$, $j \geq 0$; it is Δ_1^b-definable in S_2^1. Assume that $q(x)$ is a bounding polynomial for $\phi(a)$.*
Then

$$S_2^1 \vdash \phi(a) \equiv \tilde{a} \models_i \|\phi\|_{q(|a|)}^{|a|}(\overline{p})$$

(b) *For any proof system that is Δ_1^b-definable in S_2^1 and closed under the substitution rule, modus ponens, and that is complete for $TAUT_0$ it holds*

$$S_2^1 \vdash \, \mathrm{Con}(P) \equiv 0\text{-}RFN(P)$$

Proof. Part (a) of the lemma is proved by induction on the logical complexity of ϕ; it is analogous to Lemma 9.3.7.

For part (b) it is enough to prove the implication

$$\mathrm{Con}(P) \rightarrow \, 0\text{-}RFN(P)$$

as the opposite one is trivial. Argue in S_2^1: Assume that P proves a Σ_0^q-formula $A(\overline{p})$ but $w \models_0 \neg A$ for some assignment w.

The former implies that $P \vdash A(\overline{p}/\overline{w})$, and the latter implies $P \vdash \neg A(\overline{p}/\overline{w})$, from which $P \vdash 0$ follows. Here we use the assumption that P is closed under the substitution rule and the modus ponens. Q.E.D.

The following lemma extends Lemma 9.2.4.

Lemma 9.3.13. *Assume that $\phi(a) \in \Sigma_1^b$ and let $q(x)$ be a bounding polynomial for ϕ. Then*

$$S_2^1 \vdash \phi(a) \rightarrow \left(G_1^* \vdash \|\phi\|_{q(|a|)}^{|a|}(\tilde{a}) \right)$$

Proof. The statement is proved by induction on the logical complexity of ϕ, treating general true equations similarly to the axioms of BASIC in Lemma 9.2.4. Q.E.D.

Lemma 9.3.14. *Let ϕ be a Σ_i^b-formula, $i \geq 1$, and let $q(x)$ be a bounding polynomial for ϕ. Assume that P is a proof system Δ_1^b-defined in S_2^1.*
Then S_2^1 proves the implication

$$(i\text{-}RFN(P) \wedge \left(P \vdash \|\phi\|_{q(|x|)}^{|x|} \right) \rightarrow \forall y (|y| \leq |x| \rightarrow \phi(y))$$

Proof. The lemma follows from the following claim.

Claim. S_2^1 *proves*

$$Taut_i \left(\|\phi\|_{q(|x|)}^{|x|} \right) \wedge |y| \leq |x| \rightarrow \phi(y)$$

The claim follows from Lemma 9.3.12, part (a). Q.E.D.

The following lemma is a useful, intuitively clear property of the formula $Taut_i$.

Lemma 9.3.15. *Let $i \geq 1$ and assume that $q(x)$ is a bounding polynomial of the formula $Taut_i$. Then S_2^1 proves*

$$(A \in \Sigma_i^q \wedge n \geq |A| \wedge G_i \vdash \| Taut_i(A)\|_{q[n]}^n) \rightarrow G_i \vdash A$$

The same holds for $i = 0$ with G_1^* in place of G_0.

Proof. We present the case $i = 0$; the general case is analogous. The formula $Taut_i(A)$ is defined as

$$\forall w(|w| \leq |A|), w \models_i A$$

We think of the formula $w \models_i A$ as being defined analogously to the formulas in Definition 9.3.1 by formalizing

$$\forall u, \text{ "}u \text{ is the computation of the value of } A \text{ over } w\text{"}$$

$$\rightarrow \text{ "}u \text{ assigns to } A \text{ the value 1"}$$

The "computation" u is an assignment of values $(u)_1, \ldots, (u)_t$, each 0 or 1, to the subformulas of A, and we assume that values of the subformulas of a subformula appear in the sequence $(u)_1, \ldots, (u)_t$ sooner. Assume that $(u)_j$ corresponds to the subformula A_j of A.

Let $\text{Comp}(w, A, u)$ denote the formalization of "u is the evaluation of A over w."

Let $\overline{p}$ be the atoms of A.

Claim. S_2^1 proves

$$\forall j \leq |A|; \ Comp(\overline{p}, A, u) \rightarrow (u)_j \equiv A_j$$

and, in particular

$$Comp(\overline{p}, A, u) \rightarrow (u)_t \equiv A$$

The claim is proved by PIND on j in the formula

$$\text{Comp}(w, A, u) \rightarrow (u)_j \equiv A_j(\overline{p}/w)$$

and taking $w := \overline{p}$.

The theory S_2^1 also proves that any formula can be evaluated by induction on the length of the formula. Thus, in particular, S_2^1 proves

$$\exists u, \text{Comp}(\overline{p}, A, u)$$

and by Theorem 9.2.6 the last two formulas imply that S_2^1 proves

$$G_1^* \vdash \exists \overline{x}, \|\text{Comp}\|(\overline{p}, \tilde{A}, \overline{x})$$

and

$$G_1^* \vdash \|\text{Comp}\|(\overline{p}, \tilde{A}, \overline{q}) \rightarrow q_t \equiv A(\overline{p})$$

which entails the lemma.   Q.E.D.

Theorem 9.3.16. *For $i \geq 1$*

$$T_2^i \vdash i\text{-RFN}(G_i)$$

and

$$S_2^i \vdash i\text{-RFN}(G_i^*)$$

Proof. Using the formula $w \models_i A$ one can define a Π^b_{i+1}-formula $STaut_i(S)$ formalizing that "the sequent S consisting of $\Sigma^q_i \cup \Pi^q_i$-formulas is satisfied by all truth assignments."

In S^{i+1}_2 then by Π^b_{i+1}–PIND on the length of a G_i-proof π prove that any sequent in π is tautological. Corollary 7.2.4 then also implies that

$$T^i_2 \vdash i\text{-RFN}(G_i)$$

The proof of the second part of the theorem is more involved and we only sketch it, leaving the details to the reader. Work in S^i_2. Assume that π is a G^*_i-proof and w.l.o.g. assume that all formulas in π are Σ^q_i. Then following the proof of Theorem 7.2.3 define functions witnessing every sequent in π. This is straightforward except for the estimate of the time bound to the running time of the algorithm computing the witnessing function. To obtain the polynomial bound one needs to use the assumption that π is *treelike*.

We shall explain this on an example. If $f(w)$ is a function witnessing a sequent in a proof that is not treelike, then in the course of the construction of the witnessing functions for later sequents we may iterate f polynomially many times. That can, however, give an exponential time (e.g., a repeated squaring gives the exponentiation in polynomially many steps). If π is treelike, however, then f will never be iterated but only composed with the other witnessing functions.

The reader is invited to fill in the details, or see Krajíček and Takeuti (1992).

Q.E.D.

The following theorem is the main application of the reflection principles and it generalizes a statement from Cook (1975).

Theorem 9.3.17. *Assume that P is a proof system, Δ^b_1-defined in S^1_2, and let $0 \le j \le i$. Assume also that*

$$T^i_2 \vdash j\text{-RFN}(P)$$

Then G_i j-polynomially simulates P and, in fact, this simulation is definable in S^1_2

$$S^1_2 \vdash G_i \ge_j P$$

The same is true for S^i_2 and G^*_i in place of T^i_2 and G_i.

Proof. From the hypothesis and Theorem 9.2.5 it follows that

$$S^1_2 \vdash \left(G_i \vdash \| j\text{-RFN}(P) \|^{|a|}_{q[|a|]} \right)$$

for some bounding polynomial $q(x)$ for j-RFN(P). This implies (as G_i is, provably in S_2^1, closed under modus ponens) that

$$S_2^1 \vdash \left(G_i \vdash \|P(x, y)\|_{q(|a|)}^{|a|}(\overline{p}, \overline{q}) \wedge G_i \vdash \|y \in \Sigma_j^q\|_{q(|a|)}^{|a|}(\overline{q}) \right)$$

$$\to G_i \vdash \|\mathrm{Taut}_j(y)\|_{q(|a|)}^{|a|}(\overline{q})$$

Now argue in S_2^1: let π be a P-proof of $A \in \Sigma_j^q$: That is, $P(\pi, A)$ holds. $P(x, y)$ is a Δ_1^b-formula, so by Lemma 9.3.13 also

$$G_1^* \vdash \|P(x, y)\|(\tilde{\pi}, \tilde{A})$$

holds, and so also

$$G_i \vdash \|P\|(\tilde{\pi}, \tilde{A})$$

The same argument applies to the true Δ_1^b-formula $A \in \Sigma_j^q$, so

$$G_i \vdash \|y \in \Sigma_j^q\|(\tilde{A})$$

Hence by the previous implication

$$G_i \vdash \|\mathrm{Taut}_j(y)\|(\tilde{A})$$

By Lemma 9.3.15 then

$$G_i \vdash A$$

This concludes the proof of

$$S_2^1 \vdash G_i \geq_j P$$

That the simulation is polynomial time follows then from Theorem 7.2.3.

An identical argument, using Theorem 9.2.6 instead of 9.2.5, proves the statement for S_2^i and G_i^* in place of T_2^i and G_i.   Q.E.D.

Corollary 9.3.18. *Let P be a propositional proof system in the sense of Definition 4.1.1 that is Δ_1^b-definable in S_2^1 and closed under substitution and modus ponens. Assume*

$$S_2^1 \vdash \mathrm{Con}(P)$$

Then EF polynomially simulates P and, in fact

$$S_2^1 \vdash EF \geq P$$

Proof. By Lemma 9.3.12 the hypothesis implies that

$$S_2^1 \vdash 0\text{-RFN}(P)$$

which by Theorem 9.3.17 implies

$$S_2^1 \vdash G_1^* \geq_0 P$$

The statement then follows from Lemma 4.6.3, whose proof, as is readily veri-fiable, can be formalized in S_2^1.   Q.E.D.

The following corollary again states Lemma 4.5.5. The reason for doing so is that this time it is obtained in a simpler way (though building on nontrivial machinery). This proof was, in fact, chronologically the first one.

Corollary 9.3.19.

$$S_2^1 \vdash EF \geq SF$$

Proof. By Corollary 9.3.18 it is enough to prove Con(SF) in S_2^1, which is straight-forward: By induction on the number of steps show that every formula in it is a tautology. This requires Π_1^b–LIND available in S_2^1 through Lemma 5.2.5.

                                                            Q.E.D.

The following corollary is a very important effective version of Theorem 4.1.2. Its first version, for PV and EF, was proved by Cook (1975) .

Corollary 9.3.20. *Let $Taut_0(a)$ be a Π_1^b-formula defining in S_2^1 the set of tautolog-ical propositional (quantifier free) formulas. Assume that there is a Σ_1^b-formula $\sigma(a)$ such that*

$$T_2^i \vdash \sigma(a) \equiv Taut_0(a)$$

That is

$$T_2^i \vdash NP = coNP$$

Then tautologies $TAUT_0$ have polynomial size G_i-proofs. In fact, this is provable in S_2^1

$$S_2^1 \vdash \forall A, \ Taut_0(A) \rightarrow (G_i \vdash A)$$

The same is true for S_2^i and G_i^* (resp. S_2^1 and EF) in place of T_2^i and G_i.

Proof. Let

$$\sigma(a) = \exists x \leq t(a), \delta(a, x)$$

with $\delta \in \Delta_1^b$. Define a proof system P by

$$P(\pi, A) := \pi \leq t(A) \wedge \delta(A, \pi)$$

Then the hypothesis of the statement implies that

$$T_2^i \vdash Con(P)$$

That is by Corollary 9.3.18

$$S_2^1 \vdash G_i \geq P$$

But the hypothesis also implies that every tautology has a polynomial size
($\leq |t(A)|$) P-proof, hence also a G_i-proof.

For S_2^i and G_i^* (and S_2^1 and EF) the same argument applies.   Q.E.D.

Note that the condition of the previous statement can be strengthened as, in fact,
all Σ_i^q-tautologies will have polynomial size G_i-proofs. This is because the hypothesis implies that every Σ_∞^b-formula, and hence the formula Taut_i in particular,
is in T_2^i equivalent to a Σ_1^b-formula.

The following statement is a propositional version of the previous statement.

Theorem 9.3.21. *Assume that all Σ_1^q-tautologies (TAUT$_1$) have polynomial size
G_i-proofs.*

Then all Σ_i^q-tautologies (TAUT$_i$) have polynomial size G_i-proofs.
The same is true for G_i^* in place of G_i.

Proof. The hypothesis implies that the following $\forall \Sigma_1^b$-sentence A is true

$$A := \forall x, \mathrm{Taut}_0(x) \rightarrow \exists y, |y| \leq |x|^k \wedge \mathrm{Prf}_{G_i}(y, x)$$

where $\mathrm{Prf}_{G_i}(y, x)$ is a Δ_1^b-formalization of "y is a G_i-proof of x."
As $T_2^i \vdash i\text{-RFN}(G_i)$, by Theorem 9.3.16, we have

$$T_2^i + A \vdash \mathrm{Taut}_0(A) \equiv (\exists y, |y| \leq |A|^k \wedge \mathrm{Prf}_{G_i}(y, A))$$

That is,

$$T_2^i + A \vdash \mathrm{NP} = \mathrm{coNP}$$

Define a proof system G_i^A to be an extension of G_i by new initial sequents

$$\longrightarrow \|A\|_{q(m)}^m$$

$m = 1, 2, \ldots$, where $q(x)$ is a fixed bounding polynomial for A. Then analogously
with Theorems 9.2.5 and 9.3.16 we have that $T_2^i + A$ proves $i\text{-RFN}(G_i^A)$ and that
G_i^A simulates $T_2^i + A$-proofs of the Σ_i^b-consequences. Hence Corollary 9.3.20 also
holds for G_i^A and $T_2^i + A$ in place of G_i and T_2^i.

Thus

$$T_2^i + A \vdash \mathrm{NP} = \mathrm{coNP}$$

established previously implies that all TAUT$_i$ have polynomial size G_i^A-proofs.
But all formulas $\|A\|_{q(m)}^m \in \mathrm{TAUT}_1$ (as $A \in \Sigma_1^b$), and hence they all have by the
hypothesis of the theorem polynomial size G_i-proofs. So G_i has a polynomial
speed-up (see Definition 9.1.3) over G_i^A and the statement follows.   Q.E.D.

We shall now prove a statement about the reflection principles of G in U_2^1.

Theorem 9.3.22. *The theory U_2^1 proves that G is a sound proof system. That is, for every $i \geq 0$*

$$U_2^1 \vdash i\text{-RFN}(G)$$

Proof. The satisfaction relation for the quantified propositional formulas is $\Delta_1^{1,b}$-definable in U_2^1. A definition of

$$w \models A$$

for general $A \in \Sigma_\infty^q$ is constructed by formalizing the following:

if

$$\forall x_1 \exists y_1 \ldots \forall x_n \exists y_n \; B(\overline{x}, \overline{y}, \overline{p}) \quad \text{is a prenex normal form of } A(\overline{p})$$

then there are Skolem functions

$$F_1(x_1, \overline{p}), \ldots, F_n(x_1, \ldots, x_n, \overline{p}) \text{ s.t. } \forall \overline{x} \; B(\overline{x}, y_i/F_i, w) \text{ is true}$$

This works as the prenex normal form is Δ_1^b-definable in S_2^1, $F_1, \ldots, F_n$ can be coded by set variables of the language of U_2^1, and the last formula is $\Pi_1^b(F_1, \ldots, F_n)$.

By induction on the length of A (i.e., by $\Sigma_1^{1,b}$–PIND) U_2^1 proves the Tarski conditions for this definition of satisfaction.

This definition as stated is $\Sigma_1^{1,b}$, but it can be made $\Delta_1^{1,b}$ by choosing functions $F_1, \ldots, F_n$ in some canonical way.

Argue now in U_2^1. Let π be a G-proof of A. By induction on the length of π show that all sequents in π are satisfied by all assignments; this needs $\Pi_1^{1,b}$–LIND, which is available in U_2^1. Hence

$$\forall w, \; w \models A$$

But $w \models A$ implies $w \models_i A$ (the formula from Lemma 9.3.9); this is provable by induction on the logical complexity of A using Tarski's conditions satisfied by both $\models$ and $\models_i$.   Q.E.D.

The next corollary summarizes all corollaries of the previous theorem obtained analogously to the earlier ones for the system G_i.

Corollary 9.3.23.

1. *Let P be a propositional proof system Δ_1^b-defined in S_2^1 and assume that for some $i \geq 0$*

$$U_2^1 \vdash i\text{-RFN}(P)$$

Then G i-polynomially simulates P and, in fact

$$S_2^1 \vdash G \geq_i P$$

(after 9.3.17)

2. *If*

$$U_2^1 \vdash NP = coNP$$

then all tautologies from $TAUT_\infty := \bigcup_i TAUT_i$ *have polynomial size G-proofs and, in fact, for each* $i \geq 0$

$$S_2^1 \vdash \forall A, Taut_i(A) \rightarrow (G \vdash A)$$

(after 9.3.20)

3. *Assume that all* Σ_1^q*-tautologies have polynomial size G-proofs. Then all quantified tautologies from* $TAUT_\infty$ *have polynomial size G-proofs. (after 9.3.21)*

We conclude this section by an important corollary of Theorems 9.3.16 and 9.3.22.

Theorem 9.3.24. *Let* P *be any of the following proofs systems: EF, SF, G_i or G_i^* ($i \geq 1$), or G.*

Then there is a polynomial $p(x)$ *such that for each n there is a P-proof of*

$$\| \operatorname{Con}(P) \|_{q(n)}^n$$

of size $\leq p(n)$ *($q(x)$ is a fixed bounding polynomial of* $\operatorname{Con}(P)$*).*

Proof. For G_i, G_i^*, or G the theorem follows from Theorems 9.3.16, 9.2.5, and 9.2.6, or from 9.3.22 and 9.2.8, respectively.

For EF and SF it follows from the statement for G_1^* using Corollaries 9.3.18 and 9.3.19.   Q.E.D.

In the proof we could also use Theorems 9.3.4 and 9.1.5 for the case $P = $ EF. Note that Theorems 9.1.6 and 9.3.3 imply a similar statement for $P = $ F with $n^{(\log n)^{O(1)}}$ in place of $p(n)$.

9.4. Model-theoretic constructions

In this section we give model-theoretic proofs for Theorems 9.1.3 and 9.2.7. I believe that this side of the simulation results is important for understanding the interplay between arithmetic and propositional logic, and the fundamental problem of lower bounds for proof systems.

The following statement is a version of Theorem 9.1.3.

Theorem 9.4.1. *Let* M *be a countable model of the true arithmetic* $Th(\omega)$ *and let* $n, t \in M \setminus \omega$ *be its two nonstandard elements. Let* $\theta(a)$ *be a* $\Delta_0(R)$*-formula with* a *the only free variable.*

Assume that in M *there is no Frege proof of* $\langle \theta \rangle_{(n)}$ *of depth* $\leq t$ *and size* $\leq n^t$ *(i.e., no element* $\leq 2^{n^t}$ *codes such a proof).*

Then it is possible to define $R \subseteq M \times M$ such that

$$(n^{\omega}, R) \models I\Delta_0(R)$$

and

$$(n^{\omega}, R) \models \neg\theta(n)$$

Before we give the proof we should understand that this theorem implies Theorem 9.1.3. Assume that the formulas $\langle\theta\rangle_{(m)}$, $m < \omega$, do not have polynomial size constant-depth Frege proofs. This means that for any $k < \omega$ and any $d < \omega$ there are $m < \omega$ such that $\langle\theta\rangle_{(m)}$ does not have a depth d size $\leq m^k$ F-proof. By compactness there is a model of $\mathrm{Th}(\omega)$ and nonstandard $d, k \in M$ such that for some $m \in M$, M thinks that there is no depth d F-proof of $\langle\theta\rangle_{(m)}$ of size $\leq m^k$. Take $t := \min(d, k)$. By this theorem then there is a model of $I\Delta_0(R)$ in which $\forall x\theta(x)$ fails: That is, $\forall x\theta(x)$ is not provable in $I\Delta_0(R)$.

Proof. Let M, θ, n, and t satisfy the hypothesis of the theorem. Let Fle denote the set of propositional formulas coded in M, having a *standard* depth, built from the atoms p_{ij}, and of size $\leq n^{\omega}$. We shall form a set $T \subseteq$ Fle satisfying:

1. $\neg\langle\theta\rangle_{(n)} \in T$
2. for any $\psi \in$ Fle: $\psi \in T$ or $\neg\psi \in T$, but not both
3. if $\psi \in$ Fle has the form $\bigwedge_i \phi_i$ then $\psi \in T$ iff $\phi_i \in T$ all i
4. if $\psi \in$ Fle has the form $\bigvee_i \phi_i$ then $\psi \in T$ iff $\phi_i \in T$ some i
5. for all $\eta(x) \in \Delta_0(R)$ with parameters from n^{ω} and x the only free variable, either $\neg\langle\eta\rangle_{(0)} \in T$ or $\langle\eta\rangle_{(u)} \in T$ for all $u \in n^{\omega}$ or $\langle\eta\rangle_{(u)} \wedge \neg\langle\eta\rangle_{(u+1)} \in T$ for some $u \in n^{\omega}$

having such set T define a relation $R \subseteq n^{\omega} \times n^{\omega}$ by

$$R(i, j) \quad \text{iff} \quad p_{ij} \in T$$

Claim 1. *For any $\Delta_0(R)$-sentence ξ with parameters from n^{ω}*

$$(n^{\omega}, R) \models \xi \quad \text{iff} \quad \langle\xi\rangle \in T$$

The claim follows by conditions 2–4 posed on T.

Claim 2. $(n^{\omega}, R) \models I\Delta_0(R) + \neg\theta(n)$

This follows from conditions 1 and 5.

It remains to construct the set T having the required properties. This can be done by a completeness-type argument, but we shall cast it as a forcing-type argument.

Let $\mathcal{P}$ denote the class of subsets $S \subseteq$ Fle satisfying the conditions

(i) $\neg\langle\theta\rangle_{(n)} \in S$
(ii) for any $k < \omega$ there is no depth k size $\leq n^k$ F-proof of contradiction $(= 0)$ from formulas in S

(iii) S is definable (and hence coded) in M.

Note that for $S \in \mathcal{P}$ there is $s > \omega$ such that there is no F-proof of 0 from S of depth $\leq s$ and size $\leq n^s$; this follows by induction as it is true for all standard s. The next claim is obvious.

Claim 3. *Let $S \in \mathcal{P}$ and $\psi \in$ Fle.*
Then either $S \cup \{\psi\} \in \mathcal{P}$ or $S \cup \{\neg\psi\} \in \mathcal{P}$.

Claim 4. *Let $S \in \mathcal{P}$ and $\psi \in S$, and assume that ψ has the form $\psi := \bigvee_{j \leq r} \phi_j$.*
Then for some $j_0 \leq r$, $S \cup \{\phi_{j_0}\} \in \mathcal{P}$.

Assume otherwise: That is, for every $j \leq r$ there is a depth k_j size $\leq n^{k_j}$ F-proof of 0 from $S \cup \{\phi_j\}$, and hence depth ℓ_j size $\leq n^{\ell_j}$ F-proof π_j of $\neg\phi_j$ from S, $k_j, \ell_j < \omega$. As $\psi \in$ Fle, $r \leq |\psi| \leq n^\ell$, some $\ell \in \omega$.

Take $s > \omega$ such that there is no depth s size $\leq n^s$ F-proof of 0 from S. Each proof π_j has depth $<< s$ and size $\ll (n^s/n^\ell)$, so joining these $\leq n^\ell$ proofs gets a depth $< s$ size $< n^s$ proof of 0 from S, a contradiction.

Claim 5. *Let $S \in \mathcal{P}$ and $\psi \in S$, and let ψ have the form $\bigwedge_i \phi_i$.*
Then $S \cup \{\phi_i \mid$ all $i\} \in \mathcal{P}$.

This is seen analogously to Claim 4.

Claim 6. *Let $S \in \mathcal{P}$ and let $\eta(x)$ be a $\Delta_0(R)$ formula with the parameters from n^ω and with x the only free variable.*
Then one of the following sets is in $\mathcal{P}$ too:
(a) $S \cup \{\neg\langle\eta\rangle_{(0)}\}$
(b) $S \cup \{\langle\eta\rangle_{(u)} \mid u \in n^\omega\}$
(c) $S \cup \{\langle\eta\rangle_{(u)}\} \cup \{\neg\langle\eta\rangle_{(u+1)}\}$ some $u \in n^\omega$

To prove Claim 6 assume otherwise, so there is a depth k_0 size $\leq n^{k_0}$ proof π_{-1} of $\langle\eta\rangle_{(0)}$ from S, a depth k_u size $\leq n^{k_u}$ proof π_u of

$$\langle\eta\rangle_{(u)} \to \langle\eta\rangle_{(u+1)}$$

from S for all $u \in n^\omega$, and a depth k size $\leq n^k$ proof from S of the disjunction

$$\bigvee_{u \in X} \neg\langle\eta\rangle_{(u)}$$

for some $X \subseteq n^\omega$ of size $\leq n^k$.

For any nonstandard s, joining proofs $\pi_{-1}, \pi_0, \ldots, \pi_v$ for $v = \max(X)$ by cuts entails all $\langle\eta\rangle_{(u)}, u \in X$ by a depth s size $\leq n^s$ proofs, obtaining thus a depth s size $\leq n^s$ proof of 0 from S, contradicting $S \in \mathcal{P}$.

Now we are ready to construct the set T. Let $\psi_1, \psi_2, \ldots$ enumerate the set Fle and $\eta_1(x), \eta_2(x), \ldots$ enumerate all $\Delta_0(R)$ formulas with parameters from n^ω, and with one free variable x.

Construct a sequence $S_0, S_1, \ldots \in \mathcal{P}$ such that

(i) $S_0 := \{\neg \langle \theta \rangle_{(n)}\}$

(ii) $S_i \subseteq S_{i+1}$

(iii) $\psi_i \in S_i$ or $\neg \psi_i \in S_i$

(iv) if $\psi_i \in S_i$ and $\psi_i = \bigvee_{j \leq r} \phi_j$ then $\phi_{j_0} \in S_i$, some $j_0 \leq r$

(v) if $\psi_i \in S_i$ and $\psi_i = \bigwedge_{j \leq r} \phi_j$ then all $\phi_j \in S_i$

(vi) either $\neg \langle \eta_i \rangle_{(0)} \in S_i$ or $\langle \eta_i \rangle_{(u)} \wedge \neg \langle \eta_i \rangle_{(u+1)} \in S_i$ for some $u < n^\omega$, or $\langle \eta_i \rangle_{(u)} \in S_i$ all $u \in n^\omega$.

Having S_i satisfying the conditions, S_{i+1} exists by Claims 3–6. Set

$$T := \bigcup_i S_i$$

The set T fulfills requirements 1–5. Q.E.D.

The reason for the particular forcing-type formulation of the argument is its similarity with the following, more involved construction, which is conveniently expressed by using forcing. Another reason is a later finite version of this construction developed in Section 13.3.

Theorem 9.4.2. *Let $(M, \mathcal{X})$ be a model of V_1^1 and let $\tilde{\tau}(p_1, \ldots, p_n) \in \mathcal{X}$ be a propositional formula in $(M, \mathcal{X})$.*

Then the following two conditions are equivalent:

1. *In $(M, \mathcal{X})$ there is no EF-proof of $\tilde{\tau}$.*
2. *There is a $\Sigma_0^{1,b}$-elementary extension $(M', \mathcal{X}')$ of $(M, \mathcal{X})$ in which $\neg \tilde{\tau}$ is satisfiable.*

Proof. Assume that 1 fails, and let $\pi \in \mathcal{X}$ be an EF-proof of $\tilde{\tau}$ in $(M, \mathcal{X})$. As $(M', \mathcal{X}')$ is a $\Sigma_0^{1,b}$-elementary extension, π is an EF-proof of $\tilde{\tau}$ in $(M', \mathcal{X}')$ as well (see Definition 9.3.1). But by Theorem 9.3.4 $\tilde{\tau}$ must be tautologically true in $(M', \mathcal{X}')$; hence 2 fails.

Assume now that 1 holds and assume also that $(M, \mathcal{X})$ is countable. Construct $(M', \mathcal{X}')$ as follows.

By compactness there is a countable elementary extension $(M_0, \mathcal{X}_0)$ of $(M, \mathcal{X})$ satisfying V_1^1 such that:

(i) there is $t \in M_0$ such that for all $v \in M$, $v < t$

(ii) in $(M_0, \mathcal{X}_0)$ there is no EF-proof of $\tilde{\tau}$.

Let $(M^*, \mathcal{X}^*)$ be a substructure of $(M_0, \mathcal{X}_0)$ defined by:

(i) $M^* = \{v \in M_0 \mid \exists w \in M, v \leq w\}$

(ii) $\mathcal{X}^* = \{\tilde{\beta} \in X_0 \mid \tilde{\beta} \subseteq M^*\}$.

We define in $(M_0, \mathcal{X}_0)$ several families. Let $\{\overline{p}\}$ be the atoms of $\tilde{\tau}$ and let $\mathrm{Fle}(\overline{p}) \subseteq \mathcal{X}_0$ be the formulas with the atoms among $\{\overline{p}\}$. Further let A be the set of the atoms $\{\overline{p}\}$ plus new atoms of the form q_ψ, one for each $\psi \in \mathrm{Fle}(\overline{p})$, and let $\mathrm{Fle} \subseteq \mathcal{X}_0$ be the set of the formulas with atoms among A.

Let $C \subseteq \mathcal{X}_0$ be the family of tuples of elements of $A \cup \{0, 1\}$. Let

$$C^* := \{\beta \in C \mid |\beta| \in M^*\}$$

where $|(q_{\psi_1}, \ldots, q_{\psi_m})| = m$, and let

$$\mathrm{Fle}^* := \{\phi \in \mathrm{Fle} \mid |\phi| \in M^*\}$$

We will consider $\tilde{\beta} \in \mathcal{X}_0$ simultaneously also as an element of C: the tuple of bits of the characteristic function of $\tilde{\beta}$ (so for such $\tilde{\beta}$: $\tilde{\beta} \in C^* \equiv \beta \in \mathcal{X}^*$).

The following claim is established by induction on the logical complexity of B.

Claim 1. *Let $B(\beta)$ be a $\Sigma_0^{1,b}$-formula and let $\tilde{\beta} \in \mathcal{X}^*$.*
Then

$$(M^*, \mathcal{X}^*) \models B(\tilde{\beta}) \to \exists \pi \, Prf_F\left(\pi, \langle B \rangle (\overline{q} / \tilde{\beta})\right)$$

where $\overline{q}$ are atoms corresponding to β.

This is analogous to Lemma 9.3.13.

We will construct a set $G \subseteq \mathrm{Fle}$ satisfying the following conditions:

(1) $\neg \tilde{t} \in G$,

(2) for all $\psi \in \mathrm{Fle}^*$ exactly one of ψ, $\neg \psi$ is in G,

(3) whenever $\pi \in \mathcal{X}_0$ is an EF-proof of ψ from the assumptions $\psi_1, \ldots, \psi_r$, $|\pi| \in M^*$ and all $\psi_i \in G$, then also $\psi \in G$,

(4) if $\psi \in G$, $\psi \in \mathrm{Fle}^*$, and $\psi = \bigvee_{1 \leq i \leq r} \psi_i$, then $\psi_j \in G$ for some $1 \leq j \leq r$,

(5) for any $\Sigma_0^{1,b}$-formula $H(\phi, x)$ with the parameters from C^* and any $v \in M^*$, one of the following three conditions holds

    (a) $\neg \langle H(\phi, 0) \rangle_v(\tilde{\delta}) \in G$, all $\tilde{\delta} \in C^*$ of length $\leq t(v)$,

    (b) $\langle H(\phi, v) \rangle_v(\tilde{\delta}) \in G$, for some $\tilde{\delta} \in C^*$ of length $\leq t(v)$,

    (c) there is $v' < v$ such that

$$\langle H(\phi, v') \rangle_v(\tilde{\delta}) \in G \quad \text{and} \quad \neg \langle H(\phi, v' + 1) \rangle_v(\tilde{\varepsilon}) \in G$$

    for some $\tilde{\delta} \in C^*$ of length $\leq t(v)$ and for all $\tilde{\varepsilon} \in C^*$ of length $\leq t(v)$.

The term $t(v)$ implicitly bounds the size of the interval whose subsets can be substituted for ϕ in H for $x \leq v$.

Assume for a moment that we have such a set G. Define a structure $(M^*[G], \mathcal{X}^*[G])$ by

$$M^*[G] := M^* \quad \text{and} \quad \mathcal{X}^*[G] := C^* / \sim$$

where $\sim$ is an equivalence relation defined by

$$\tilde{\beta}_1 \sim \tilde{\beta}_2 \text{ iff } \langle \beta_1 = \beta_2 \rangle_u(\tilde{\beta}_1, \tilde{\beta}_2) \in G$$

(u the maximum of the lengths of $\tilde{\beta}_1$, $\tilde{\beta}_2$). Note that $(M^*[G], \mathcal{X}^*[G])$ is an extension of $(M^*, \mathcal{X}^*)$ and hence of $(M, \mathcal{X})$ too.

Claim 2. *Let $B(\beta)$ be any $\Sigma_0^{1,b}$-formula with parameters from C^* and $\tilde{\beta} \in C^*$. Then we have for all sufficiently large u*

$$(M^*[G], \mathcal{X}^*[G]) \models B(\tilde{\beta}/ \sim) \text{ iff } \langle B \rangle_u(\tilde{\beta}) \in G$$

In particular, $(M^*[G], \mathcal{X}^*[G])$ is a $\Sigma_0^{1,b}$-elementary extension of $(M, \mathcal{X})$.

The claim follows from conditions (2)–(4) posed on G. For example, that all $\Sigma_0^{1,b}$-sentences true in $(M, \mathcal{X})$ also hold in $(M^*[G], \mathcal{X}^*[G])$ follows from condition (3) and Claim 1.

Claim 3. *Structure $(M^*[G], \mathcal{X}^*[G])$ is a model of V_1^1.*

Condition (5) posed on G guarantees that the induction for every $\Sigma_1^{1,b}$-formula $\exists \phi H(\phi, x)$ holds up to every $v \in M^*[G]$. The other axioms hold in $(M^*[G], \mathcal{X}^*[G])$ obviously. In particular, the $\Sigma_0^{1,b}$–CA is guaranteed by Claim 2.

Claim 4. *There is $\tilde{\alpha} \in \mathcal{X}^*[G]$ such that*

$$(M^*[G], \mathcal{X}^*[G]) \models (\tilde{\alpha} \models \neg \tilde{\tau})$$

By condition (1) posed on G and Claim 2, $\tilde{\alpha}$ is a satisfying assignment for $\neg \tilde{\tau}$ in $(M^*[G], \mathcal{X}^*[G])$, where

$$\tilde{\alpha} := \overline{p}^{\alpha}/ \sim$$

It remains to construct the set G satisfying the preceding five requirements. We shall use two simple technical properties of system EF.

Form a set $T \subseteq$ Fle consisting of all formulas:

  (i) $q_{p_i} \equiv p_i$, whenever $p_i \in \{\overline{p}\}$,
  (ii) $q_{\neg \psi} \equiv (\neg q_\psi)$, whenever $\psi \in \text{Fle}(\overline{p})$,
  (iii) $q_{\psi_1 \circ \psi_2} \equiv (q_{\psi_1} \circ q_{\psi_2})$, whenever $\psi_1, \psi_2 \in \text{Fle}(\overline{p})$ and $\circ = \vee, \wedge$.

A set of formulas $S \subseteq$ Fle is said to ℓ-*entail* formula ψ iff there is an F-proof of size at most ℓ of ψ with the axioms from $S \cup T$. A set S is called ℓ-*consistent* iff S does not ℓ-entail 0.

Claim 5. *Let $S \subseteq$ Fle be a $\Delta_1^{1,b}$-definable in $(M_0, \mathcal{X}_0)$, and assume that ψ has an EF-proof from S of size ℓ in $(M_0, \mathcal{X}_0)$.*
  Then S also $O(\ell^2)$-entails ψ in $(M_0, \mathcal{X}_0)$.

This follows as every extension axiom of size t in the EF-proof can be proved (after suitably renaming the extension atoms) from T by an F-proof of size $O(t^2)$.

Claim 6. *Let $S \subseteq$ Fle be a $\Delta_1^{1,b}$- definable in $(M_0, \mathcal{X}_0)$ and assume that S is ℓ-consistent in $(M_0, \mathcal{X}_0)$, where ℓ is nonstandard.*

Then for every formula ψ of size at most $\ell^{2^{-1}}$ one of the sets $S \cup \{\psi\}$ or $S \cup \{\neg\psi\}$ is $\ell^{2^{-1}}$-consistent.

Also, for every disjunction $\bigvee_{i \le r} \psi_i \in Fle$ of size at most $\ell^{3^{-1}}$ one of the sets $S \cup \{\bigwedge_{i \le r} \neg\psi_i\}$ or $S \cup \{\bigvee_{i \le r} \psi_i\} \cup \{\psi_j\}$, some $j \le r$, is $\ell^{3^{-1}}$-consistent.

The first part is obvious. For the second part, assuming that all $r + 2$ sets are $\ell^{3^{-1}}$-inconsistent would allow us to construct in an obvious way a proof of 0 from S of size at most

$$(r + 2)\ell^{3-1} + O\left(|\bigvee_{i \le r} \psi_i|^2\right) \le \ell$$

This is a contradiction.

Claim 7. *Let $S \subseteq Fle$ be a $\Delta_1^{1,b}$-definable family of formulas in (M_0, X_0), and assume that S is ℓ-consistent in (M_0, X_0), where $\ell \in (M_0 \setminus M^*)$.*

Let $H(\phi, x)$ be a $\Sigma_0^{1,b}$-formula with parameters from C^* and M^*. Let $v \in M^*$, and assume that a term $t(v)$ bounds the size of the interval whose subsets can be substituted for ϕ in H for all $x < v$.

Then one of the following sets is $\ell^{3^{-1}}$-consistent:

(i) $S \cup \{\neg\langle H(\phi, 0)\rangle_v(\tilde{\delta}) \mid \tilde{\delta} \in C^*, |\tilde{\delta}| \le t(v)\}$,

(ii) $S \cup \{\langle H(\phi, v)\rangle_v(\tilde{\delta})\}$, some $\tilde{\delta} \in C^*$ of length $\le t(v)$,

(iii) $S \cup \{\langle H(\phi, v')\rangle_v(\tilde{\delta})\} \cup \{\neg\langle H(\phi, v' + 1)\rangle_v(\tilde{\rho}) \mid \tilde{\rho} \in C^*, |\tilde{\rho}| \le t(v)\}$, some $\tilde{\delta} \in C^*$ of size $\le t(v)$ and $v' < v$.

To prove Claim 7 take the formula $D(u)$

$$\forall w \le u \exists \bar{r}_w \in C, S \; \ell^{3^{-1}}\text{-entails formula } \langle H(\phi, w)\rangle(\bar{r}_w)$$

The formula $D(u)$ is a $\Sigma_1^{1,b}$-formula and witnesses $\bar{r}_w$ are actually from C^* (using the bound $|\bar{r}_w| \le t(v)$).

As $(M_0, X_0) \models V_1^1$ one of the two cases must occur:

(a) $D(v)$ holds in (M_0, X_0)

(b) there exists minimal $u \le v$ for which $D(u)$ fails in (M_0, X_0).

In case (a) define

$$S' := S \cup \{\langle H(\phi, v)\rangle(\bar{r}_v)\}$$

where $\bar{r}_v$ is a witness to the existential quantifier of $D(v)$. The set S' is $\ell/2$-consistent as otherwise one could $\ell/2 + \ell^{3^{-1}} \le \ell$-entail 0 from S, which would be a contradiction.

In case (b) let $u \le v$ be the first u such that $D(u)$ fails. Take a set

$$S' := S \cup \{\langle H(\phi, u - 1)\rangle(\bar{r}_{u-1})\} \cup \{\neg\langle H(\phi, u)\rangle(\bar{q}) \mid \bar{q} \in C, |\bar{q}| \le t(v)\}$$

for $u \geq 1$ (and again $\bar{r}_{u-1}$ the relevant witness) or

$$S' := S \cup \{\neg\langle H(\phi, 0)\rangle(\bar{q}) \mid \bar{q} \in C, \ |\bar{q}| \leq t(v)\}$$

for $u = 0$.

We claim that S' is $\ell^{3^{-1}}$-consistent. Assume otherwise and w.l.o.g. let $u \geq 1$. The set $S + \langle H(\phi, u - 1)\rangle(\bar{r}_{u-1})$ then $O(\ell^{2/3})$-entails some disjunction of the form

$$\bigvee_{\bar{q} \in I} \langle H(\phi, u)\rangle(\bar{q})$$

where $I \subseteq C^*$. But then $\langle H(\phi, u)\rangle(\bar{r})$ can also be $O(\ell^{2/3})$-entailed from $S + \langle H(\phi, u - 1)\rangle(\bar{r}_{u-1})$, where $\bar{r}$ is a new tuple defined by extension atoms using a case distinction considering which disjunct in the disjunction is true (cf. Claim 5). Note that $|\bar{r}| \leq t(v)$. This contradicts the assumption that $D(u)$ fails; hence S' is $\ell^{3^{-1}}$-consistent.

Define now the family $\mathcal{P}$ of all $H \subseteq \mathrm{Fle}$ that are $\Delta_1^{1,b}$-definable in $(M_0, \mathcal{X}_0)$ and that are ℓ-consistent for some $\ell \in (M_0 \backslash M^*)$; such ℓ exists by our assumption about $(M_0, \mathcal{X}_0)$. Note that $\{\neg\tilde{t}\} \in \mathcal{P}$.

Family $\mathcal{P}$ is partially ordered by the inclusion relation $\subseteq$. Class $\mathcal{Q} \subseteq \mathcal{P}$ is *dense* if

$$\forall H \in \mathcal{P} \exists H' \in \mathcal{Q}, H \subseteq H'$$

Class $\mathcal{Q}$ is *definable* if there is a formula $\Psi(X)$ in the language of V_1^1 augmented by new metavariable X such that

$$\mathcal{Q} = \{H \in \mathcal{P} \mid (M, \mathcal{X}, H) \models \Psi(H)\}$$

Class $\mathcal{G} \subseteq \mathcal{P}$ is *generic* if it satisfies the following conditions:
  (i) if $H \in \mathcal{G}$ and $H' \subseteq H$ then $H' \in \mathcal{G}$,
  (ii) $\mathcal{G}$ intersects every dense, definable subclass of $\mathcal{P}$.

Claim 8. *Let $\mathcal{G} \subseteq \mathcal{P}$ be a generic class and assume that $\{\neg\tilde{t}\} \in \mathcal{G}$. Put*

$$G := \bigcup \mathcal{G}$$

Then G satisfies conditions (1)–(5) and hence $(M^*[G], \mathcal{X}^*[G])$ is a model of V_1^1 in which the formula $\neg\tilde{t}$ is satisfiable.

As model $(M_0, \mathcal{X}_0)$ is countable there are only countably many dense definable subclasses of $\mathcal{P}$; hence by the standard argument a generic class $\mathcal{G}$ exists. By Claims 5, 6, 7 the classes of those $K \in \mathcal{P}$ that fulfill condition (2) for $\psi \in \mathrm{Fle}^*$

$$\psi \in K \quad \text{or} \quad \neg\psi \in K$$

are clearly definable and dense, as are the classes of $K \in \mathcal{P}$ that fulfill condition

(4) for $\psi = \bigvee_{i \le r} \psi_i \in \mathrm{Fle}^*$, that is

$$\bigwedge_{i \le r} \neg \psi_i \in K \quad or \quad \{\psi, \psi_j\} \subseteq K, \; some \; j \le r$$

and the classes of $K \in \mathcal{P}$ that fulfill condition (5) for $H(\phi, x)$ and $v \in M^*$, that is

$\{\neg \langle H(\phi, 0)\rangle_v(\tilde{\delta}) \mid \tilde{\delta} \in C, |\tilde{\delta}| \le t(v)\} \subseteq K$ or

$\{\langle H(\phi, v)\rangle_v(\tilde{\delta})\} \subseteq K$,   some $\tilde{\delta} \in C$ of length $\le t(v)$ or

$\{\langle H(\phi, v')\rangle_v(\tilde{\delta})\} \cup \{\neg \langle H(\phi, v' + 1)\rangle_v(\tilde{\rho}) \mid \tilde{\rho} \in C, |\tilde{\rho}| \le t(v)\} \subseteq K$,
some $\tilde{\delta} \in C$ of length $\le t(v)$ and $v' < v$

Hence any G defined from a generic $\mathcal{G}$ satisfies conditions (1)–(5).

This concludes the description of the forcing construction of the model

$$(M', \mathcal{X}') = (M^*[G], \mathcal{X}^*[G])$$

Q.E.D.

We leave it as an exercise for the reader to give proofs for the preceding two theorems by modifying the proofs of Theorems 9.1.3 and 9.2.7 and by working with $I\Delta_0(R)$ (resp. V_1^1) plus the $\Sigma_0^{1,b}$-diagram of the original model.

9.5. Witnessing and test trees

A class of search problems comes from tasks to witness an existential quantifier in a first order sentence valid over all finite structures. Let $\exists \overline{x} \phi(\overline{x})$ be a first order sentence in a relational language consisting, for simplicity, of one binary relation R. Assume that $\exists \overline{x} \phi(\overline{x})$ is valid in every finite structure (i.e., in every digraph in the case of the language $\{R\}$). The search task is, Given a digraph find its vertices $\overline{v}$ such that $\phi(\overline{v})$ holds in the digraph.

A particular model for solving search problems is the *test tree*, defined as a decision tree (cf. Definition 3.1.12) but allowing labeling of the inner nodes by arbitrary formulas and leaves by tuples $\overline{v}$.

A bounded formula of the form $\exists \overline{x} \le t(a)\theta(a, \overline{x})$ defines for every natural number n a sentence $\exists \overline{x} \le t(n)\theta(n, \overline{x})$ whose validity is determined in a finite structure: an initial interval of N of the form $[0, s(n)]$, $s(a)$ a suitable term.

If $\exists \overline{x} \le t(n)\theta(n, \overline{x})$ is valid for all n we obtain a sequence of search problems, and if the formula $\exists \overline{x} \le t(a)\theta(a, \overline{x})$ is actually provable in bounded arithmetic, then by the results of Chapter 7 we have information in terms of the computational complexity of the witnessing function. In this section we translate this into the terms of the complexity of the test trees solving the associated search problems.

Call a propositional formula $\Sigma_i^{S,t}$ iff it satisfies conditions (a)–(d) of part 1 of Definition 10.4.8. For example, $\Sigma_1^{S,t}$-formulas are disjunctions of conjunctions of size $\le t$.

Lemma 9.5.1. *Let $i \geq 1$. Let $\exists u \leq a\theta(a, u)$ be a Σ^b_{i+1}-formula, $\theta \in \Pi^b_i$. Assume that*

$$T^i_2(R) \vdash \forall x \exists u \leq x\, \theta(x, u)$$

Then for every n there is a test tree T_n finding, given R, a valid formula from the set

$$\{\langle \theta \rangle_{n,u} \mid u \leq t(n)\}$$

and satisfying the following conditions:

1. *the height of T_n is $\log(n)^{O(1)}$*
2. *every test formula in T_n is $\Sigma^{S,t}_i$, where $S = 2^t$ and $t = \log(n)^{O(1)}$*

Proof. The proof of Theorem 9.1.3 (based on Theorem 7.1.4) shows how to translate the $T^i_2(R)$-proof of the formula $\exists u \leq t(a)\theta(a, u)$ using the translation $\langle \ldots \rangle$ of Definition 9.1.1 into a constant-depth LK-proof π' (cf. Definition 4.3.10). The size of π' is now $2^{\log(n)^{O(1)}}$ as the term may contain the # function (cf. Lemma 9.1.2).

As the cut-elimination was applied first to the original $T^i_2(R)$- proof all formulas in π' or their negations are equivalent to a $\Sigma^{S,t}_i$-formula (S and t as earlier), or equal to $\langle \exists u \leq t(a)\theta(a, u) \rangle_n$. This is because $\Sigma^b_0(R)$-formulas translate into formulas of size $\log(n)^{O(1)}$ (Lemma 9.1.2) and can be written both as disjunction of conjunctions of size at most t and as conjunctions of disjunctions of size at most t.

The IND-rule of $T^i_2(R)$ is simulated in π' by $\leq S$ cuts that can be arranged in a binary tree of height $\leq t$. The $\forall \leq :$ *right* and $\exists \leq :$ *left* are simulated by $\bigwedge :$ *right* and $\bigvee :$ *left* with $\bigwedge, \bigvee$ of unbounded ($\leq S$) arity.

The proof π' is easily turned into a constant-depth LK refutation π of the set of sequents

$$\{\longrightarrow \langle \neg\theta(a, u) \rangle_{n,u} \mid u \leq t(n)\}$$

Having the truth evaluation for the atoms (i.e., the relation R) construct a path $P(R) = \xi_1, \ldots, \xi_r$ of sequents of π satisfying

1. ξ_1 is the end-sequent
2. all sequents are false for R
3. ξ_{i+1} is a hypothesis of the inference of π giving ξ_i
4. ξ_r is an initial sequent

Conditions 2 and 4 imply that ξ_r must have the form

$$\longrightarrow \langle \neg\theta \rangle_{n,u}, \qquad \text{some } u \leq t(n)$$

Hence $\theta(n, u)$ is true for R. Thus the path $P(R)$ contains an answer to the search problem.

We need to define ξ_{i+1} knowing ξ_i. This is trivial in all inferences except the cut-rule, $\bigwedge :$ *right* and $\bigvee :$ *left*.

In case of the cut-rule let ξ_{i+1} be the unique false hypothesis; to decide which one it is one needs to query the truth value of the cut formula.

In case of $\bigwedge$: *right* with the principal formula $\bigwedge_{i \leq m} A_i, m \leq S$, it is necessary to find A_i false for R. This is done by binary search querying $\log(m) \leq t$ truth values of formulas of the form $\bigwedge_{s \leq i \leq s'} A_i$.

The case of $\bigvee$: *left* is dual to $\bigwedge$: *right*. As the number of $\bigvee$: *left*s and $\bigwedge$: *right*s on any path through π is bounded by an independent constant (e.g. by the number of $\exists \leq$: *left* and $\forall \leq$: *right* inferences in the original $T_2^i(R)$-proof that is independent of m) the total number of queries is $O(t)$ on any path $P(R)$.

This shows that the test tree can be constructed from π by turning it upside down and simulating the $\bigvee$: *left* and $\bigwedge$: *right* inferences as earlier (cf. Theorem 4.2.3).

Q.E.D.

Note that for the formula $\exists u \leq t(a)\theta(a, u)$ that is $\Sigma_i^b(R)$ this gives no information as then the binary search for u works as well.

We shall state here without a proof a lemma relevant to the test trees but based on ideas developed later in Chapter 11. The proof of the lemma is almost identical with the proof of Theorem 11.3.2 (based on the proof of Theorem 11.2.5).

Lemma 9.5.2. *Let $\exists x\theta(x)$ be a sentence in a relational language L' and assume that $\neg\exists x\theta(x)$ has an infinite model. Assume also that $\exists x\theta(x)$ is valid in every finite structure.*

Denote by X_n the search problem to find in a given structure with n elements some u satisfying $\theta(u)$.

Then there is no constant k such that each X_n could be solved by a test tree of height $\leq \log(n)^k$ querying the validity of the $\Sigma_1^{S,t}$-formulas $t = \log(n)^k$, $S = 2^t$.

Note that such test trees for X_n are guaranteed to exist if $\theta(x)$ has the form $\forall y\phi(x, y)$ and the formula $\exists x < a\forall y < a\phi(x, y)$ is provable in $T_2^1(L')$.

Next we reexamine Lemma 9.5.1 for the theory $S_2^i(R)$ in place of $T_2^i(R)$. We shall consider *witnessing test trees* instead of ordinary test trees. An ordinary test tree in a node querying $\exists x \leq n\psi(x)$ branches into two paths, one for the affirmative answer and one for the negative one. The witnessing test tree will allow, besides these nodes, nodes that branch into $n + 2$ paths: one for the negative answer and $n + 1$ for all possible witnesses $u \leq n$ for the affirmative answer. Hence we must know a witness if we want to proceed. Such nodes are called *witnessing*.

Lemma 9.5.3. *Let $i \geq 1$. Let $\exists u \leq a\theta(a, u)$ be a Σ_{i+1}^b-formula, $\theta \in \Pi_i^b$. Assume*

$$S_2^i(R) \vdash \forall x\exists u \leq a\theta(a, u)$$

Then for every n there is a witnessing test tree T_n finding, given R, a valid formula from the set

$$\{\langle\theta\rangle_{n,u} \mid u \leq t(n)\}$$

and satisfying the following conditions:
1. the height of T_n is $O(\log(\log(n)))$
2. on every path in T_n there are only $O(1)$ witnessing nodes
3. every test formula in T_n is $\Sigma_i^{S,t}$, where $S = 2^t$ and $t = \log(n)^{O(1)}$

Proof. The proof of the lemma is similar to the proof of Lemma 9.5.1. We need to notice only that as the PIND-rule is used instead of the IND-rule there are only $\leq t$ cuts (arranged in a binary tree of height $O(\log(t))$) needed for the simulation of the PIND-rule.

The $\bigwedge$: *right* and $\bigvee$: *left* inferences are treated this time with the new witnessing nodes of the tree. Q.E.D.

The reader may observe that Lemmas 9.5.1 and 9.5.3 are versions of the witnessing Theorems 7.2.3 (for T_2^{i-1} in place of S_2^i there) and 7.3.5.

9.6. Bibliographical and other remarks

The translation of Section 9.1 was first considered in Paris and Wilkie (1985) for $I\Delta_0(R)$ in a model-theoretic context (Theorem 9.4.1). The proof-theoretic presentation (and the extension to U_1^1 and V_1^1) follows Krajíček (1995a).

The translation of Σ_∞^b-formulas in Section 9.2 generalizes the translation defined in Cook (1975) for PV-equations and was developed in Krajíček and Pudlák (1990a) and somewhat differently in Dowd (unpublished). Propositions 9.2.2–9.2.6 are from Krajíček and Pudlák (1990a). Corollary 9.2.7 is from Cook (1975). The relation between G and U_2^1 was proved in Krajíček and Takeuti (1990). Using a different translation a relation between G and an equational system PSA was proved in Dowd (1979).

The idea to use the reflection principles for simulations is due to Cook (1975) and was extended to systems G_i, G_i^*, G, and subsystems of F and EF in Krajíček and Pudlák (1990a), Krajíček and Takeuti (1990, 1992), and Krajíček (1995a). The underlying idea is always the same, but different systems often present some nontrivial technical problems.

Corollary 9.3.19 was first observed in Dowd (unpublished). Corollary 9.3.20 is from Krajíček and Pudlák (1990a), as well as Theorem 9.3.21.

In view of the remark after the proof of Theorem 9.3.24 it would be interesting to have a theory satisfying Theorem 9.3.3 and Theorem 9.2.6 with a polynomial bound instead of bound $n^{(\log n)^{O(1)}}$. This would entail (by the same proof) Theorem 9.3.24 for $P = $ F. In fact, Theorem 9.3.24 for $P = $ F was proved by a direct construction in Buss (1991), and analyzing that construction Arai (1991) suggested a subtheory of U_1^1 fulfilling the preceding requirements: It is an extension of $I\Sigma_0^{1,b}$ by a form of inductive definition. Another approach would be using the equational theory ALV proposed in Clote (1992) relating to F similarly as PV relates to EF.

However, Clote (1992) does not verify that the soundness of F is provable in the theory.

The proof of Theorem 9.4.1 is a variant of the original proof of Paris and Wilkie (1985) to allow generalization to V_1^1 (Theorem 9.4.2), which is from Krajíček (1995a), and generalizes an unpublished construction of Wilkie.

The content of Section 9.5 expands a bit on a remark in Krajíček (1994).

10

Finite axiomatizability problem

This chapter surveys the known facts about the

Fundamental problem. *Is bounded arithmetic S_2 finitely axiomatizable?*

As we shall see (Theorem 10.2.4), this question is equivalent to the question whether there is a model of S_2 in which the polynomial time hierarchy PH does not collapse.

10.1. Finite axiomatizability of S_2^i and T_2^i

In this section we summarize the information about the fundamental problem that we have on the grounds of the knowledge obtained in the previous chapters.

Theorem 10.1.1. *Each of the theories S_2^i and T_2^i is finitely axiomatizable for $i \geq 1$.*

Proof. By Lemma 6.1.4, for $i \geq 1$ there is a Σ_i^b-formula $\mathrm{UNIV}_i(x, y, z)$ that is a *universal* Σ_i^b-formula (provably in S_2^1). This implies that T_2^i and S_2^{i+1}, $i \geq 1$, are finitely axiomatizable over S_2^1.

To see that S_2^1 is also finitely axiomatizable, verify that only a finite part of S_2^1 is needed in the proof of Lemma 6.1.4. Q.E.D.

The next statement generalizes this theorem.

Theorem 10.1.2. *Let $1 \leq i$ and $2 \leq j$. Then the set of the $\forall \Sigma_j^b$-consequences of T_2^i*

$$\forall \Sigma_j^b(T_2^i) := \left\{ \phi \in \Sigma_j^b \mid T_2^i \vdash \phi \right\}$$

is finitely axiomatizable.

The sets $\forall \Sigma_j^b(S_2^i)$ and $\forall \Sigma_j^b(U_2^1)$ are also finitely axiomatizable.

Proof. Let $\phi(a) \in \Sigma_j^b$. Assume

$$T_2^i \vdash \phi(a)$$

Then by Theorem 9.2.5

$$S_2^1 \vdash \left(\forall x \ G_i \vdash \|\phi\|_{q(|x|)}^{|x|} \right)$$

for some bounding polynomial $q(y)$ for ϕ, and hence also

$$S_2^1 + j\text{-RFN}(G_i) \vdash \left(\forall x \ \text{Taut}_j(\|\phi\|_{q(|x|)}^{|x|}) \right)$$

This implies by Lemma 9.3.12

$$S_2^1 + j\text{-RFN}(G_i) \vdash \forall x \ \phi(x)$$

Hence every $\forall\Sigma_j^b$-consequence of T_2^i follows from $S_2^1 + j\text{-RFN}(G_i)$. Note that also $j\text{-RFN}(G_i) \in \forall\Sigma_j^b(T_2^i)$ (by Theorem 9.3.16) and that S_2^1 has a finite $\forall\Sigma_2^b$-axiomatization.

An analogous argument applies to $\forall\Sigma_j^b(S_2^i)$, using G_i^* in place of G_i, and to $\forall\Sigma_j^b(U_2^1)$, using G in place of G_i.   Q.E.D.

The following statement is a separation of the theories S_2^i and T_2^i conditioned on a combinatorial property.

Theorem 10.1.3. *Let* $i \geq 1$, $i \geq j \geq 0$ *and assume that the proof system* G_i^* *does not* j-*polynomially simulate the system* G_i.

Then $S_2^i \neq T_2^i$ *and, in fact,* T_2^i *is not* $\forall\Sigma_j^b$-*conservative over* S_2^i.

Proof. The $\forall\Sigma_j^b$-formula $j\text{-RFN}(G_i)$ is provable in T_2^i (by Theorem 9.3.16) and (by Theorem 9.3.17) is provable in S_2^i if and only if G_i^* j-polynomially simulates G_i (provably in S_2^1).   Q.E.D.

10.2. T_2^i versus S_2^{i+1}

Inspired by Theorem 7.4.1 we consider the following computational complexity principle.

The principle is associated with an optimization problem (see Section 6.3) of the type: $R(x, y)$ is a binary relation satisfying

$$\forall x \ R(x, 0) \ \wedge \ \forall x, y \ (R(x, y) \rightarrow |y| \leq |x|)$$

and the task is, given a find b such that

$$R(a, b) \text{ and } |b| \text{ is maximal}$$

In Section 6.4 we considered only polynomial time predicates $R(x, y)$, but now we allow higher complexity.

Definition 10.2.1. *For $i \geq 0$, the principle Ω_i is the following computational principle:*

For any relation $R(x, y) \in \Pi_i^p$ there are $\square_{i+1}^p$-functions

$$f_1(a), f_2(a, b_1) \ldots, f_k(a, b_1, \ldots, b_{k-1})$$

that solve the preceding optimization problem in the counterexample interactive way described at the beginning of Section 6.3. That is, the following is true:
either $\forall z R^*(a, f_1(a), z)$ is true,   or   if b_1 is s.t. $\neg R^*(a, f_1(a), b_1)$
then $\forall z R^*(a, f_2(a, b_1), z)$ is true,   or   if b_2 is s.t. $\neg R^*(a, f_2(a, b_1), b_2)$
then ...

...

...

then $\forall z R^*(a, f_k(a, b_1, \ldots, b_{k-1}), z)$ is true
where the relation $R^*(x, y, z)$ is defined by

$$R(x, y) \wedge (|y| < |z| \leq |x| \to \neg R(x, z))$$

Recall the convention that $\Pi_0^p = \Delta_1^p = P$.

Lemma 10.2.2. *The principle Ω_i is implied by $\Sigma_{i+1}^p = \Delta_{i+1}^p$, and Ω_i implies $\Sigma_{i+1}^p \subseteq \Delta_{i+1}^p/poly$ and hence also $\Sigma_{i+2}^p = \Pi_{i+2}^p$.*

Proof. That Ω_i is implied by $\Sigma_{i+1}^p = \Delta_{i+1}^p$ follows from the fact that the binary search for an optimal b can be performed (under the hypothesis $\Sigma_{i+1}^p = \Delta_{i+1}^p$) by a $\square_{i+1}^p$-function. The principle Ω_i then holds with $k = 1$, in fact.

The proof of the second part is more involved. We show that the hypothesis that Ω_i is valid implies $\Sigma_{i+1}^p \subseteq \Delta_{i+1}^p/poly$. That the latter implies $\Sigma_{i+2}^p = \Pi_{i+2}^p$ is by Theorem 3.1.6 straightforwardly generalized to $i > 0$.

We shall treat only the case $i = 0$; the general case is completely analogous.

Let $A(v)$ be a Σ_1^p-predicate and assume $A(v)$ is defined

$$A(v) := \exists w \leq v \, B(v, w)$$

with $B(v, w)$ a Δ_1^p-predicate.

We want to show that there is a polynomially bounded *advice function* $h(n)$ (cf. Section 2.2) such that for some $\square_1^p$-function $g(u, v)$

$$\forall v, \ (\exists w \leq v \, B(v, w)) \to (B(v, g(h(|v|), v)))$$

holds. Call w *a witness for* v if $w \leq v \wedge B(v, w)$, and define a Δ_1^p-relation

$$R(a, b) := \text{ if } a = \langle v_1, \ldots, v_r \rangle \quad \text{and} \quad b = \langle w_1, \ldots, w_s \rangle \quad \text{then}$$

$$s \leq r \quad \text{and} \quad \forall \ell \leq s: \text{``}w_\ell \text{ is a witness for } v_\ell\text{''}$$

Now we apply the principle Ω_0: We assume that there are $\Box_1^p$-functions $f_1(a)$, $f_2(a, b_1), \ldots, f_k(a, b_1, \ldots, b_{k-1})$ interactively computing, given a, an optimal b such that $R(a, b)$ holds.

Fix length n. We shall describe how the advice $h(n)$ is constructed. Let $V_1 = \{|v| = n \mid \exists w \leq v \, B(v, w)\}$ and let $w(v)$ denote a witness for $v \in V_1$. For a k-tuple (the same k as is the number of functions $f_1, \ldots, f_k$) $a = \langle v_1, \ldots, v_k \rangle$ of distinct elements from V_1 consider the following algorithm constructing a pair $\langle \ell, w \rangle$:

Step 1 Compute $f_1(a)$.

Step 2
If $f_1(a) = \langle w_1', \ldots, w_j' \rangle$, $j \geq 1$, and $R(a, f_1(a))$ holds
then put $\langle \ell, w \rangle := \langle 1, w_1' \rangle$ and STOP.
Else compute $f_2(a, \langle w(v_1) \rangle)$ and GO TO Step 3.

Step m ($2 < m \leq k$):
If $f_{m-1}(a, \langle w(v_1) \rangle, \ldots, \langle w(v_1), \ldots, w(v_{m-2}) \rangle) = \langle w_1', \ldots, w_j' \rangle$, $j \geq m - 1$, and $R(a, \langle w_1', \ldots, w_j' \rangle)$ holds
then put $\langle \ell, w \rangle := \langle m - 1, w_{m-1}' \rangle$ and STOP.
Else compute $f_m(a, \langle w(v_1) \rangle, \ldots, \langle w(v_1), \ldots, w(v_{m-1}) \rangle)$ and GO TO Step $m + 1$.

Step $k + 1$ If the algorithm reaches this step then it necessarily holds that

$$f_k(a, \langle w(v_1) \rangle, \ldots, \langle w(v_1), \ldots, w(v_{k-1}) \rangle) = \langle w_1', \ldots, w_k' \rangle \quad \text{and}$$

$$R(a, \langle w_1', \ldots, w_k' \rangle) \text{ is true}$$

Put $\langle \ell, w \rangle := \langle k, w_k' \rangle$ and STOP.

The idea of the algorithm is that from the witnesses $w(v_1), \ldots, w(v_{\ell-1})$ it is possible to compute by a $\Box_1^p$-algorithm a witness (namely w) for $v_\ell \in V$. Now we utilize this for a construction of the advice $h(n)$.

For Q a $(k - 1)$-element subset of V_1 and $v \in V_1 \setminus Q$ we say that Q *helps* v if and only if there is an ordering $\{v_1, \ldots, v_{\ell-1}, v_{\ell+1}, \ldots, v_k\}$ of Q such that the previous algorithm assigns to the k-tuple

$$\langle v_1, \ldots, v_{\ell-1}, v, v_{\ell+1}, \ldots, v_k \rangle$$

the pair $\langle \ell, w \rangle$, where w is a witness for v.

Put $N_1 = |V_1|$. The previous algorithm produces the pair $\langle \ell, w \rangle$ from any k-tuple of elements of V_1; this shows that the number of pairs (Q, v) such that Q helps v is at least $\binom{N_1}{k}$.

On the other hand, there are only $\binom{N_1}{k-1}$ $(k - 1)$-element subsets Q of V_1, and so there must be one subset $Q_1 \subseteq V_1$ such that Q_1 helps to at least

$$\frac{\binom{N_1}{k}}{\binom{N_1}{k-1}} = \frac{N_1 - k + 1}{k}$$

elements $v \in V_1$.

Put

$$V_2 := V_1 \setminus \{v \in V_1 \mid Q_1 \quad \text{helps } v\}$$

and $N_2 = |V_2|$. Analogously there must be a $(k-1)$-element subset $Q_2 \subseteq V_2$ that helps at least

$$\frac{\binom{N_2}{k}}{\binom{N_2}{k-1}} = \frac{N_2 - k + 1}{k}$$

elements $v \in V_2$.

Iterating this procedure we obtain a decreasing chain

$$V_1 \supseteq V_2 \supseteq \dots$$

and a sequence

$$Q_1 \subseteq V_1, Q_2 \subseteq V_2, \dots$$

of $(k-1)$-element subsets Q_j of V_j, such that Q_j helps all elements $v \in V_j \setminus V_{j+1}$.

A simple computation shows

$$N_{j+1} < \left(\frac{k-1}{k}\right)^j \cdot N_1 + k$$

and so we get $N_t \le k$ after at most t steps of the procedure for

$$t := O\left(\frac{1}{\log_2(k/k-1)} \cdot \log_2(N_1)\right) = O(n)$$

Define the advice $h(n)$ to be the sequence of all pairs

$$\langle v, w(v) \rangle$$

for all elements

$$v \in Q_1 \cup \dots \cup Q_{t-1} \cup V_t$$

The function $g(u, v)$ for $u = h(|v|)$ is then defined by the algorithm

1. try whether $v \in Q_1 \cup \dots \cup Q_{t-1} \cup V_t$ and if so output $w(v)$ (which is a part of $h(|v|)$).
2. if 1 fails try consecutively whether Q_j helps v, $j = 1, \dots, t - 1$. One of them must, giving a witness for v (which $g(u, v)$ outputs).

The second part of the lemma then follows, observing that, for example, the problem of computing the maximal size of a clique in a graph, that is, optimizing $|b|$ in

$$R(a, b) := \text{"}b \text{ is a clique in graph } a\text{"}$$

is an NP-complete problem.   Q.E.D.

The disadvantage of the construction of the advice in the proof of the previous lemma is that it is not apparent how to formalize it within the theory S_2. If one had a construction that could be formalized in S_2, then Theorem 10.2.4 could also be formalized in S_2. The next lemma offers such a construction, albeit for a bit weaker statement.

Lemma 10.2.3. *Let $i \geq 0$ and assume that for every Π_i^b-predicate $R(x, y)$ there are $\Box_{i+1}^p$-functions $f_1(a), \ldots, f_k(a, b_1, \ldots, b_{k-1})$ such that for $R^*(x, y, z)$ defined from $R(x, y)$ as in Definition 10.2.1 T_2^i proves*

$$\forall x, y_1, \ldots, y_k; \ R^*(x, f_1(x), y_1) \vee \ldots \vee R^*(x, f_k(x, y_1, \ldots, y_{k-1}), y_k)$$

Then

$$T_2^i \vdash \Sigma_{i+1}^p \subseteq \Pi_{i+1}^p/\text{poly}$$

This means that for every Σ_{i+1}^b-formula $A(v)$ there is a Π_{i+1}^b-formula $C(v, u)$ for which

$$T_2^i \vdash \forall x \exists u \forall v; \ |x| = |v| \to A(v) \equiv C(v, u)$$

Proof. Let $A(v)$ be a Σ_{i+1}^b-formula of the form $\exists w \leq v \ B(v, w)$ with $B \in \Pi_i^b$, and let $R(x, y)$ be defined for B as in the proof of Lemma 10.2.2, where functions $f_1, \ldots, f_k$ fulfill the hypothesis of the lemma.

Fix length n and consider a formula $E(t, u)$ that is the conjunction of the conditions

(i) $u = \langle v_{t+1}, \ldots, v_k \rangle \wedge t \geq 1 \wedge |v_{t+1}| = \cdots = |v_k| = n$

(ii)

$$\forall v_1, \ldots, v_t, w_1, \ldots, w_t \ \forall \ell, w; \ \bigwedge_{1 \leq i \leq t} (w_i \leq v_i \wedge B(v_i, w_i))$$
$$\wedge \langle \ell, w \rangle = D(\langle v_1, \ldots, v_k \rangle) \to \ell \leq t$$

where $D(a)$ is the algorithm described in the proof of Lemma 10.2.2 with $w(v_i) := w_i, i = 1, \ldots, t$.

Let t_0 be the minimal $t < k$ such that for some u, $E(t_0, u)$ holds. Such t_0 exists as there are only finitely many possibilities for t_0; in particular, no induction axioms are needed to assure the existence of t_0.

Claim. *T_2^i proves that for $|v| = n$ and*

$$u = \langle v_{t_0+1}, \ldots, v_k \rangle$$

such that $E(t_0, u)$ holds the following conditions are equivalent

1. *$A(v)$ holds*

2. $C(v, u)$ holds where $C(v, u)$ is the formula

$$\forall v_1, \ldots, v_{t_0-1} \, \forall w_1, \ldots, w_{t_0-1} \, \forall \ell, w,$$

$$\left(\bigwedge_{0 \leq i < t_0} (|v_i| = n \wedge w_i \leq v_i \wedge B(v_i, w_i)) \right.$$

$$\wedge \langle \ell, w \rangle = D(\langle v_0, \ldots, v_{t_0-1}, v, v_{t_0+1}, \ldots, v_k \rangle))$$

$$\left. \longrightarrow ((\ell < t_0 \wedge w \leq v_\ell \wedge B(v_\ell, w)) \vee (\ell = t_0 \wedge w \leq v \wedge B(v, w))) \right)$$

To see the claim, reason in T_2^i and assume first that $A(v)$ holds: that is, that for some $w \leq v$ the formula $B(v, w)$ holds. Then the second condition is satisfied by Ω_i, that is, by the definition of the algorithm $D(a)$ from the functions $f_1, \ldots, f_k$, and by the choice of t_0 and u.

Assume now that condition 2 is satisfied but that $A(v)$ does not hold. That means that $\langle \ell, w \rangle$ from 2 must satisfy $\ell < t_0$, but then $u' = \langle v, v_{t_0+1}, \ldots, v_k \rangle$ would satisfy $E(t_0 - 1, u')$, contradicting the previous choice of t_0 as the minimal t for which $\exists x \, E(t, x)$ is valid.

The definition of the algorithm $D(a)$ is Δ_{i+1}^b and the formula $C(v, u)$ from condition 2 is Π_{i+1}^b.   Q.E.D.

We can now employ these two lemmas to get a strong theorem about the relation of theories T_2^i and S_2^{i+1}.

Theorem 10.2.4 (Krajíček, Pudlák and Takeuti 1991). *Let $i \geq 1$. Then the theories T_2^i and S_2^{i+1} are distinct, assuming that $\Sigma_{i+1}^p \not\subseteq \Delta_{i+1}^p/poly$, or assuming that the polynomial time hierarchy PH does not collapse to its $(i + 2)^{nd}$ level: $\Sigma_{i+2}^p \neq \Pi_{i+2}^p$.*

In fact, the theories T_2^i and S_2^{i+1} are distinct, assuming that T_2^i does not prove that $\Sigma_{i+1}^p \subseteq \Pi_{i+1}^p/poly$.

The same is true for $i = 0$ with PV_1 in place of T_2^i.

Proof. We shall show that $T_2^i = S_2^{i+1}$ implies that the principle Ω_i is provable in T_2^i, that is, that for every Π_i^b-formula $R(x, y)$ there are $\square_{i+1}^p$-functions $f_1(a), \ldots, f_k(a, b_1, \ldots, b_{k-1})$ Δ_{i+1}^b-definable in T_2^i such that T_2^i proves the disjunction

$$R^*(a, f_1(a), b_1) \vee \ldots \vee R^*(a, f_k(a, b_1, \ldots, b_{k-1}), b_k)$$

with $R^*(x, y, z)$ defined from R as in Definition 10.2.1.

Consider the formula

$$\psi(x, y) := |y| \leq |x| \wedge R(x, y)$$

It is a Π_i^b-formula and $\psi(x, 0)$ holds (by the assumption that $\forall x \, R(x, 0)$ holds).

By Lemma 5.2.7 S_2^{i+1} admits the Π_i^b–LENGTH–MAX principle: That is, S_2^{i+1} proves that there is y of the maximal length satisfying $\psi(x, y)$

$$\exists y(|y| \leq |x|) \, \forall z(|z| \leq |x|), \, R(x, y) \wedge (|y| < |z| \rightarrow \neg R(x, z))$$

That is

$$\forall x \, \exists y \, \forall z \, R^*(x, y, z)$$

(with the bounds $\leq x$ to y and z implicit).

Now use the hypothesis of the theorem to get

$$T_2^i \vdash \forall x \exists y \forall z \, R^*(x, y, z)$$

Theorem 7.4.1 then applies (with $\phi = R^*$) and guarantees the existence of the $\square_{i+1}^p$-functions $f_1, \ldots, f_k$ with the required property

$$T_2^i \vdash R^*(a, f_1(a), b_1) \vee \ldots \vee R^*(a, f_k(a, b_1, \ldots, b_{k-1}), b_k)$$

By Lemma 10.2.2 this implies that $\Sigma_{i+1}^p \subseteq \Delta_{i+1}^p/\text{poly}$ and hence $\Sigma_{i+2}^p = \Pi_{i+2}^p$, while Lemma 10.2.3 implies that T_2^i proves that $\Sigma_{i+1}^p \subseteq \Pi_{i+1}^p/\text{poly}$.   Q.E.D.

Corollary 10.2.5. *Let $i \geq 1$ and assume that $T_2^i = S_2^{i+1}$. Then $T_2^i = S_2$. The same holds for $i = 0$ with PV_1 in place of T_2^i.*

Proof. We use the second part of Theorem 10.2.4 by which the hypothesis of the lemma implies that T_2^i proves $\Sigma_{i+1}^p \subseteq \Pi_{i+1}^p/\text{poly}$.

First we show that $T_2^i = T_2^{i+1}$. Let $\sigma(x)$ be a Σ_{i+1}^p-formula and, working in T_2^i, assume that

$$\sigma(0) \wedge \forall x < v(\sigma(x) \rightarrow \sigma(x+1)) \wedge \neg \sigma(v)$$

By $\Sigma_{i+1}^p \subseteq \Pi_{i+1}^p/\text{poly}$ we know that for every length n there is an advice u_n such that

$$\forall x, |x| = n \rightarrow \sigma(x) \equiv C(x, u_n)$$

where $C(x, y)$ is a fixed Π_{i+1}^p-formula. Joining the advice $u_0, \ldots, u_{|v|}$ into one u we have

$$\forall x \leq v, \sigma(x) \equiv C'(x, u)$$

where

$$C'(x, u) := C(x, u_{|x|})$$

is a Π_{i+1}^p-formula.

But then $\sigma(x)$ is provably Δ_{i+1}^b on the interval $[0, v]$ and as S_2^{i+1} admits Δ_{i+1}^b- IND (by Lemma 5.2.9), $T_2^i (= S_2^{i+1})$ satisfies the induction for $\sigma(x)$ on any interval $[0, v]$. This shows that $T_2^i = T_2^{i+1}$.

To show that, in fact, $T_2^i = S_2$, observe that

$$T_2^i \vdash \Sigma_{i+1}^p \subseteq \Pi_{i+1}^p/\text{poly}$$

implies that any Σ_∞^b-formula $\eta(x)$ is in T_2^i on any interval $[0, v]$ equivalent to a Π_{i+1}^b-formula (with parameters depending on v); this is proved by induction on the logical complexity of η. Hence the induction for any $\eta(x)$ follows from Π_{i+1}^b–IND, and hence from T_2^i as we have already established that $T_2^i = T_2^{i+1}$.

Q.E.D.

Corollary 10.2.6. *To show that S_2 is not finitely axiomatizable it is necessary and sufficient to show the existence of a model M of S_2 in which the polynomial time hierarchy PH does not collapse, that is, in which for every i the class $\Sigma_{i+1}^b(M)$ of subsets of M definable by Σ_{i+1}^b-formulas with parameters is strictly larger than the class $\Sigma_i^b(M)$.*

Proof. S_2 is finitely axiomatizable iff (by Theorem 10.1.1) $S_2 = T_2^i$ for some i iff (by Theorem 10.2.5) $T_2^i = S_2^{i+1}$ for some i. This implies (by Theorem 10.2.4)

$$T_2^i \vdash \Sigma_{i+1}^p \subseteq \Pi_{i+1}^p/\text{poly}$$

for some i.

We shall show that the last condition implies that

$$T_2^i \vdash \Sigma_{i+3}^p = \Pi_{i+3}^p$$

which trivially implies $T_2^{i+3} = S_2$, completing thus the proof of the corollary.

The inclusion $\Sigma_{i+1}^p \subseteq \Pi_{i+1}^p/\text{poly}$ obviously implies (in T_2^i)

$$\Sigma_{i+1}^p/\text{poly} \subseteq \Pi_{i+1}^p/\text{poly}$$

and in particular $PH \subseteq \Pi_{i+1}^p/\text{poly}$. By analogy with Theorem 3.1.6 (for $i + 2$ in place of 1) this gives (again in T_2^i)

$$PH = \Sigma_{i+3}^p = \Pi_{i+3}^p$$

Q.E.D.

We note that Buss (1993b) showed that $\Sigma_{i+1}^p \subseteq \Pi_{i+1}^p/\text{poly}$ implies even that $PH \subseteq \mathcal{B}(\Sigma_{i+2}^p)$: that every set in PH is a Boolean combination of Σ_{i+2}^p-sets. This is not needed, however, for the previous corollary.

10.3. S_2^i versus T_2^i

This section presents a conditional separation of theories S_2^i and T_2^i.

Theorem 10.3.1 (Krajíček 1993). *Let $i \geq 1$. Then the theories S_2^i and T_2^i are distinct, assuming that the classes $P^{\Sigma_i^p}[O(\log n)]$ and Δ_{i+1}^p are distinct. In fact, the assumption can be weakened to the assumption that S_2^i does not prove*
$$P^{\Sigma_i^p}[O(\log n)] = \Delta_{i+1}^p.$$

Proof. Let $i \geq 1$. By Theorem 6.1.2 T_2^i can Σ_{i+1}^b-define all $\square_{i+1}^p$-functions and, in particular, all Δ_{i+1}^p-predicates. By Corollary 7.3.6 the predicates Σ_{i+1}^b-definable in S_2^i are exactly those from the class $P^{\Sigma_i^p}[O(\log n)]$ and, in fact, Theorem 7.3.5 shows that the equivalence of a $P^{\Sigma_i^p}[O(\log n)]$-predicate to a Σ_{i+1}^b-predicate is provable in S_2^i.

Hence $S_2^i = T_2^i$ implies that any Σ_{i+1}^b-definable predicate in T_2^i (i.e., any Δ_{i+1}^p-predicate) is in S_2^i equivalent to a $P^{\Sigma_i^p}[O(\log n)]$-predicate.   Q.E.D.

Note that it is open whether $P^{\Sigma_i^p}[O(\log n)] = \Delta_{i+1}^p$ implies the collapse of the polynomial time hierarchy PH. We thus cannot obtain a statement analogous to Corollary 10.2.5.

Corollary 10.3.2. *Assume that there is a model M of S_2 in which the classes $\Sigma_0^b(\Sigma_i^b)(M)$ and $\Delta_{i+1}^b(M)$ are distinct for all $i \geq 1$.*
Then all theories S_2^i and T_2^i are mutually distinct.

Proof. The class $\Sigma_0^b(\Sigma_i^b)$ is the class of formulas that are obtained by logical connectives and sharply bounded quantification from Σ_i^b-formulas. These define exactly the $P^{\Sigma_i^p}[O(\log n)]$-predicates (cf. Theorem 6.2.3). Hence the hypothesis implies, via Theorem 10.3.1, that $S_2^i \neq T_2^i$, all $i \geq 1$.

The inequality $P^{\Sigma_i^p}[O(\log n)] \neq \Delta_{i+1}^p$ implies, in particular, that $\Sigma_i^p \neq \Sigma_{i+1}^p$, and hence by Corollary 10.2.6 also that $\mathrm{PV}_1 \neq S_2^1$ and $T_2^i \neq S_2^{i+1}$, $i \geq 1$.
                    Q.E.D.

10.4. Relativized cases

In this section we prove, using the results of previous two sections, that all theories $S_2^i(R)$ and $T_2^i(R)$ are distinct.

Recall that these are the theories defined as S_2^i and T_2^i but in the language $L \cup \{R\}$ extending L by a new binary relation symbol R. Equivalently we could work with the theories S_2^i and T_2^i in the language L_2 of the second order bounded arithmetic.

Lemma 10.4.1. *Assume that there is an oracle A such that the relativized polynomial hierarchy PH^A does not collapse. Then the theories $T_2^i(R)$ and $S_2^{i+1}(R)$ are distinct.*

Assuming that there is an oracle A such that all relativized classes $L^{\Sigma_i^p(A)}$, $\Delta_{i+1}^p(A)$ are different, then the theories $S_2^i(R)$ and $T_2^i(R)$ are distinct.

Proof. The lemma is a corollary of the proofs of Theorems 10.2.4 and 10.3.1, which both "relativize" to theories $S_2^i(R)$ and $T_2^i(R)$. This is because the definability and the witnessing theorems from Chapters 6 and 7 readily relativize.   Q.E.D.

A construction of an oracle separating the levels of the polynomial time hierarchy was proposed by Furst, Saxe, and Sipser (1984). It reduced the existence of such an oracle to the existence of exponential lower bounds on the size of depth $d - 1$ circuits computing Sipser's function $S_d(x_{i_1,...,i_d})$ (cf. Section 3.1). Such bounds were announced by Yao (1985) and proved by Hastad (1989). In the following theorem we use similar combinatorics to construct oracle A for which even

$$\leq_{tt} \left(\Sigma_i^P(A) \right) \neq \Delta_{i+1}^P(A)$$

for all $i \geq 1$. Recall the class $\leq_{tt} (\Sigma_i^P)$ from Theorem 6.2.3.

Definition 10.4.2.

   (a) For $i \geq 1$ define the $\Pi_{i-1}^b(\alpha)$-formulas

     (1) $\psi_1(x, y_1) := y_1 = 0 \vee \alpha(\langle i, x, y_1 \rangle)$

     (2) $\psi_2(x, y_1) := y_1 = 0 \vee \forall y_2 < (x \log(x))^{1/2} \; \alpha(\langle i, x, y_1, y_2 \rangle)$

     (3)

$$\psi_i(x, y_1) := y_1 = 0$$

$$\vee \forall y_2 < x \exists y_3 < x \ldots Q_{i-1} y_{i-1} < x \; Q_i y_i < \left(\frac{i \cdot x \log(x)}{2} \right)^{1/2}$$

$$\times \alpha(\langle i, x, y_1, \ldots, y_i \rangle)$$

     where Q_{i-1} (resp. Q_i) is $\forall$ (resp. $\exists$) iff i is odd.

   (b) For $i \geq 1$ define the predicate

$$P_i(x, \alpha) := \text{"the maximal } y_1 < x \text{ satisfying } \psi_i(x, y_1) \text{ is odd"}$$

The predicate P_i generalizes the ODDMAXSAT problem.

Lemma 10.4.3. *The predicate $P_i(x, A)$ is in the class $\Delta_{i+1}^P(A)$, for all $i \geq 1$ and $A \subseteq \omega$.*

Proof. The algorithm finding maximal $y_1 < x$ such that $\psi_i(x, y_1)$ based on the binary search defines a $\Delta_{i+1}^P(A)$-function.   Q.E.D.

The crucial idea allowing later combinatorial arguments is to fix in $\psi_i(x, y_1)$ values of x and y_1 and to think of such $\psi_i(m, u)$ as of a predicate defined for sets $A \subseteq \omega$. Next we shall define the Boolean circuits $\overline{\psi}_i(m, u)$ computing such a predicate.

Definition 10.4.4.

(a) *The input variables of circuit* $\overline{\psi}_i(m, u)$ *are of the form*

$$p_{u, y_2, \ldots, y_{i-1}, t}$$

for every $(i - 2)$-*tuple* $y_2, \ldots, y_{i-1} < m$ *and* $t < (i \cdot m \log(m)/2)^{1/2}$.

(b) *The circuit* $\overline{\psi}_i(m, u)$ *is the propositional translation* $\langle \psi_i(x, y_1) \rangle(m, u)$ *with the atomic formula* $\alpha(\langle i, x, y_1, \ldots, y_i \rangle)$ *translated as* $p_{y_1, \ldots, y_i}$ *(cf. Definition 9.1.1).*

(c) *The circuit* C_i^m *is a disjunction of* $\lceil m/2 \rceil$ *conjunctions*

$$\overline{\psi}_i(m, u) \wedge \bigwedge_{u < v < m} \neg \overline{\psi}_i(m, v)$$

one for each odd $u < m$.

The following lemma is straightforward.

Lemma 10.4.5. *For any* $i \geq 1$ *and* m:

1. *The circuit* $\overline{\psi}_i(m, u)$ *computes the truth value of* $\psi_i(m, u)$ *under the assignment*

$$p_{y_1, \ldots, y_i} := 1 \quad \text{iff} \quad \langle i, m, y_1, \ldots, y_i \rangle \in A$$

2. *Under the same assignment the circuit* C_i^m *computes the truth value of the predicate* $P_i(m, A)$.

The next definition and lemma are crucial technical concepts in the method of random restrictions in Boolean complexity.

Definition 10.4.6.

(a) *Let* $(B_j)_j$ *be a partition of the atoms of the circuit* C_i^m *into* m^{i-1} *classes of the form*

$$\left\{ p_{y_1, \ldots, y_{i-1}, t} \mid t < \left(\frac{i \cdot m \log(m)}{2} \right)^{1/2} \right\}$$

one for every choice of $y_1, \ldots, y_{i-1} < m$.

(b) *Let* $0 < q < 1$ *be a real number. A probability space* R_q^+ *of random restrictions (cf. Section 3.1) is a space of restrictions* ρ *determined by the following process*

(a)

$$s_j := \begin{cases} *, & \text{with probability } q \\ 0, & \text{with probability } 1 - q \end{cases}$$

(b) for every atom $p \in B_j$

$$\rho(p) := \begin{cases} s_j, & \text{with probability } q \\ 1, & \text{with probability } 1 - q \end{cases}$$

(c) A probability space R_q^- is defined as R_q^+ interchanging the roles of 0 and 1.

(d) For any $\rho \in R_q^+$, $g(\rho)$ is a further restriction and renaming of the atoms defined as follows (for all j):

(a) for j s.t. $s_j = *$ let $p_j = p_{y_1,...,y_{i-1},t}$ be an atom from B_j given value $*$ by ρ with the least t,

(b) $g(\rho)$ gives value 1 to all $p \in B_j$, $p \neq p_j$ s.t. $\rho(p) = *$,

(c) $g(\rho)$ renames p_j to atom $p_{y_1,...,y_{i-1}}$.

(e) For $\rho \in R_q^-$, $g(\rho)$ is defined as in (d) but again interchanging 0 with 1.

(f) For G a circuit with the atoms among those of the circuit C_i^m, G^ρ denotes the circuit obtained from G by first performing the restriction ρ and then also the restriction $g(\rho)$. Note that the atoms of G^ρ are among those of C_{i-1}^m.

The following lemma is crucial. We refer to Hastad (1989) for full details of the proof.

Lemma 10.4.7. *Fix a value $q := (2 \cdot i \cdot \log(m)/m)^{1/2}$ and assume that m is sufficiently large. Then the following three conditions hold.*

1. *Let G be a depth 2 subcircuit of C_i^m: That is, G is either an OR of ANDs of size $< (i \cdot m \log(m)/2)^{1/2}$ or an AND of ORs of size $< (i \cdot m \log(m)/2)^{1/2}$. Pick a random ρ from R_q^+ if G is an OR of ANDs and from R_q^- if it is an AND of ORs.*
    Then with the probability at least

    $$1 - \frac{1}{3} m^{-i+1}$$

    G^ρ is an OR (resp. an AND) of at least $((i-1) \cdot m \log(m)/2)^{1/2}$ different atoms.

2. *Let $i \geq 3$. Pick ρ random from R_q^+ for i even or from R_q^- for i odd. Then with the probability at least*

    $$\frac{2}{3}$$

    the circuit $(C_i^m)^\rho$ contains C_{i-1}^m. That is, after some renaming of the atoms $(C_i^m)^\rho$ becomes C_{i-1}^m.

3. *Let $i = 2$. Pick ρ from R_q^+ randomly. Then with a probability at least*

$$\frac{2}{3}$$

the circuit $(C_2^m)^\rho$ contains circuit C_1^n, for $n = (m \log(m)/2)^{1/2}$.

Proof. We sketch the proofs of the parts of the lemma.

(1) Assume that G is an OR of ANDs and ρ is drawn from R_q^+ (the case when G is an AND of ORs and $\rho \in R_q^-$ is treated similarly).

An AND gate of G corresponds to class B_j of atoms and takes after ρ the value s_j with the probability at least

$$1 - (1-q)^{|B_j|} = \left(1 - \left(\frac{2i \log(m)}{m}\right)^{1/2}\right)^{\left(\frac{im \log(m)}{2}\right)^{1/2}} > 1 - \frac{1}{6}m^{-i}$$

Hence with probability at least

$$1 - \frac{1}{6}m^{-i+1}$$

this is true for all ANDs in G.

The expected number of ANDs assigned the values s_j in the definition of ρ and not the value 0 is $m \cdot q = (2im \log(m))^{1/2}$, and thus with probability at least

$$1 - \frac{1}{6}m^{-i}$$

we can get at least $((i-1)m \log(m)/2)^{1/2}$ s_j's assigned.

Hence with probability at least

$$1 - \frac{1}{3}m^{-i+1}$$

the circuit G^ρ is an OR of at least $((i-1)m \log(m)/2)^{1/2}$ different atoms.

(2) In C_i^m there are m^{i-2} different subcircuits G of depth 2. From (1) we have that with the probability at least

$$1 - \frac{1}{3}m^{-1} \geq \frac{2}{3}$$

all of them are restricted by ρ as required in (1). So, after renaming the atoms, $(C_i^m)^\rho$ becomes the circuit C_{i-1}^m.

(3) For $i = 2$ the circuit $\overline{\psi}_i(m, u)$ is just an AND of size $(m \log(m))^{1/2}$ corresponding to the classes B_j, and there are m of them. Hence (1) implies that with the probability at least $(5/6)$ they are all assigned the value s_j, which is again with the probability at least $(5/6)$ equal to $*$ for at least $(m \log(m)/2)^{1/2}$ of these ANDs.

Q.E.D.

Definition 10.4.8.

1. *A Boolean circuit is called* $\Sigma_{i,m}^{S,t}$ *if*

    (a) *it has depth* $i + 1$

    (b) *its top gate is an OR and ORs and ANDs alternate in levels*

    (c) *it has at most S gates in levels* $2, 3, \ldots, i + 1$

    (d) *its bottom (depth 1) gates have arity at most* t

    (e) *its atoms are among those of* C_i^m.

2. *A tt-reducibility* $D = \langle f; E_1, \ldots, E_r \rangle$ *of type* (i, m, k) *is a Boolean function* $f(w_1, \ldots, w_r)$ *in* $r \leq \log(m)^k$ *variables and with* r $\Sigma_{i,m}^{S,t}$-*circuits* $E_1, \ldots, E_r$, *where* $S = 2^{\log(m)^k}$ *and* $t = \log(S)$.

3. *A tt-reducibility D of type* (i, m, k) *computes a function of the atoms of* C_i^m: *First evaluate* $w_j := E_j$ *on the atoms and then evaluate* $f(w_1, \ldots, w_r)$ *on* w_j's.

Lemma 10.4.9. *Let G be an AND of ORs of size* $\leq t$ *with the atoms among those of* C_i^m. *Pick* ρ *randomly from* R_q^+ *or from* R_q^-.
Then with probability at least

$$1 - (6qt)^s$$

the circuit G^ρ *can be written as an OR of ANDs of size* $< s$.

The same is valid for the probability of the switching an OR of ANDs into an AND of ORs.

This is one of Hastad's two switching lemmas. For the proof consult Hastad (1989).

Lemma 10.4.10. *Let D be a tt-reducibility of type* (i, m, k) *and let* $q := (2i \log(m)/m)^{1/2}$. *Pick* ρ *at random from* R_q^+ *or from* R_q^-.
Then with the probability at least

$$1/2$$

is

$$D^\rho := \langle f; E_1^\rho, \ldots, E_r^\rho \rangle$$

a tt-reducibility of type $(i - 1, m, k)$.

Proof. Put $t := s := \log(m)^k$ and apply Lemma 10.4.9. Then the probability of a failure to switch one depth 2 subcircuit of any E_j is at most

$$(6qt)^t = \left(6 \left(\frac{2i \log(m)}{m} \right)^{1/2} \log(m)^k \right)^{\log(m)^k} < 2^{-2 \cdot \log(m)^k}$$

for sufficiently large m.

There are $\leq \log(m)^k \, 2^{\log(m)^k}$ such depth 2 subcircuits, so with probability at least

$$1 - \log(m)^k 2^{-\log(m)^k} > 1/2$$

all of them are switched. The switched subcircuits can then be merged with the gates at level 3, to decrease the depth of circuits E_j by 1.   Q.E.D.

The following lemma shows the nonexistence of a particular tt-reducibility.

Lemma 10.4.11. *Let $i \geq 1$ and k be fixed. Then for m sufficiently large there is no tt-reducibility of type (i, m, k) correctly computing the predicate $P_i(m, A)$ for all $A \subseteq \omega$.*

Proof. The proof of the lemma divides into two claims.

Claim 1. *Let D be a tt-reducibility of type (i, m, k) computing the predicate $P_i(m, A)$ for all $A \subseteq \omega$.*

Then there is a tt-reducibility D_1 of type $(1, m, k)$ computing correctly the predicate $P_1((m \log(m)/2)^{1/2}, B)$ for every $B \subseteq \omega$.

Claim 2. *For k fixed and m sufficiently large there is no tt-reducibility of type $(1, m, k)$ computing correctly $P_1((m \log(m)/2)^{1/2}, B)$ for every $B \subseteq \omega$.*

We prove Claim 1 first. The predicate $P_i(m, A)$ is computed by the circuit C_i^m (Lemma 10.4.5). By Lemmas 10.4.7 and 10.4.10 (and the probability q defined there) a random restriction ρ (drawn from R_q^+ for i even and from R_q^- for i odd) converts with probability $\geq 1/6$ simultaneously C_i^m into C_{i-1}^m and $D_i = D$ into a tt-reducibility D_{i-1} of the type $(i - 1, m, k)$. Hence there exists such a restriction ρ and clearly C_{i-1}^m and D_{i-1} compute the same predicate, that is, $P_{i-1}(m, A)$.

Apply this conversion $(i - 1)$-times (with part 3 of Lemma 10.4.7 in the last step) to prove the claim.

Now we prove Claim 2 by a direct diagonalization of any type $(1, m, k)$ tt-reducibility D.

Put $n := (m \log(m)/2)^{1/2}$ and $t := \log(m)^k$. Let $D = \langle f; E_1, \ldots, E_r \rangle$ be a type $(1, m, k)$ tt-reducibility and for simplicity of notation denote C_1^m by C.

We shall construct a sequence of sets of numbers A_s^+, A_s^-, I_s satisfying

1. $A_s^+ \cap A_s^- = \emptyset$ and any number in $A_s^+ \cup A_s^-$ is smaller than t.
2. $|A_s^+| \leq s$ and $|A_s^+ \cup A_s^-| \leq st$.
3. at least half of the numbers $\leq \max(A_s^+)$ belong in $A_s^+ \cup A_s^-$.
4. $I_s \subseteq \{1, \ldots, r\}$ and $|I_s| = s$.
5. for every $B \subseteq \omega$ for which

$$A_s^+ \subseteq B \ \wedge \ A_s^- \cap B = \emptyset$$

and for every $j \in I_s$ it holds that

$$E_j^B = 1$$

E_j^B denotes the circuit E_j evaluated according to B.

Put $A_0^+ := A_0^- := I_0 := \emptyset$. Assume that we have A_s^+, A_s^-, I_s satisfying the conditions stated.

Put $B := A_s^+$ (hence $E_j^B = 1$ for all $j \in I_s$) and consider three cases:

(a) $D^B = 1$ but $\max(B)$ is even, or $D^B = 0$ but $\max(B)$ is odd. In this case STOP.

(b) $D^B = 1$ and $\max(B)$ is odd. Consider a set

$$S = \{x < 2^t \mid \max(A_s^+) < x, x \text{ even}, x \notin A_s^+\}.$$

By conditions 1, 2, and 3, the set S is nonempty. Two subcases may occur:

(b1) We may add some $x \in S$ to B to form $B' := B \cup \{x\}$ such that $D^{B'} = D^B = 1$.

In this subcase put $A_{s+1}^+ := A_s^+ \cup \{x\}$ and $A_{s+1}^- := A_s^-$ and STOP.

(b2) We cannot add any $x \in S$ to B with the property (b1).

Take $x := \min(S)$ and put $A_{s+1}^+ := A_s^+ \cup \{x\}$. As the circuit D changes value, some E_{j_0} for $j_0 \notin I_s$ had to receive new value 1. Take an AND of E_{j_0} (containing x), which becomes true, and add indices of all atoms negatively occurring there to A_s^- to form A_{s+1}^-. Note that this is correct as none of them is in A_s^+.

Put $I_{s+1} := I_s \cup \{j_0\}$ and GO TO step $s + 2$.

(Note that the new sets A_{s+1}^+, A_{s+1}^-, and I_{s+1} fulfill the conditions 1–5. For example, 3 holds as x was chosen to be the minimal element of S.)

(c) $D^B = 0$ and $\max(A_a^+)$ is even. Take a set

$$S = \{x < 2^t \mid \max(A_s^+) < x, x \text{ odd}, x \notin A_s^+\}$$

and proceed analogously with case (b).

If the sequence is not completed at step s, necessarily $I_s \not\subseteq I_{s+1}$ and hence the construction must eventually halt.

Take $B := A_s^+$ for the final s. Clearly D^B does not agree with C^B; this proves Claim 2 and hence also the lemma. Q.E.D.

Theorem 10.4.12. *There is an oracle A such that for all i*

$$\leq_{tt} (\Sigma_i^p(A)) \neq \Delta_{i+1}^p(A)$$

and hence also

$$\Sigma_i^p(A) \neq \Pi_i^p(A)$$

Proof. We construct the oracle $A \subseteq \omega$ such that for all $i \geq 1$ the predicate $P_i(x, A)$ is not in $\leq_{tt} (\Sigma_i^p(A))$. This is enough by Lemma 10.4.3.

Let M_j, $j = 0, 1, \ldots$ enumerate all polynomial time machines. We shall consider successively pairs (i, j) and build A in stages to assure that M_j is not a tt-reducibility of $P_i(x, A)$ to $\Sigma_i^p(A)$ (to some canonical $\Sigma_i^p(A)$-complete set).

Let A_s be the approximation of A constructed in the first s stages and let (i, j) be a pair to be considered next. Choose number $m := m_{s+1}$ so large that all numbers considered in the first s stages are small w.r.t. m. The machine M_j outputs on m a Boolean function $f(w_1, \ldots, w_r)$ and queries $z_1, \ldots, z_r$ to $\Sigma_i^p(A)$. Such queries naturally correspond to the evaluations of a $\Sigma_{i,m}^{S,\log(S)}$-circuit (a propositional translation of the instances of the $\Sigma_i^b(A)$-formula defining the $\Sigma_i^p(A)$ complete set), with $S = 2^{\log(m)^k}$.

In these circuits first evaluate the atoms "$n \in \alpha$" according to A_s and set to 0 all atoms "$n \in \alpha$" for n not of the form $\langle i, m, y_1, \ldots, y_i \rangle$.

This leaves us with a tt-reducibility of type (i, m, k), and by Lemma 10.4.11 it cannot compute the predicate $P_i(x, A)$ correctly for all $A \subseteq \omega$.

Take a finite $A_{s+1} \supseteq A_s$ in such a way that the reducibility fails, and hence the machine M_j does also.

Proceed to the next pair (i, j).

This completes the proof of the theorem.   Q.E.D.

Corollary 10.4.13. *For all* i

$$S_2^i(R) \neq T_2^i(R) \neq S_2^{i+1}(R)$$

and

$$PV_1(R) \neq S_2^1(PV_1, R)$$

Proof. The proof follows from Theorem 10.4.12 and Lemma 10.4.11.   Q.E.D.

10.5. Consistency notions

This section is devoted to examining what can be achieved by Gödel-type arguments in the context of the problem of the finite axiomatizability of bounded arithmetic.

That *Peano arithmetic* PA is not finitely axiomatizable was proved by Ryll-Nardzewski (1952). Rabin (1961) strengthened this to show that PA is not axiomatizable by any set of axioms of a bounded quantifier complexity. PA has subtheories $I\Delta_0 \subseteq I\Sigma_1^0 \subseteq I\Sigma_2^0 \subseteq \ldots$, based on induction for the Σ_i^0-formulas, and each of them is finitely axiomatizable (utilizing universal Σ_i^0-formulas, analogously to Section 10.1). Every theory $I\Sigma_{i+1}^0$ proves the consistency $\mathrm{Con}(I\Sigma_i^0)$ of theory $I\Sigma_i^0$, which is, by Gödel's theorem, not provable in $I\Sigma_i^0$ itself. Hence $I\Sigma_{i+1}^0 \neq I\Sigma_i^0$. We refer the reader to Hájek and Pudlák (1993) for information on this topic.

Of course, we would like to apply the same idea in bounded arithmetic but there are serious obstacles. First is a theorem of Paris and Wilkie (1987b) (for the definition of Q see Section 2.1; Exp is a Π_2^0 axiom $\forall x\,\exists y,\ |y| = x$).

Theorem 10.5.1. *The theory S_2+ Exp does not prove the consistency of Robinson's arithmetic Q.*

It follows that we cannot separate the theories S_2^i and T_2^i by statements of the form $\mathrm{Con}(S_2^i)$.

A natural notion to consider then is the notion of *bounded consistency*, $\mathrm{BdCon}(T)$, formalizing *"there is no T-proof of $0 = 1$ using only bounded formulas."* But Pudlák (1990) showed that S_2 does not prove $\mathrm{BdCon}(S_2^1)$ either.

However, we do not have to guess various modifications of the consistency statements as Theorem 10.1.2 tells us that $\mathrm{Con}(G_i)$, defined in Section 9.3, is the strongest consistency statement (over S_2^1) provable in T_2^i: That is, every $\forall\Pi_1^b$-consequence of T_2^i follows in S_2^1 from $\mathrm{Con}(G_i)$. Thus the problem is to prove that S_2^i does not prove $\mathrm{Con}(G_i)$. A problem in carrying out the usual diagonalization argument is that the *metanotion* of provability (provability in S_2^i) is different from the notion (the provability in G_i) to which the diagonalization is applied.

This obstacle can be removed to some extent. In particular, we define a notion of *regular provability* in (fragments of) S_2 and show that $\mathrm{RCon}(T_1^i)$, a regular consistency of T_1^i, is in S_2^1 equivalent to $\mathrm{Con}(G_i)$. However, we are unable to prove that S_2^i does not prove $\mathrm{RCon}(T_1^i)$.

I include this material primarily to illustrate what cannot be achieved by Gödel-type arguments. Some corollaries are nevertheless obtained.

Definition 10.5.2. *An i-regular proof of a #-free sequent*

$$\Gamma \to \Delta$$

of Σ_∞^b-formulas is a triple

$$\langle\, \pi, \bar{t}, \bar{d} \,\rangle$$

satisfying

1. *π is an LKB proof of $\Gamma \to \Delta$ using the Σ_i^b–IND-rule*
2. *the function symbol # does not occur in π*
3. *all formulas in π are strictΣ_i^b*

4. *proof π is in a* free variable normal form*: the eigenvariables (the variables eliminated in $\forall \leq$: right and $\exists \leq$: left) are all mutually distinct and are also distinct from all parameters*

5. *if $\overline{a}$ are all parameters (i.e., the free variables in $\Gamma \rightarrow \Delta$) and $\overline{b} = (b_0, \ldots, b_k)$ all other free variables in π then*

   (i) *if the elimination inference of b_u is below the elimination inference of b_v then $u < v$*

   (ii) *the elimination rule of any b_u is one of*

   Σ_i^b–IND

   $$\frac{A(b_u), \Pi \longrightarrow \Sigma, A(b_u + 1)}{A(0), \Pi \longrightarrow \Sigma, A(r(\overline{a}, b_0, \ldots, b_{u-1}))}$$

   $\exists \leq$: *left*

   $$\frac{b_u \leq r(\overline{a}, b_0, \ldots, b_{u-1}), A(b_u), \Pi \longrightarrow \Sigma}{\exists x \leq r(\overline{a}, b_0, \ldots, b_{u-1})\, A(x), \Pi \longrightarrow \Sigma}$$

   $\forall \leq$: *right*

   $$\frac{b_u \leq r(\overline{a}, b_0, \ldots, b_{u-1}), \Pi \longrightarrow \Sigma, A(b_u)}{\Pi \longrightarrow \Sigma, \forall x \leq r(\overline{a}, b_0, \ldots, b_{u-1})\, A(x)}$$

6. *$\overline{t}$ is a tuple of $k + 1$ terms $t_u(\overline{a})$ and it holds that*

   $$b_0 \leq t_0(\overline{a}), \ldots, b_{u-1} \leq t_{u-1}(\overline{a}) \longrightarrow r(\overline{a}, b_0, \ldots, b_{u-1}) \leq t_u(\overline{a})$$

   where $r(\overline{a}, b_0, \ldots, b_{u-1})$ is the term from the elimination inference of b_u in 5.

7. *$\overline{d}$ is a $(k+1)$-tuple of proofs d_u such that each d_u is a proof of the sequent from 6 that is without the IND-rule, is quantifier-free and #-free, and the only variables in d_u are $\overline{a}, b_0, \ldots, b_{u-1}$.*

$RCon(T_1^i)$ is a $\forall\Pi_1^b$-*formula formalizing that* "there is no i-regular proof of the empty sequent."

We only sketch the idea of the following statement; the details are in Krajíček and Takeuti (1992).

Theorem 10.5.3. *For $i \geq 1$, T_2^i is not $\forall\Pi_1^b$-conservative over T_1^i. In particular, the formula $RCon(T_1^i)$ is provable in T_2^i but not in T_1^i.*

Proof (sketch). First we shall discuss why $RCon(T_1^i)$ is provable in T_2^i. The idea is to use a partial truth definition for the #-free strictΣ_i^b formulas. Such a definition is a Σ_i^b-formula $Tr_i(x, y)$ in two variables that has the property that for any #-free strictΣ_i^b formula $\phi(\overline{a})$ and any tuple $n = \langle n_1, \ldots, n_k \rangle$

$$Tr_i(\lceil \phi \rceil, n) \equiv \phi(\overline{n})$$

where $\lceil \phi \rceil$ is the Gödel number of ϕ: that is, a number coding formula ϕ as a syntactic object. Such a definition is constructed by induction on i, and for the induction step as well as for the case $i = 0$ a Δ_1^b-definition of a value of a #-free term

$$\mathrm{val}(\lceil t \rceil, n) = t(n_1, \ldots, n_k)$$

for any term $t(\overline{a})$ is needed.

The function $\mathrm{val}(x, y)$ is defined straightforwardly using a sequence of the values of subterms (which is unique and hence the function is Δ_1^b-definable) but we have to, in advance, bound $\mathrm{val}(x, y)$ by a term of L.

Assume that the coding of terms and formulas is defined in a natural way (cf. Section 5.4) such that, in particular

$$|\lceil t \rceil| = O(|t|)$$

By induction on the syntactic complexity of the term $t(\overline{a})$ one shows

$$t(\overline{n}) \leq (\max(\overline{n}) + 2)^{O(|t|)}$$

and hence the bound

$$\mathrm{val}(x, y) \leq (x \# y)^{O(1)}$$

is sufficient.

The partial truth definition $\mathrm{Tr}_i(x, y)$ we obtain in this way is a Σ_i^b-formula, and S_2^1 proves Tarski's conditions. We may extend Tr_i to a partial truth definition STr_i for sequents consisting only of strict Σ_i^b-formulas. The formula STr_i is then Δ_{i+1}^b in S_2^1.

Now let the triple $\langle \pi, \overline{t}, \overline{d} \rangle$ be an i-regular proof, where π is a sequence of the sequents $S_1, \ldots, S_\ell$. Let $\overline{a}$ be the parameter variables and $b_0, \ldots, b_k$ all other free variables. Consider the Π_{i+1}^b-formula $\phi(s)$

$$\forall r \leq s \forall b_0 \leq t_0(\overline{a}), \ldots, b_k \leq t_k(\overline{a}) \, \mathrm{STr}_i(S_r)$$

where $t_0, \ldots, t_k$ are the terms $\overline{t}$ provided by the i-regular proof.

The formula $\phi(1)$ holds as any initial sequent is trivially true, and the implication $\phi(s) \rightarrow \phi(s + 1)$ holds as all inference rules are sound (and using Tarski's conditions S_2^1 can prove it) and the terms $t_0, \ldots, t_k$ are correctly chosen, with proofs of the correctness provided by the proofs $d_0, \ldots, d_k$, which are a part of the i-regular proof as well.

Hence Π_{i+1}^b–LIND implies $\phi(\ell)$ and, in particular, that S_ℓ is true and so S_ℓ cannot be the empty sequent. Hence S_2^{i+1} proves $\mathrm{RCon}(T_1^i)$ and the first part of the theorem follows from Corollary 7.2.4.

Take a diagonal formula $A(a)$ such that the equivalence

$$A(a) \equiv (\forall w \leq a^u, \ w \text{ is not an } i\text{-regular proof of } A(\tilde{a}))$$

is provable in T_1^i, where $A(a)$ is a #-free Π_1^b-formula and $\tilde{a}$ formalizes the dyadic numeral, and u is a constant to be specified later.

By the same argument T_2^i proves that "$A(\tilde{a})$ *is true*," that is, $A(a)$. Assume for the sake of contradiction that $A(a)$ is provable in T_1^i and we may take D to be some i-regular proof of $A(a)$.

Let m be any number and $\underline{m}$ its dyadic numeral. As

$$a = \underline{m}, A(a) \to A(\underline{m})$$

can be proved substituting $\underline{m}$ for b in some fixed i-regular proof of

$$a = b, A(a) \to A(b)$$

and hence (by cut with $\to A(a)$)

$$a = \underline{m} \to A(\underline{m})$$

and thus also

$$\exists x \le \underline{m}, x = \underline{m} \to A(\underline{m})$$

has an i-regular proof of size $O(|m|) = O(\log m)$. Trivially all sequents

$$\to \exists x \le \underline{m}, x = \underline{m}$$

have i-regular proofs of size $O(\log m)$ and consequently also all

$$\to A(\underline{m})$$

have i-regular proofs of size $O(\log m)$, say $\le v \cdot \log(m)$, for all sufficiently large m (the constant v is independent of m, u in A and D). We may assume that we have $v \le u$. Then for m large enough there is an i-regular proof $w_m \le 2^{v \cdot \log(m)} \le m^v \le m^u$ of $A(\underline{m})$. But that implies $\neg A(\underline{m})$: a contradiction. Hence the formula $A(a)$ is not provable in T_1^i.

This proves that T_2^i is not $\forall \Pi_1^b$-conservative over T_1^i. To prove that, in fact, $\mathrm{RCon}(T_1^i)$ is not provable in T_1^i one shows that its provability would imply that the diagonal formula A is provable. However, this requires slightly more estimates to the lengths of the proofs as we do not have a priori all Löb's conditions sufficient for the derivability of Gödel's theorem (cf. Smoryńsky 1977). In particular, the condition

$$\mathrm{Prf}_i(a, b) \to \exists x\, \mathrm{Prf}_i(x, \ulcorner \mathrm{Prf}_i[\tilde{a}, \tilde{b}] \urcorner)$$

is not generally valid in T_1^i, where $\mathrm{Prf}_i(a, b)$ formalizes that "a *is an i-regular proof of b.*"

We shall skip these estimates as we have no further use for them.   Q.E.D.

Recall that the theory T_k^i is defined as T_2^i but with the function symbols $\omega_1(x), \ldots, \omega_{k-1}(x)$ in the language (cf. Theorem 5.1.6) and with the additional axioms

$$|\omega_{j+1}(x)| = \omega_j(|x|)$$

$j = 1, \ldots, k - 2$, in BASIC. The theories S_k^i are defined similarly.

Corollary 10.5.4. *For $i \geq 1$ and $k \geq 1$ it holds that*

   1. *The theory T_{k+1}^i is not $\forall\Pi_1^b$-conservative over T_k^i.*

   2. *The theory S_{k+2}^{i+1} is not $\forall\Pi_1^b$-conservative over S_{k+1}^{i+1}.*

Proof. Let $k > 1$. Say that a triple w is a *restricted T_k^i-proof* of a #-free strictΣ_i^b-formula $A(a)$ iff w is an i-regular proof of a sequent of the form

$$2 \leq |c|^{(k)}, |a|^{(k)} \cdot |a|^{(k)} \cdot \ldots \cdot |a|^{(k)} \leq |c|^{(k)} \longrightarrow A(a)$$

where $|a|^{(k)}$ appears j times and

$$j \leq |w|^{(k+1)}$$

Symbol $|a|^{(k)}$ denotes the k-times iterated function $|x|$ on a.

The number j grows slowly but exceeds any standard number and thus the compactness argument implies that any #-free strictΣ_i^b-formula $A(a)$ provable in T_k^i actually has a restricted T_k^i-proof (note that the antecedent of the sequent stands for "$\omega_k^{(j)}(a)$ exists").

Given a and w the number $c := \omega_k(a \cdot w)$ satisfies the antecedent of the preceding sequent, and as T_2^i proves that every i-regular proof is also sound formula $A(a)$ has to be true. Hence T_{k+1}^i proves that every *restricted T_k^i proof* is sound.

We show that T_k^i itself does not prove that every restricted T_k^i proof is sound. Take a #-free Π_1^b-formula $A(a)$ obtained by a diagonalization such that

$$S_2^1 \vdash A(a) \equiv \left(\forall w \leq a^u, \text{"w is not restricted T_k^i proof of $A(a)$"}\right)$$

By an argument similar to the end of the proof of the previous statement $T_k^i \nvdash A(a)$, but $T_{k+1}^i \vdash A(a)$.

Part 2 of the corollary follows from the first part, as a simple consequence of Corollary 7.2.4 is that S_{k+2}^{i+1} is $\forall\Sigma_{i+1}^b$-conservative over T_{k+2}^i.   Q.E.D.

We may use a similar provability notion to show that the theory U_2^1 is not $\forall\Pi_1^b$-conservative over $I\Delta_0$ (and U_3^1 over S_2, etc.). First note, however, that we already know that U_2^1 is different from $I\Sigma_0^{1,b}$; for example, U_2^1 can prove that every Boolean formula can be evaluated, whereas $I\Sigma_0^{1,b}$ cannot prove that (cf. Section 9.3, the discussion after Theorem 9.3.4).

Call a triple $\langle \pi, \bar{t}, \bar{d} \rangle$ a *regular proof* if it is i-regular for some i, and let $\mathrm{RCon}(T_1)$ denotes a #-free $\forall \Pi_1^b$-formula (in the first order language L) formalizing that "*there is no regular proof of the empty sequent.*"

Theorem 10.5.5. *The theory U_2^1 is not $\forall \Pi_1^b$-conservative over T_1. In particular, U_2^1 proves formula $\mathrm{RCon}(T_1)$ whereas T_1 does not prove this formula.*

Proof. The idea of a proof of $\mathrm{RCon}(T_1)$ in U_2^1 is that one can find a $\Delta_1^{1,b}$-definition of truth for the #-free Σ_∞^b-formulas. This is somewhat similar to the truth definition for the quantified propositional formulas (cf. Section 9.3, the proof of Theorem 9.3.22), and we leave the details to the reader.

The rest is shown similarly to the argument in the proof of Theorem 10.5.3.

Q.E.D.

We conclude this section by a technical lemma that we state for the completeness of the presentation but shall not prove (the proof would follow the idea of the proof of Theorem 9.2.5).

Lemma 10.5.6. *Let $i \geq 1$. Then*

$$S_2^1 \vdash \mathrm{Con}(G_i) \equiv \mathrm{RCon}(T_1^i)$$

and

$$S_2^1 \vdash \mathrm{Con}(G) \equiv \mathrm{RCon}(T_1)$$

10.6. Bibliographical and other remarks

Theorems 10.1.2 and 10.1.3 are from Krajíček and Pudlák (1990a). Definition 10.2.1, Lemma 10.2.2, and the first part of Theorem 10.2.4 are from Krajíček et al. (1991). Lemma 10.2.3 is from Buss (1993b) and Zambella (1994) (based on Kadin 1988), and it implies the last part of Theorem 10.2.4 and Corollaries 10.2.5 and 10.2.6.

The content of Sections 10.3 and 10.4 is from Krajíček (1993) and the presentation of 10.4.2–10.4.12 closely follows that paper. Corollary 10.4.13 is from Krajíček et al. (1991) and Krajíček (1993).

The content of Section 10.5 is from Krajíček and Takeuti (1992), with the exception of Theorem 10.5.1, which is from Paris and Wilkie (1987b), and Theorem 10.5.5, which is from Krajíček and Takeuti (1990). Paris and Wilkie (1987b) have considered restricted provability notions.

There are several topics related to questions considered in Section 10.5, although they do not seem to have some immediate implications for bounded arithmetic (or, more accurately, for the questions about bounded arithmetic we study here). But these topics appear to me quite important so I shall at least mention three of them: the truth definition for bounded formulas, in Lessan (1978), Dimitracopoulos (1980), Paris and Dimitracopoulos (1988), and Takeuti (1988); interpretability in theories of arithmetic in Pudlák (1985) and Wilkie (1986); and restricted consistency statements in Pudlák (1986, 1987) and Paris and Wilkie (1987b).

11

Direct independence proofs

From Section 10.4 we know that all theories $S_2^i(R)$ and $T_2^i(R)$ are distinct. In this chapter we examine specific, more direct independence proofs for theories $S_2^1(R)$, $T_2^1(R)$, and $S_2^2(R)$, and we strengthen Corollary 10.4.3.

11.1. Herbrandization of induction axioms

In this section we shall examine the following idea for independence proofs: Take an induction axiom for a $\Sigma_i^b(\alpha)$-formula. It has the complexity $\forall \Sigma_{i+1}^b(\alpha)$. Introduce a new function symbol to obtain a *Herbrand form* of the axiom, as at the beginning of Section 7.3. But this time we reduce the axiom to an existential formula. This allows us to use a simpler witnessing theorem (Theorem 7.2.3) than the original form of the axiom would require.

Consider first the simplest case (which will turn out to be the only one for which the idea works). Let $\alpha(x, y)$ be a binary predicate. Then the herbrandization of the induction axiom for the formula $A(a) := \exists u \leq a, \alpha(u, a)$

$$A(0) \wedge \forall x \leq a, (A(x) \rightarrow A(x + 1)) \rightarrow A(a)$$

is the formula

$$\alpha(0, 0) \wedge (\forall x, y \leq a, (\alpha(x, y) \wedge x \leq y)$$
$$\rightarrow (\alpha(f(x, y), y + 1) \wedge f(x, y) \leq y + 1)) \rightarrow \exists u \leq a, \alpha(u, a)$$

Denote this formula $\text{IND}_H(A(a))$.

Theorem 11.1.1. *The formula $\text{IND}_H(A(a))$ is provable in $T_2^1(\alpha, f)$ but not in $S_2^1(\alpha, f)$. Hence $T_2^1(\alpha, f)$ is not $\forall \Sigma_1^b(\alpha, f)$-conservative over $S_2^1(\alpha, f)$.*

Proof. The antecedent of the formula $\text{IND}_H(A(a))$ implies the antecedent of the induction axiom for $A(x) := \exists y \leq x, \alpha(y, x)$; hence $T_2^1(\alpha, f)$ implies the

210

formula $\exists u \leq a, \alpha(u, a)$. This shows

$$T_2^1(\alpha, f) \vdash \text{IND}_H(A(a)).$$

Assume now that also $S_2^1(\alpha, f) \vdash \text{IND}_H(A(a))$. By (the relativization of) Theorem 7.2.3 there must be a polynomial time oracle machine $M^{\alpha, f}$ computing on an input a a witness for $\text{IND}_H(A(a))$, for every $\alpha \subseteq \omega \times \omega$ and $f : \omega^2 \to \omega$ of a polynomial growth. We shall show, however, that for any such machine there are α, f, and a for which $M^{\alpha, f}(a)$ is not a witness to $\text{IND}_H(A(a))$.

Let $M^{\alpha, f}$ be any machine and fix a sufficiently large (depending on the running time of $M^{\alpha, f}$). Start a computation of $M^{\alpha, f}$ on the input a. When the machine will ask queries $\alpha(x, y)$ or ask for values of $f(x, y)$, we shall determine the answer to the query by the following rules:

1. Assign $\alpha(0, 0)$ the value TRUE.
2. To the query $[\alpha(x, y)?]$ assign TRUE (and answer YES) if for all $t \leq y, t \neq x$ query $\alpha(t, y)$ was already assigned FALSE. Otherwise assign to $\alpha(x, y)$ FALSE (and answer NO).
3. If the machine requests the value $[f(x, y) =?]$ consider three cases:

    (a) $\alpha(x, y)$ is assigned TRUE. Then choose $t \leq y + 1$ s.t. $\alpha(t, y + 1)$ is already assigned TRUE if such t exists; otherwise choose arbitrary $t \leq y + 1$ such that $\alpha(t, y + 1)$ was not assigned a truth value yet. Answer $f(x, y) = t$ and assign to $\alpha(t, y + 1)$ the value TRUE.

    (b) $\alpha(x, y)$ is assigned FALSE. Choose some $t \leq y + 1$ s.t. $\alpha(t, y + 1)$ is already assigned FALSE if such t exists; otherwise pick arbitrary $t \leq y + 1$ such that $\alpha(t, y + 1)$ has no truth value yet. Answer $f(x, y) = t$ and assign to $\alpha(t, y + 1)$ the value FALSE.

    (c) $\alpha(x, y)$ has no truth value yet. Assign to it some value following rule 2 and then define $f(x, y)$ according to (a) or (b).

Claim. *$\alpha(x, y)$ cannot receive the truth value TRUE during the answers to the first y oracle queries.*

The claim is verified by induction on y.

Now assume that a is large enough that for the running time n^k, $|a|^k < a$. Then no $\alpha(x, a)$ could receive the value TRUE.

Assume $M^{\alpha, f}(a)$ claims to output a witness to $\text{IND}_H(A(a))$. If it is a pair $(x, y), x \leq y \leq a$ such that $\alpha(x, y)$ is assigned TRUE, then either $f(x, y)$ is defined and $\leq y + 1$ and $\alpha(f(x, y), y + 1)$ is also assigned TRUE, or undefined (and then define $f(x, y)$ following 3).

Take α_0 to be the relation consisting of those pairs (x, y) such that $\alpha(x, y)$ received value TRUE and f_0 any total function extending the partial function f. Clearly then the output of $M^{\alpha_0, f_0}(a)$ is not a witness to $\text{IND}_H(A(a))$.

Q.E.D.

A natural thing is to try to apply the same idea to the induction axioms for formulas of the form

$$A(a) = \exists x_1 \le a \forall y_1 \le a \dots Q x_i \le a, \ \alpha(\overline{x}, \overline{y}, a)$$

(with i alternating quantifiers). A herbrandization $\text{IND}_H(A(a))$ of the induction axiom for the formula $A(a)$ has the form

$$\alpha(\overline{0}, \overline{0}, 0) \wedge (\forall b, x_1, \dots, x_m, t_1, \dots, t_n \le a,$$

$$\tilde{\alpha}(\overline{x}, y_j/f_j, b) \to \tilde{\alpha}(z_k/g_k, \overline{t}, b+1)) \to \exists u_1, \dots, u_m \le a, \ \tilde{\alpha}(\overline{u}, v_\ell/h_\ell, a)$$

where

(i) $m = \lceil i/2 \rceil$ and $n = \lfloor i/2 \rfloor$ and $\tilde{\alpha}$ is the formula

$$x_1 \le b \wedge (y_1 \le b \to (x_2 \le b \wedge (\dots (\alpha(\overline{x}, \overline{y}, b)) \dots)))$$

(ii) the function symbol f_j has the arguments $b, x_1, \dots, x_j, t_1, \dots, t_{j-1}$
(iii) the function symbol g_k has the arguments $b, x_1, \dots, x_k, t_1, \dots, t_{k-1}$
(iv) the function symbol h_ℓ has the arguments $a, u_1, \dots, u_\ell$.

Lemma 11.1.2. *Let $i \ge 1$ and let $A(a)$ be a $\Sigma_i^b(\alpha)$-formula of the form*

$$\exists x_1 \le a \forall y_1 \le a \dots \alpha(\overline{x}, \overline{y}, a)$$

α an $(i+1)$-ary predicate.

Then the formula $\text{IND}_H(A(a))$ defined earlier is provable in $T_2^1(\alpha, \overline{f}, \overline{g}, \overline{h})$.

Proof. Consider the formula $B(b)$

$$B(b) = \exists u_1, \dots, u_m \le b, \tilde{\alpha}(\overline{u}, v_\ell/h_\ell, b)$$

with h_ℓ a function symbol depending on $b, u_1, \dots, u_\ell, \ell = 1, \dots, n$.

Assume the conjunction in the antecedent of the formula $\text{IND}_H(A(a))$ is valid. Then $B(0)$ is valid as it follows from $\alpha(0, \dots, 0)$. Also the implication

$$B(b) \to B(b+1)$$

is valid for $b \le a$, which is seen as follows. Assume $B(b)$, that is

$$\exists u_1, \dots, u_m \le b, \ \tilde{\alpha}(\overline{u}, v_\ell/h_\ell, b)$$

which by the second conjunct of the antecedent of $\text{IND}_H(A(a))$ implies

$$\forall \overline{u}, \overline{t}, \ \tilde{\alpha}(z_k/g_k, \overline{t}, b+1)$$

where $g_k = g_k(b, u_1, \dots, u_k, t_1, \dots, t_{k-1})$. Substitute for t_ℓ the function h_ℓ and existentially quantify the terms $g_k(t_\ell/h_\ell)$ on the places of z_k's. This gives $B(b+1)$.

By $\Sigma_1^b(\alpha, \overline{h})$–IND then $B(a)$, which is the desired succedent of $\text{IND}_H(A(a))$, follows.   Q.E.D.

11.2. Weak pigeonhole principle

We have already considered the weak pigeonhole principle in Section 4.2 (Theorem 4.2.4) and in Section 7.3 (Theorem 7.3.7) but in both cases in slightly different versions, so now we define the principle again, as it will be used in this and further chapters.

Definition 11.2.1. $PHP(R)_n^m$ is the following bounded $L_{PA}(R)$ formula

$$\neg\, (\forall x < m \exists! y < n\, R(x, y) \,\wedge\, \forall y < n \exists! x < m\, R(x, y))$$

It says that the relation R is not a graph of a 1–1 function from $m = \{0, 1, \ldots, m - 1\}$ onto $n = \{0, 1, \ldots, n - 1\}$.

In the case when $m > n+1$ the formula is called *the weak pigeonhole principle*; most prominent cases of the weak pigeonhole principle are $m = 2n, n^2$, and 2^n. Later we shall also consider a stronger version of the principle saying that there is not an injective map from m into n.

In the rest of the section we omit explicit reference to the relation R in the formula $PHP(R)_n^m$.

Lemma 11.2.2. *The theory* $S_2^1(R)$ *proves the weak pigeonhole principle* $PHP_n^{2^n}$. *In other words*

$$S_2^1(R) \vdash PHP_{|a|}^a$$

Proof. Assume $\neg\, PHP_{|a|}^a$ for $R \subseteq 2^n \times n$, where $n = |a|$. Define the formula

$$\text{``}j \in R^{-1}(i)\text{''} \quad \text{iff} \quad \exists x \leq 2^n, \mathrm{bit}(x, j) = 1 \wedge R(x, i)$$

which is $\Delta_1^b(R)$ in $S_2^1(R)$ as the witness x is unique, and consider a $\Delta_1^b(R)$-formula

$$\text{``}i \notin R^{-1}(i)\text{''}$$

As the bounded comprehension axiom CA is available in $S_2^1(R)$ for $\Delta_1^b(R)$-formulas (cf. Lemma 5.4.2), we have

$$\exists y \leq 2^n \forall i < n, (\mathrm{bit}(y, i) = 1) \equiv \text{``}i \notin R^{-1}(i)\text{''}$$

and the usual diagonal argument applies to y.   Q.E.D.

Theorem 11.2.3. *For all* k

$$S_1^3(R) + \exists y, y = a^{|a|^{(k)}} \vdash PHP(R)_a^{a^2}$$

where $|a|^{(k)}$ *is k-times iterated function* $|x|$.

Proof. We shall work in $S_2(R)$ first and later see how much induction we used and how much we used the function #. Assume $\neg PHP_a^{a^2}$ with $R \subseteq a^2 \times a$, and let

$F : a \times a \mapsto a$ be the bijection whose graph is R, that is

$$F(x, y) = z \quad \text{iff} \quad R(a \cdot x + y, z)$$

Consider a complete binary tree of height t. We identify its nodes with the 0–1 sequences of length at most t and consequently with the numbers $< 2^t$.

Given $i < a$, we shall label the nodes of the tree by numbers $< a$ by induction on the height:

1. label $\ell(i, \epsilon) = i$, where ϵ the empty sequence (i.e., ϵ is the root)
2. if $\ell(i, x) = u$ and $x0, x1$ are the left and the right sons of x, and $F(v, w) = u$, then set

$$\ell(i, x0) = v \quad \text{and} \quad \ell(i, x1) = w$$

In this sense $i < a$ codes a sequence of numbers $< a$ of length 2^t: the labels of the leaves ordered lexicographically. Note that $\ell(i, x)$ is $\Delta_1^b(R)$-definable from i and in the definition one needs to code sequences of numbers $< a$ of the length $O(|x|) = O(t)$: That is, one needs that the number a^t exists.

On the other side let $g : 2^t \to a$ be any map definable by a Δ_1^b-formula. Then there is $i < a$ such that

$$g(x) = \ell(i, x)$$

for all $x < 2^t$. This is proved by induction as follows. Consider the formula

$$A_g(r) := r \le t \to \forall x \in \{0, 1\}^{t-r} \, \exists i < a \, \forall y \in \{0, 1\}^r, \ell(i, y) = g(x \frown y)$$

Then $A_g(0)$ holds as for $x \in \{0, 1\}^t$ we can take $i := g(x)$. Assume that $A_g(r)$ holds and let $x \in \{0, 1\}^{t-r-1}$ be arbitrary. By the induction hypothesis there are $i_0, i_1 < a$ such that for all $y \in \{0, 1\}^r$

$$\ell(i_0, y) = g(x \frown 0 \frown y) \quad \text{and} \quad \ell(i_1, y) = g(x \frown 1 \frown y)$$

Take $i := F(i_0, i_1)$; hence $i < a$. It is straightforward to verify that for all $y \in \{0, 1\}^{r+1}$

$$\ell(i, y) = g(x \frown y)$$

The formula $A_g(r)$ is Π_3^b so $S_1^3(R)$ implies that $A_g(t)$ holds, for $t \le |a|$. Take $i < a$ witnessing $A_g(t)$ (the quantifier $\forall x$ is void). Then

$$\forall y \in \{0, 1\}^t, \ell(i, y) = g(y)$$

We shall summarize the preceding arguments in a claim.

Claim. *The theory $S_1^3(R) + \neg PHP_a^{a^2}$ proves that if a^t exists, $t \le |a|$, then there is a Δ_1^b-definable map*

$$\ell(i, x) : a \times 2^t \to a$$

such that for any Δ_1^b-*definable map*

$$g(x) : 2^t \to a$$

there is $i_g < a$ *such that*

$$\forall x < 2^t, \ell(i_g, x) = g(x)$$

Now we are ready to obtain the contradiction from the existence of the map F. Take $t := |a|$ and consider the Δ_1^b-definable map $h : 2^t = a \to a$ defined by

$$h(i) = \begin{cases} i, & \text{if } \ell(i, i) \neq i \\ 0, & \text{if } \ell(i, i) = i \quad \text{and} \quad i \neq 0 \\ 1, & \text{if } \ell(i, i) = i \quad \text{and} \quad i = 0 \end{cases}$$

By the previous observation above there is $i_h < a$ such that

$$\forall y < a, \ell(i_h, y) = h(y)$$

and, in particular

$$\ell(i_h, i_h) = h(i_h)$$

which is impossible by the definition.

This shows that $S_1^3(R)$ proves $\neg \, \mathrm{PHP}_a^{a^2}$ assuming a^t exists for $t = |a|$.

Assume now that a^t exists for $t = ||a||$. The claim implies that there is a Δ_1^b-definable map $\ell(i, x)$ enumerating for $i < a$ all Δ_1^b-definable maps from 2^t to a, that is, from $|a|$ to a. Hence using this map we can encode sequences of numbers $< a$ of length $< |a|$ needed in the definition of the function ℓ' enumerating all Δ_1^b-definable functions from $2^{|a|}$ to a, and the contradiction is obtained as before.

Generally, if $t = |a|^{(k)}$ then we use the claim k-times to obtain the contradiction. As all maps $\ell(i, x), \ell'(i, x), \ldots$ constructed in the repeated use of the claim are $\Delta_1^b(R)$, the formula $A_g(r)$ used in its proof is $\Pi_3^b(R)$ and hence the whole argument can be formalized in $S_1^3(R)$.   Q.E.D.

Theorem 11.2.4 (Paris, Wilkie, and Woods 1988).

$$T_2^2(R) \vdash \mathrm{PHP}(R)_a^{a^2}$$

and

$$T_2^2(R) \vdash \mathrm{PHP}(R)_a^{2a}$$

Proof. The first part of the theorem follows from Theorem 11.2.3 and Corollary 7.2.4. For the second part let G be an injection of $2a$ into a. If we iterate G k-times we get a map

$$G^{(k)} : 2^k \cdot a \mapsto a$$

which is $\Delta_1^b(G)$-definable.

For $k = |a|$ this gives an injection

$$F : a^2 \mapsto a$$

and Theorem 11.2.3 applies.   Q.E.D.

Theorem 11.2.5 (Krajíček 1992).

$$S_2^2(R) \nvdash \mathrm{PHP}(R)_a^{a^2}$$

In fact, let f be a function such that a is eventually majorized by $2^{f(a)^\epsilon}$, for any $\epsilon > 0$. Then

$$S_2^2(R) \nvdash \mathrm{PHP}(R)_{f(a)}^a$$

Proof. Assume for the sake of contradiction that

$$S_2^2(R) \vdash \mathrm{PHP}(R)_a^{a^2}$$

The formula $\mathrm{PHP}(R)_a^{a^2}$ is $\Sigma_2^b(R)$, and hence by Theorem 7.2.3 there is a polynomial time machine M^B with access to a $\Sigma_1^b(R)$-oracle B that on input a outputs one of the following elements:

    (a) $x < a^2$ such that $\forall y < a, \neg R(x, y)$
    (b) $x_1 < x_2 < a^2$ and $y < a$ such that $R(x_1, y) \wedge R(x_2, y)$
    (c) $y < a$ such that $\forall x < a^2, \neg R(x, y)$
    (d) $x < a^2$ and $y_1 < y_2 < a$ such that $R(x, y_1) \wedge R(x, y_2)$.

We shall contradict this hypothesis by proving the following claim.

Claim. *Let M^B be a polynomial time oracle Turing machine with access to a $\Sigma_1^b(R)$-oracle B.*

Then there is $R \subseteq a^2 \times a$ such that M^B on input a with the oracle B does not output a witness to $PHP(R)_a^{a^2}$.

To prove the claim assume that the oracle $B(b)$ has the form

$$B(b) := \exists w \le t(b), N^R(w, b)$$

where $N^R(w, b)$ is a $\Delta_1^b(R)$-formula formalizing (cf. Section 6.1)

    "w is an accepting computation of machine N^R with oracle R on input b"
where N^R is a polynomial time oracle Turing machine.

Fix a large enough (we shall analyze the precise magnitude of a needed later). Start the computation of M^B on a. As in the proof of Theorem 11.1.1 we shall answer the oracle queries and construct the approximations $R_0^+, R_1^+, \ldots$ to R and $R_0^-, R_1^-, \ldots$ to $(a^2 \times a) \setminus R$. Put $R_0^+ := R_0^- := \emptyset$. Assume that after i queries $[B(b_1)?], \ldots, [B(b_i)?]$ we have $R_0^+ \subseteq R_1^+ \subseteq \ldots \subseteq R_i^+$ and $R_0^- \subseteq R_1^- \subseteq \cdots \subseteq R_i^-$ satisfying for $j \le i$ the conditions
    1. $R_j^+ \cap R_j^- = \emptyset$

2. R_j^+ is a graph of a partial 1–1 function from a^2 to a

3. $|R_j^+ \cup R_j^-| \le j \cdot t_M(t_N(|a|))$, where $t_M(n)$ and $t_N(n)$ are the time bounds of the machines M and N.

4. For any $S \supseteq R_i^+$, $S \cap R_i^- = \emptyset$ that is a graph of partial 1–1 function from a^2 to a, the oracle answer to the queries $[B^S(b_j)?]$, $j = 1, \dots, i$, are fixed as given in the construction.

Let $[B(b_{i+1})?]$ be the $(i + 1)$-st query. Consider two cases:

(i) There is $S \supseteq R_i^+$, $S \cap R_i^- = \emptyset$ that is a graph of partial 1–1 function from a^2 to a, and $w \le t(b_{i+1})$ such that $N^S(w, b_{i+1})$ holds.

(ii) otherwise.

In case (i) answer YES. Extend R_i^+ (resp. R_i^-) to R_{i+1}^+ (resp. to R_{i+1}^-) by adding to it all pairs $(x, y) \in a^2 \times a$ such that $[S(x, y)?]$ is a query in the computation w and is in w answered affirmatively (resp. negatively). As there are at most $t_M(t_N(|a|))$ such queries, the sets R_{i+1}^+, R_{i+1}^- will satisfy condition 3.

In case (ii) answer NO and put $R_{i+1}^+ := R_i^+$ and $R_{i+1}^- := R_i^-$.

Clearly R_{i+1}^+, R_{i+1}^- satisfy all four conditions 1–4.

Let $R := R_{i_0}^+$, for i_0 the last query, and run the machine M^R on input a. It cannot output $x_1 < x_2 < a^2$ and $y < a$ such that

$$R(x_1, y) \wedge R(x_2, y)$$

as R is a partial 1–1 function. Assume it outputs $x_0 < a^2$, which should be a witness to

$$\forall y < a, \; \neg R(x, y)$$

But by condition 3

$$|R^+ \cup R^-| \le t_M(|a|) \cdot t_N(t_M(|a|)) = |a|^{O(1)} \ll a$$

so we can always find $y_0 < a$ such that

$$R' := R \cup \{(x_0, y_0)\}$$

is a partial 1–1 map disjoint with $R_{i_0}^-$. This means, by condition 4, that $M^{B(R')}$ would on input a output the same $x_0 < a^2$, but now

$$\exists y < a, \; R'(x_0, y)$$

This shows that no output of such a machine can satisfy **(a)** or **(b)**. Analogously the output cannot satisfy **(c)** or **(d)** either. Hence the machine does not witness $\text{PHP}(R)_a^{a^2}$.

For the second part of the theorem note that earlier we needed only that

$$t_M(n) \cdot t_N(t_M(n)) = n^{O(1)} < f(a)$$

where $n = |a|$, that is, that for any $k < \omega$ we can choose a large enough that

$$|a|^k < f(a)$$

That is

$$a < 2^{f(a)^{1/k}}$$

Q.E.D.

As the formula $PHP(R)_a^{a^2}$ is $\forall \Sigma_2^b(R)$, last two theorems imply the next corollary.

Corollary 11.2.6. *The theory $T_2^2(R)$ is not $\forall \Sigma_2^b(R)$-conservative over the theory $S_2^2(R)$.*

We shall improve upon this corollary. Let us consider now the weak pigeonhole principle formalized with function symbols f for the bijection whose graph is the relation R and g for its inverse function. The formula $WPHP(a, f, g)$ is

$$\exists x < 2a \ f(x) \geq a \ \vee \ \exists y < a \ g(y) \geq 2a$$
$$\vee \ \exists x < 2a \ g(f(x)) \neq x \ \vee \ \exists y < a \ f(g(y)) \neq y$$

Note that the formula $WPHP(a, f, g)$ is $\Box_2^p(f, g)$-witnessed as the binary search looking for appropriate x or y asks only $\Sigma_1^p(f, g)$-questions. Recall Definition 7.5.1.

Theorem 11.2.7. *The formula $WPHP(a, f, g)$ is not witnessed by a PLS-problem. In particular, $WPHP(a, f, g)$ is not provable in $S_2^2(f, g)$.*

Proof. The second sentence follows from Theorem 7.5.3 and Corollary 7.2.4.

For the first part assume that $WPHP(a, f, g)$ is witnessed by a PLS-problem $P = P^{f,g}$: That is, the locally optimal solutions s from the solution space $F_P(a)$ are witnesses to $WPHP(a, f, g)$. W.l.o.g. assume that the solution space is $F_P(a) := \{0, 1\}^{|a|^k}$ and let $C^{f,g}(a, s)$ and $N^{f,g}(a, s)$ be the polynomial-time algorithms with the oracle for the functions f and g computing the cost and the neighborhood function.

Assume for the sake of contradiction that the local optima do provide witnesses for $WPHP(a, f, g)$: Whenever $s = \langle w, t_1, \ldots \rangle \in F_P(a)$ is locally optimal, then w witnesses one of the four existential quantifiers in the formula $WPHP(a, f, g)$.

We shall show that for any $C^{f,g}$ and $N^{f,g}$ there are f, g for which a local optimum fails to provide a witness.

Fix a large enough. Call a computation of a polynomial time machine with the oracle for f, g *good* if the oracle answers form a *partial 1–1 map from $\leq 2a$ to $\leq a$*.

Let $m \in N$ be the minimal number such that there are $s \in F_P(a)$ and a good computation of $C^{f,g}$ yielding $C^{f,g}(a, s) = m$. Set s_m to be one such s and f_m to be the oracle answers in some good computation of $C^{f,g}(a, s_m) = m$. Note that f_m is a partial 1–1 map and $|f_m| = \log(a)^{O(1)}$.

Extend f_m to h by adding to f_m the oracle answers in some good computation of $N^{f,g}(a, s_m)$ consistent with f_m. Hence $|h| = \log(a)^{O(1)}$ and by our assumption

$$N^h(a, s_m) = s_m$$

as s_m is even globally optimal.

It follows that s_m is locally optimal in every problem $P^{f,g}$ for all $f \supseteq h$ and $g \supseteq h^{(-1)}$, and thus it has either the form

$$s_m = \langle w, t_1, \ldots \rangle$$

in which case $f(w) \geq a$ should hold for all $f \supseteq h$ or $g(w) \geq 2a$ for all $g \supseteq h^{(-1)}$, or the form

$$s_m = \langle \langle x, y \rangle, t_1, \ldots \rangle$$

in which case $x \neq y$ and $f(x) = f(y)$ should hold for all $f \supseteq h$ or $g(x) = g(y)$ for all $g \supseteq h^{(-1)}$.

In both cases it is, however, easy to find $f \supseteq h$ and $g \supseteq h^{(-1)}$ not obeying the required property, as $|h| = \log(a)^{O(1)} \ll a$.   Q.E.D.

From this theorem and Theorem 11.2.4 we get the following corollary.

Corollary 11.2.8. *The theory $T_2^2(R)$ is not $\forall \Sigma_1^b(R)$-conservative over the theory $S_2^2(R)$.*

Proof. The formula WPHP(a, f, g) is $\forall \Sigma_1^b(f, g)$ but in a language with the function symbols f, g. If we translate the formula naively into a language with a relation symbol R interpreting R as the graph of f we get only a $\forall \Sigma_2^b(R)$-formula. However, a better translation into a relational language is derived by interpreting $R \subseteq 2a \times (|a| + 1)$ as

$$R(x, i) \text{ true} \quad \text{iff} \quad \text{bit}(f(x), i) = 1$$

and $S \subseteq a \times (|a|)$ as

$$S(y, j) \text{ true} \quad \text{iff} \quad \text{bit}(g(y), j) = 1$$

The relation R is called *the bit-graph* of f.

Then the formula translates into a $\forall \Sigma_1^b(R, S)$-formula. Moreover, two relations R and S can be easily Δ_1^b-coded by one relation.

It is a trivial corollary of Theorem 11.2.7 that this formula is not PLS^{R,S}-witnessed either, and hence the corollary follows.   Q.E.D.

11.3. An independence criterion

S. Riis (1993a) obtained a nice sufficient condition (11.3.2) for a combinatorial principle to be unprovable in $S_2^2(R)$. His original proof is model-theoretic but we give another proof of a slightly stronger statement.

Theorem 11.3.1. *Assume that* $\Phi = \forall \overline{x} \exists \overline{y} \phi(\overline{x}, \overline{y})$, ϕ *open, is a sentence in a relational language* $\{R\}$ *consisting of one binary relation (the case of a general relational language is completely analogous). Assume that* Φ *has an* infinite *model.*

Then there are no polynomial-time machine M and NP(R)-oracle A(u) (i.e., a $\Sigma_1^b(R)$-*formula) such that for every finite structure* $([0, a], R)$ *given as an input to M the machine queries A evaluated in* $([0, a], R)$ *and eventually outputs* $\overline{x} \in [0, a]$ *satisfying*

$$([0, a], R) \models \forall \overline{y} \neg \phi(\overline{x}, \overline{y})$$

In other words, the formula

$$\exists \overline{x} \leq a \forall \overline{y} \leq a, \ \neg \phi(\overline{x}, \overline{y})$$

is not witnessed by a polynomial time machine with an NP(R) oracle.

Proof. We shall follow the idea of the proof of Theorem 11.2.5. Let (K, R_K) be an infinite model of Φ.

Let M be a machine running in time $\leq n^\ell, n = |a|$; let $A(u)$ have the form

$$A(u) = \exists w(|w| \leq |u|^t \wedge N^R(u, w))$$

where $N^R(u, w)$ formalizes that

> "*w is a computation of the polynomial-time machine N on the input u working with the oracle R.*"

For any w, $|w| \leq |u|^t$, there are $D_w \subseteq [0, a]$ and $R_w \subseteq D_w \times D_w$ such that for every $R \subseteq [0, a]^2$ satisfying

$$R_w \subseteq R \wedge (D_w^2 \setminus R_w) \cap R = \emptyset$$

the truth value of $N^R(u, w)$ over the structure $([0, a], R)$ is fixed. We say that D_w, R_w forces that truth value. Note that $|D_w| \leq n^t$.

Start a computation of M on a, a sufficiently large. We claim that there is a sequence of $D_i \subseteq [0, a]$ $R_i \subseteq D_i \times D_i$ and

$$F_i : (D_i, R_i) \mapsto (K, R_K)$$

for $i = 0, 1, \ldots$, satisfying
1. $|D_i| \leq i \cdot n^t$
2. $D_i \subseteq D_{i+1}, R_i \subseteq R_{i+1}, (D_i^2 \setminus R_i) \cap R_{i+1} = \emptyset$
3. $F_{i+1} \downarrow D_i = F_i$

4. F_i is an embedding and

$$\forall x, y \in D_i, R_i(x, y) \equiv R_K(F_i(x), F_i(y))$$

5. for all $R \subseteq [0, a]^2$, $R_i \subseteq R$, and $(D_i^2 \setminus R_i) \cap R = \emptyset$, the truth values of first i queries $A(u_1), \ldots, A(u_i)$ in $([0, a], R)$ are fixed.

Set $D_0 = R_0 = F_0 = \emptyset$. Assume that we have (D_i, R_i, F_i) and that $[A(u)?]$ is the $(i + 1)$-st query. Consider two cases.

(i) There is w, $|w| \leq |u|^t$ such that

    (a) (D_w, R_w) forces $A(u)$ be true

    (b) $R = R_i \cup R_w$ satisfies $(D_i^2 \setminus R_i) \cap R = \emptyset$

    (c) there is an embedding

$$F : (D_i \cup D_w, R_i \cup R_w) \mapsto (K, R_K)$$

    such that $F \downarrow D_i = F_i$

(ii) otherwise

In Case (i) set $D_{i+1} := D_i \cup D_w$, $R_{i+1} := R_i \cup R_w$ and $F_{i+1} := F$, answer the query affirmatively, and resume the computation of the machine.

In Case (ii) set

$$(D_{i+1}, R_{i+1}, F_{i+1}) := (D_i, R_i, F_i)$$

answer the query negatively, and resume the computation.

Conditions 1–4 are satisfied trivially; 5 is true as we prefer, by Case (i), the affirmative answer to the negative one.

Let $(D, R, F) = (D_{i_0}, R_{i_0}, F_{i_0})$ be the last triple in this sequence, so $i_0 \leq n^t$. Let $\bar{x}$ be the output of M; w.l.o.g. we assume that $\bar{x}$ is disjoint from D. Let $\bar{x}'$ be any tuple from K disjoint with the range $\mathrm{Rng}(F)$ of F.

As $(K, R_K) \models \Phi$ there is a tuple $\bar{y}'$ such that

$$(M, R_M) \models \phi(\bar{x}', \bar{y}')$$

Again w.l.o.g. assume $\bar{y}'$ is disjoint with $\{\bar{x}'\} \cup \mathrm{Rng}(F)$. Pick any $\bar{y}$ disjoint with $\{\bar{x}\} \cup D$ (this is possible as $|D| \leq n^{\ell+t} \ll a$) and extend R to R' such that the embedding

$$((\bar{x}, \bar{y}, D), R') \mapsto ((\bar{x}', \bar{y}', \mathrm{Rng}(F)), R_K)$$

extends F.

This R' fulfills the hypothesis of condition 5 and so M outputs $\bar{x}$ when the queries are evaluated over $([0, a], R')$. But $([0, a], R') \models \phi(\bar{x}, \bar{y})$, so $\bar{x}$ does not witness $\exists \bar{x} \forall \bar{y} \neg \phi(\bar{x}, \bar{y})$.   Q.E.D.

Note that the same proof shows a slightly stronger (a nonuniform version) statement with the Turing machine M replaced by a witnessing test tree satisfying the conditions of Lemma 9.5.3. Examples of the weak pigeonhole principle formulated with a relation symbol and with function symbols show that the theorem is not valid for formulas in language with function symbols (cf. the discussion before Theorem 11.2.7).

Theorem 11.3.2 (Riis 1993a). *Let $\Phi = \forall x \exists y \, \phi(x, y)$ be a first order sentence in a language L' (not necessarily relational) disjoint with L and assume that it has an infinite model (we do not require ϕ to be open).*
   Then

$$S_2^2(L') \nvdash \exists x < a \forall y < a, \; \neg\phi^{<a}(x, y)$$

where $\phi^{<a}$ denotes ϕ with all quantifiers bounded by a.

Proof. Assume $\phi(x, y)$ has the form

$$\forall u_1 \exists v_1 \ldots \forall u_k \exists v_k, \; \psi(x, y, \overline{u}, \overline{v})$$

Let Φ_S be the skolemization of Φ

$$\Phi_S := \forall x, u_1, \ldots, u_k, \; \psi\left(x, y/G(a, x), \overline{u}, v_i/H_i(a, x, u_1, \ldots, u_i)\right)$$

For every r-ary function symbol $f(z_1, \ldots, z_r)$ in Φ_S consider a new $(r+1)$-ary predicate symbol R_f and a formula $\mathrm{Def}(R_f)$

$$\forall z_1, \ldots, z_r \exists t, \, R(z_1, \ldots, z_r, t)$$
$$\wedge \, \forall z_1, \ldots, z_r, t_1, t_2, \; R(z_1, \ldots, z_r, t_1) \wedge R(z_1, \ldots, z_r, t_2) \to t_1 = t_2$$

and replace in Φ_S the occurrences of f using R_f without an increase of the quantifier complexity to get a formula Φ_S'. For example, replace a negative occurrence of a formula like $P(f(z))$ in Φ_S by $\exists t, R_f(z, t) \wedge P(t)$ and a positive one by $\forall t, R_f(z, t) \to P(t)$. Hence Φ_S' is also universal.
   Take a formula Φ_{rel} to be the conjunction

$$\Phi_S' \wedge \bigwedge_f \mathrm{Def}(R_f)$$

f ranging over the function symbols of Φ_S. The formula Φ_{rel} is thus $\forall\exists$.
   If Φ has an infinite model, so does Φ_S' and Φ_{rel}. By Theorem 11.3.1 the formula $(\neg\Phi_{\mathrm{rel}})^{<a}$ is not witnessed by a machine with an $NP(L'')$ oracle, where L'' lists all relations of Φ_{rel}. Hence by Theorem 7.2.3 $(\neg\Phi_{\mathrm{rel}})^{<a}$ is not provable in $S_2^2(L'')$.
   Take a model M of $S_2^2(L'')$ in which

$$(M, L'') \models (\Phi_{\mathrm{rel}})^{<m}$$

holds for some $m \in M$. The relation symbols R_f of L'' satisfy in M the conditions $\mathrm{Def}(R_f)$. Hence R_f is a graph of a unique function; denote it f.

Expand M to M' by these function symbols. Clearly then

$$M' \models (\Phi_S)^{<m}$$

and hence also

$$M' \models (\Phi)^{<m}$$

We have to show, however, that M' is still a model of S_2^2 in the language L'' extended by the function symbols f. This follows from a simple observation that any Σ_2^b-formula in this extended language is equivalent in M' to the $\Sigma_2^b(L'')$-formula because an atomic formula $f(\bar{z}) = t$ is equivalent to a $\Delta_1^b(L'')$-formula (same as previously).   Q.E.D.

Corollary 11.3.3. *None of the following principles is provable in $S_2^2(L')$:*
  1. *the weak pigeonhole principle $PHP(R)_a^{a^2}$*
  2. *the modular counting principle $MOD_k(R, S)_a$ formalizing (for fixed standard k) that for each a at least one of the following properties must fail:*

     (a) *R is a partition of $a = \{0, \ldots, a - 1\}$ into k-element classes*

     (b) *S is a partition of $a \setminus \{0\}$ into k-element classes*

  3. *no linear ordering R of $[0, a]$ can be dense*
  4. *every linear ordering R of $[0, a]$ has a least element*

Proof. By Theorem 11.3.2 it is enough to construct infinite models for the negations of the principles. This is trivial for all of them (interpreting a^2 as $a \times a$).
                                                                            Q.E.D.

Before the next theorem recall that in S_2 we can code sets of size $\leq |a|^k$ (cf. Section 5.4).

Theorem 11.3.4. *Let $\phi(S)$ be a first order sentence in the language L' disjoint with L, and let L' contain a unary predicate S. Assume that there is an infinite structure K for $L' \setminus \{S\}$ such that for every finite $S \subseteq K$*

$$(K, S) \models \phi(S)$$

Then for every k

$$S_2^2(L') \nvdash \exists S \subseteq [0, a], |S| \leq |a|^k \wedge \neg\phi^{<a}(S)$$

Proof. We shall indicate how the proof of Theorem 11.3.1 can be extended and leave the extension of the proof of Theorem 11.3.2 to this case to the reader.

We want to show that no polynomial time machine M with access to an NP(L') oracle evaluated over an $(L' \setminus \{S\})$-structure with the universe $[0, a]$ can, given as

input a, output $S \subseteq [0, a]$ such that
   (i) $|S| \leq \log(a)^k$
   (ii) $\neg\phi^{<a}(S)$ holds in $[0, a]$.

Choose a large enough and construct $(L' \setminus \{S\})$-structure on $[0, a]$ as in the proof of Theorem 11.3.1, utilizing the embeddings into the given infinite $(L' \setminus \{S\})$-structure K.

Assume that M outputs S, that is, a list of $\leq n^k$ elements of $[0, a]$, $|a| = n$. Then $\phi^{<a}(S)$ cannot hold in $[0, a]$ as the image $F''S$ of S in the embedding into K would be a finite set satisfying ϕ, which contradicts the assumption.   Q.E.D.

Corollary 11.3.5. *None of the following true principles is provable in* $S_2^2(L')$:
   1. every binary tree $R \subseteq [0, a]$ *has a path of length* $\leq \lceil \log(a) \rceil$
   2. every vector space $V \subseteq [0, a]$ *over* $\mathbf{F}_2$ *has a basis of size* $\leq \lceil \log(a) \rceil$

Proof. The corollary follows from Theorem 11.3.4 as there is an infinite binary tree without a finite path and no infinite vector space over $\mathbf{F}_2$ has a finite basis.
Q.E.D.

The following statement is a bit surprising.

Theorem 11.3.6. *Let* $\phi(S)$ *be a first order sentence in the language* L' *disjoint with* L *and assume that for some* ℓ

$$S_2^2(L') \vdash \exists S \subseteq [0, a], |S| \leq |a|^{(\ell)} \wedge \neg\phi^{<a}(S)$$

Then, in fact, for every k:

$$M \models \exists S \subseteq [0, a], |S| \leq |a|^{(k)} \wedge \neg\phi(S)$$

is true in every finite $(L' \setminus \{S\})$-*structure* M *on* $[0, a]$, $a < \omega$, *where* $|a|^{(k)}$ *is the length function k-times iterated.*

Proof. Assume that the conclusion is not valid: That is, for arbitrary large $a < \omega$ there is an $(L' \setminus \{S\})$-structure on $[0, a]$ such that

$$M \models \forall S \subseteq [0, a], |S| \leq |a|^{(k)} \rightarrow \phi(S)$$

As $|a|^{(k)}$ eventually majorizes any constant, by compactness there is an infinite $(L' \setminus \{S\})$-structure K such that for any finite set $S \subseteq K$

$$(K, S) \models \phi(S)$$

The theorem then follows as in the proof of Theorem 11.3.4.   Q.E.D.

11.4. Lifting independence results

In this section we develop a method, utilizing *Sipser's functions* (cf. Section 3.1), to lift the independence results for the theories $S_2^i(\alpha)$, $T_2^i(\alpha)$, $i = 1, 2$ to higher $i > 2$. In this way we shall lift Theorem 11.2.5, a version of the weak pigeonhole principle, but similar arguments work for the principles from Section 11.3.

The idea is the following. Formula $\text{PHP}(R)_a^{a^2}$ is a $\forall \Sigma_2^b(R)$-formula provable in $T_2^2(R)$ but not in $S_2^2(R)$ (Theorems 11.2.4 and 11.2.5). For technical reasons we shall represent the numbers $x < a^2$ by pairs of numbers $x_1, x_2 < a$, and we take the weak pigeonhole principle $\text{WPHP}(a, R(x, y, z))$ in the form

$$\neg\,(\forall u_1, u_2 < a \exists! w < a R(u_1, u_2, w) \,\wedge\, \forall w < a \exists! u_1 < u_2 < a R(u_1, u_2, w))$$

formalizing that $\exists u_1, u_2 < a$, $a \cdot u_1 + u_2 = x \wedge R(u_1, u_2, y)$ is not a graph of a bijection between $a^2 = a \times a$ and a.

If we replace the predicate $R(x_1, x_2, y)$ by a formula $\psi_d^\ell(a, x, y)$

$$\psi_d^\ell(a, x, y) := \forall z_1 < a \exists z_2 < a \ldots Q z_{d-1} < a$$
$$Q' z_d < \left(\frac{\ell a \log a}{2}\right)^{1/2}, \; \alpha(x_1, x_2, y, \bar{z})$$

where α is a $(d + 3)$-ary predicate (and ℓ a constant to be specified later), the formula $\text{PHP}(R)_a^{a^2}$ becomes a $\Sigma_{2+d}^b(\alpha)$-formula and just substituting $\psi_d^\ell(a, x, y)$ for $R(x, y)$ in a $T_2^2(R)$-proof of $\text{PHP}(R)_a^{a^2}$ shows that it is provable in $T_2^{2+d}(\alpha)$ (as $\Delta_1^b(R)$-formulas transform into $\Delta_{1+d}^b(\alpha)$-formulas).

We want to argue that $\text{WPHP}(\psi_d^\ell(a, x, y))_a^{a^2}$ is not provable in $S_2^{2+d}(\alpha)$. Assuming otherwise, Theorem 7.2.3 implies that $\text{WPHP}(\psi_d^\ell(a, x, y))_a^{a^2}$ is witnessed by a $\Box_{2+d}^p(\alpha)$-function: by a polynomial time Turing machine M querying a $\Sigma_{1+d}^p(\alpha)$-oracle. For a fixed sufficiently large input a we want to find α such that M fails to output the required witness. Using the partial restrictions from Section 10.4 we collapse all possible $\Sigma_{1+d}^p(\alpha)$-queries to something analogous to $\Sigma_1^p(\alpha)$-queries, and at the same time the formula $\text{WPHP}(\psi_d^\ell(a, x, y))_a^{a^2}$ will almost collapse to the original formula $\text{WPHP}(R)_a^{a^2}$. finally an argument analogous to the proof of Theorem 11.2.5 should give the wanted contradiction.

Now we present these ideas formally. We shall work with Boolean circuits with atoms of the form

$$p_{x_1, x_2, y, z_1, \ldots, z_i}, \quad \text{for} \quad x_1, x_2, y, z_1, \ldots, z_{i-1} < m, \; z_i < \left(\frac{\ell m \log(m)}{2}\right)^{1/2}$$

The set of all these variables is denoted $B^{i,\ell}(m)$.

Recall from Definition 10.4.6 the space of random restrictions R_q^+, R_q^- with $B^{i,\ell}(m)$ partitioned into classes of the form

$$\left\{ p_{x_1,x_2,y,z_1,\ldots,z_{i-1},t} \mid t < \left(\frac{\ell m \log(m)}{2} \right)^{1/2} \right\}$$

and the definition of the restricted circuit C^ρ, for $\rho \in R_q^+$ or $\rho \in R_q^-$. Definition 10.4.8 of $\Sigma_{i,m}^{S,t}$-circuits is used in the same form except that atoms of the circuit are from $B^{i,\ell}(m)$ now.

Hastad's two Lemmas 10.4.7 (converting a depth d Sipser function into depth $d-1$ one) and 10.4.9 (switching a $\Sigma_{i,m}^{S,t}$ circuit into a $\Sigma_{i-1,m}^{S,t}$ circuit) hold literally, with the same choice of the probability q.

Definition 11.4.1. *Fix i and ℓ. A circuit-oracle is a function C assigning to any natural number u a Boolean circuit C_u with variables from $B^{i,\ell}(m)$, $m = m(u)$ also a function of u.*

For any $\alpha \subseteq \omega^{2+i}$, the circuit oracle C defines a particular set C^α

$$C_\alpha := \{u \mid C_u^\alpha = 1\}$$

where C_u^α is the evaluation of C_u under the evaluation of atoms

$$p_{x_1,x_2,y,z_1,\ldots,z_i} = 1 \quad \textit{iff} \quad (x_1, x_2, y, z_1, \ldots, z_i) \in \alpha$$

For S, t, and m functions of u, a circuit oracle is called $\Sigma_{i,m}^{S,t}$ iff $C_u \in \Sigma_{i,m(u)}^{S(u),t(u)}$ for all u.

Note that the propositional translation $\langle \ldots \rangle$ (Definition 9.1.1) of bounded formulas into Boolean formulas provides a correspondence between a $\Sigma_i^B(\alpha)$-oracle and a $\Sigma_{i,m}^{S,t}$-circuit-oracle, with the functions $S = 2^{\log(m)^{O(1)}}$, $t = \log(S)$, and $m = 2^{\log(m)^{O(1)}}$.

Definition 11.4.2. *Fix m and k, and let $[m]^k$ denote the set of k-tuples of elements of $\{0, \ldots m-1\}$.*

A $(k-u)$-dimensional cylinder in $[m]^k$ is any set of the form

$$\{(x_1, \ldots, x_k) \in [m]^k \mid x_{i_1} = a_1, \ldots, x_{i_u} = a_u\}$$

for any fixed $i_1 < \cdots < i_u$ and $a_1, \ldots, a_u < m$.

The notion of a cylinder is a technical one and its relevance comes from the next theorem. Note that there are $\binom{k}{r} m^{k-r}$ many r-dimensional cylinders in $[m]^k$.

Theorem 11.4.3. *Let $d \geq 1$, let α be a $(3+d)$-ary predicate symbol, and let $\psi_d^\ell(a, x, y)$ be a $\Pi_d^b(\alpha)$-formula defined at the beginning of the section.*

Let $A(a, R)$ be any $\Sigma_\infty^b(R)$-formula with all quantifiers bounded by $< a$, in which the only free variable is a, in which there are no function symbols, and in which all variables in all occurrences of α are bounded.

Assume that M is a polynomial-time oracle machine with Σ_{1+d}^p-oracle B such that M^B computes on input a some witness to the formula $A(a, \psi_d^\ell(a, x, y))$.

Then there is a constant $c \geq 1$ such that for any sufficiently large m there is $Q \subseteq N^3$ and a $\Sigma_{1,m}^{S,t}$-circuit oracle C^1 with variables from $B^{0,\ell-d}(m)$ satisfying the conditions

1. *for all u: $m(u) = m$, $S(u) = 2^{\log(m)^c}$, $t(u) = \log(m)^c$*
2. *for any $r = 1, 2, 3$ and any r-dimensional cylinder U in $[m]^3$*

$$|U \setminus Q| \geq m^{r-1/2}$$

3. *for every $R^0 \subseteq N^3$ for which $R^0 \cap Q = \emptyset$ the machine M with the oracle $C_{R^0}^1$ computes on input m a witness to the formula $B(m, R^0[x_1, x_2, y])$.*

Proof. Choose m large enough so that Lemmas 10.4.7 and 10.4.9 hold with $\ell \geq d + 4$, and fix $a := m$. W.l.o.g. we forbid as an element into any α any $(3 + d)$-tuple $(x_1, x_2, y, z_1, \ldots, z_d)$ for which one of $x_1, x_2, y, z_1, \ldots, z_{d-1}$ is $\geq m$ or for which $z_d \geq (\ell m \log(m)/2)^{1/2}$. Identify the formula $\psi_d^\ell(a, x_1, x_2, y)$ with a depth d circuit $\langle \psi_d^\ell(a, x_1, x_2, y) \rangle_{(m)}$ computing $\psi_d^\ell(a, x_1, x_2, y)$ from the atoms of $B^{d,\ell}(m)$.

Let B be the $\Sigma_{d+1}^b(\alpha)$-oracle. Since M runs in polynomial time any u, for which B is queried is bounded by $\leq 2^{\log(m)^c}$, some $c < \omega$. Moreover, all quantifiers in all such instances $B(u)$ are bounded by numbers of the same magnitude. For any $u \leq 2^{\log(m)^c}$ $B(u)$ is computed by a circuit $\langle B \rangle_{(u)}$ built from the atoms of $B^{d,\ell}(m)$; each such $\langle B \rangle_{(u)}$ is a $\Sigma_{d+1,m}^{S,t}$-circuit with $m(u) = m$, $S(u) = 2^{\log(m)^c}$ and $t(u) = \log(S(u))$. Hence we may think of B as a $\Sigma_{d+1,m}^{S,t}$-circuit oracle rather than a $\Sigma_{d+1}^b(\alpha)$-oracle.

Choose randomly partial restrictions ρ_i from $R_{i,m}^+(q_i)$ for i odd and from $R_{i,m}^-(q_i)$ for i even, $i = d, d-1, \ldots, 1$, where the probability q_i is set to be

$$q_i := \left(\frac{2(\ell - d + i) \log(m)}{m} \right)^{1/2}$$

Denote by C^η the circuit $(\cdots ((C^{\rho_d})^{\rho_{d-1}}) \ldots)^{\rho_1}$.

By Lemma 10.4.9 any $\Sigma_{d+1,m}^{S,t}$-circuit collapses to a $\Sigma_{1,m}^{S,t}$-circuit with probability at least

$$1 - S \cdot (6t)^t \cdot \sum_{1 \leq i \leq d} q_i^t \geq 1 - S \cdot (6t)^t \cdot d \cdot q_d^t$$

(as $q_d \geq \cdots \geq q_1$) and hence with probability at least

$$1 - S^2 \cdot d \cdot (6q_d t)^t \geq 1 - 2^{-\frac{1}{3} \log(m)^{c+1}}$$

this happens for all S circuits $\langle B \rangle_{(u)}$ of the circuit oracle (for m large enough w.r.t. c and ℓ).

To see what happens with the circuits $\psi_d^\ell(a, x_1, x_2, y)$ we apply Lemma 10.4.7. Restricting the circuit by random $\rho_d, \ldots, \rho_1$ produces circuits

$$\left(\psi_d^\ell(a, x_1, x_2, y) \right)^{\rho_d}, \; \left((\psi_d^\ell(a, x_1, x_2, y))^{\rho_d} \right)^{\rho_{d-1}}, \ldots$$

which with the probabilities at least $1 - (1/3)m^{-\ell+d-1}$ contain $\psi_{d-1}^{\ell-1}(a, x_1, x_2, y)$, $\psi_{d-2}^{\ell-2}(a, x_1, x_2, y), \ldots$. There is m^3 of such circuits; hence after the restrictions $\rho_d, \ldots, \rho_2$ all these circuits contain

$$\psi_{d-d+1}^{\ell-d+1}(a, x_1, x_2, y)$$

with the probability at least

$$\geq 1 - \frac{1}{3}(d-1)m^{-\ell+d+2}$$

We have to analyze what happens with these circuits $\psi_1^{\ell-d+1}(a, x_1, x_2, y)$ after the random restriction ρ_1 more closely to be able to establish condition 2 of the theorem.

Assume U is an r-dimensional cylinder, $r = 1, 2, 3$. Analogously with the proof of Lemma 10.4.7 (part 2) we have that with the probability at least

$$1 - \frac{1}{6}m^{-\ell+d-1+r}$$

there are at least

$$m^r \left(\frac{(\ell-d)\log(m)}{m} \right)^{1/2} \geq m^{r-(1/2)}$$

many $*$'s assigned to the m^r circuits corresponding to $(x_1, x_2, y) \in U$. But at the same time, with probability at least

$$1 - \frac{1}{6}m^{-\ell+d-1+r}$$

none of these circuits collapses to 1. Summing up these two probabilities, with the probability at least

$$1 - \frac{1}{3}m^{-\ell+d-1+r}$$

all M^r circuits corresponding to U collapse to either $*$ (i.e., to atom $p_{x_1,x_2,y}$) or to 0, and at most $m - m^{r-(1/2)}$ of them collapse to 0.

There are $3m^2$ 1-dimensional cylinders, $3m$ 2-dimensional ones, and 1 3-dimensional, so the preceding holds for all of them with probability at least

$$\geq 1 - m^{-\ell+d+1}$$

By the earlier computations the $\Sigma_{d+1,m}^{S,t}$-circuit oracle B collapses to $\Sigma_{1,m}^{S,t}$-circuit oracle with probability at least

$$1 - 2^{-\frac{1}{3}\log(m)^{c+1}}$$

and by the preceding with probability at least

$$\geq 1 - m^{-\ell+d+1} - \frac{1}{3}(d-1)m^{-\ell+d+1} \geq 1 - \frac{4}{3}m^{-\ell+d+1}$$

all circuits $\psi_d^\ell(a, x_1, x_2, y)$ collapse to $p_{x_1,x_2,y}$ or 0 with the *"cylinder property"* 2 of the theorem holding.

Hence for m large enough this probability is at least $1/2$ and hence there is a combined restriction η satisfying all these conditions. Finally, for some such fixed η put

$$Q := \left\{ (x_1, x_2, y) \mid (\psi_d^\ell(x_1, x_2, y))^\eta = 0 \right\}$$

and define the $\Sigma_1^{S,t}$-circuit oracle C^1 by

$$C_u^1 := \left(\langle B \rangle_{(u)} \right)^\eta$$

Q.E.D.

Now we apply the previous general theorem to a particular formula.

Theorem 11.4.4. *For all $d \geq 0$, the $\Sigma_{2+d}^b(\alpha)$-formula*

$$WPHP\left(a, \psi_d^\ell(x_1, x_2, y)\right)$$

where $\psi_d^\ell(x_1, x_2, y)$ (and ℓ) were defined previously, is provable in the theory $T_2^{2+d}(\alpha)$ but not in $S_2^{2+d}(\alpha)$.

Proof. By Theorem 11.2.4 WPHP(a, R) is provable in $T_2^2(R)$. If we replace R in the $T_2^2(R)$-proof by $\psi_d^\ell(x_1, x_2, y)$, the $\Sigma_2^b(R)$-induction formulas become $\Sigma_{2+d}^b(\alpha)$ formulas, and hence the proof becomes a $T_{2+d}^2(\alpha)$-proof of WPHP$(a, \psi_d^\ell(x_1, x_2, y))$.

Let $d \geq 1$ and assume for the sake of contradiction that $S_{2+d}^2(\alpha)$ does prove the formula. By Theorem 7.2.3 there must by a polynomial time oracle machine M with access to a $\Sigma_{1+d}^b(\alpha)$-oracle witnessing the formula WPHP$(a, \psi_d^\ell(x_1, x_2, y))$. By Theorem 11.4.3 for any sufficiently large m there are
  (i) $Q \subseteq [m]^3$
  (ii) $\Sigma_1^{S,t}$-circuit oracle C^1
with the properties stated there, such that M with access to C_R^1 witnesses WPHP(m, R) for all $R \subseteq [m]^3$, $R \cap Q = \emptyset$.

Now construct a sequence $R_0^+, R_0^-, R_1^+, R_1^-, \ldots \subseteq [m]^3$ such that
  1. $R_0^+ := \emptyset$ and $R_0^- := Q$

2. $R_i^+ \cap R_i^- = \emptyset$
3. R_i^+ is a graph of a partial 1–1 function from $m \times m$ to m
4. $|R_i^+ \cup R_i^-| \leq |Q| + j \cdot \log(m)^c$, where c is the constant from Theorem 11.4.3
5. For any $S \subseteq [m]^3$, $S \supseteq R_i^+$, and $S \cap R_i^- = \emptyset$ that is a graph of a partial 1–1 function from $m \times m$ to m, the oracle answers to first i queries $[C_{u_j}^1 ?]$, $j = 1, \ldots, i$, are fixed.

This sequence is constructed identically to the sequence in the proof of Theorem 11.2.5, noting that if a $\Sigma_{1,m}^{S,t}$-circuit oracle is true, its truth can be forced by $\leq t$ values of some atoms.

Let $R^+ := R_{i_0}^+$ and $R^- := R_{i_0}^-$ be the last pair in the construction, so in particular:

$$|(R^+ \cup R^-) \setminus Q| \leq i_0 \cdot \log(m)^c \leq \log(m)^{2c}$$

Hence the cylinder condition 2 of Theorem 11.4.3 implies that for every cylinder U

$$|U \setminus (R^+ \cup R^-)| \geq m^{1/2} - \log(m)^{2c} \geq 1$$

Now assume that M with the oracle $C_{R^+}^1$ on the input m outputs $u_1, u_2 < m$ that should witness the formula; that is, it should hold that

$$\forall v < m, \ \neg R(u_1, u_2, v)$$

Take the cylinder U

$$U := \left\{ (x_1, x_2, y) \in [m]^3 \mid x_1 = u_1, \ x_2 = u_2 \right\}$$

By the previous discussion there is at least one $v < m$ such that

$$(u_1, u_2, v) \notin R^-$$

Hence for

$$R := R^+ \cup \{(u_1, u_2, v)\}$$

machine M with oracle C_R^1 on input m also outputs pair (u_1, u_2), but now

$$\exists v < m, (u_1, u_2, v) \in R$$

Candidates of witnesses for the other quantifiers of the formula WPHP are treated similarly. This concludes the proof.   Q.E.D.

Corollary 11.4.5 (Buss and Krajíček 1994). *For any $i \geq 1$, the theory $T_2^i(\alpha)$ is not $\forall \Sigma_i^b(\alpha)$-conservative over $S_2^i(\alpha)$.*

Corollary 11.4.5 reduces the case $i > 2$ via Theorem 11.4.3 to the base case $i = 2$. We could, however, also reduce it to the base case $i = 1$ and argue in the base case in analogy with Theorem 11.1.1. Or to use a more elegant version of the herbrandization of the induction axioms for a $\Sigma_1^b(\alpha)$-formula, the *iteration principle* from Buss and Krajíček (1994)

$$(0 < f(0) \wedge \forall x < a, x < f(x) \rightarrow f(x) < f(f(x)))) \rightarrow \exists x < a, f(x) \geq a$$

Theorem 11.4.3 has to be modified to be applicable to formulas with function symbols. The main difficulty is that after the series of the restrictions the bit-graph of a function is determined to such an extent that the values of the function are determined for most arguments as well. That leaves no room for a diagonalization.

This technical problem can be overcome; we refer the reader to Chiari and Krajíček (1994). Here we confine ourselves to stating without proof the main application of the modified lifting method.

Theorem 11.4.6 (Chiari and Krajíček 1994). *For any $i \geq 2$, the theory $T_2^i(\alpha)$ is not $\forall \Sigma_{i-1}^b(\alpha)$-conservative over $S_2^i(\alpha)$.*

11.5. Bibliographical and other remarks

Section 11.1 is based on Krajíček (1992); earlier Buss (1986) contained a separation of $S_2^1(f)$ and $T_2^2(f)$. Paris et al. (1988) prove WPHP in $I\Delta_0 + \Omega_1$; the analysis of the proof gives Theorems 11.2.3 and 11.2.4. Theorem 11.2.5 is from Krajíček (1992); Corollary 11.2.6 was obtained independently by Pudlák (1992a). Theorem 11.2.7 and Corollary 11.2.8 are from Chiari and Krajíček (1994).

The content of 11.3 is essentially that of Riis (1993a), generalized and with new proofs (the original ones were model-theoretic and applied to a language without function symbols). One corollary of his approach that we do not derive in this way was a construction of a model of $T_2^1(R)$ in which $S_2^2(R)$ does not hold.

The content of Section 11.4 is based on Buss and Krajíček (1994), which extended a construction from Krajíček (1993). Theorem 11.4.6 is from Chiari and Krajíček (1994).

12

Bounds for constant-depth Frege systems

Constant depth *Frege* systems, or equivalently constant-depth LK systems (recall Definition 4.3.11), are the strongest systems for which some nontrivial lower bounds are known.

We present these bounds in this chapter. We shall start with the known upper bounds, however, to get an idea of the strength of the systems.

12.1. Upper bounds

Theorem 9.1.3 and Corollary 9.1.4 can be used as sources of the upper bounds for constant-depth *Frege* proofs: To assure the existence of polynomial size constant-depth F-proofs of instances of a combinatorial principle (and generally of any translation $\langle A \rangle_n$ of a bounded formula), one only needs to prove the principle in bounded arithmetic $I\Delta_0(R)$, or perhaps in $I\Delta_0(R)$ augmented by an extra function. The results of Section 11.2 then imply the following theorems.

Theorem 12.1.1. *There are constant-depth Frege proofs of size $n^{O(\log(n))}$ of the weak pigeonhole principle PHP_n^{2n}:*

$$\bigvee_{i < 2n} \bigwedge_{j < n} \neg p_{ij} \vee \bigvee_{i_1 < i_2 < 2n} \bigvee_{j < n} (p_{i_1 j} \wedge p_{i_2 j})$$
$$\vee \bigvee_{j < n} \bigwedge_{i < 2n} \neg p_{ij} \vee \bigvee_{j_1 < j_2 < n} \bigvee_{i < 2n} (p_{ij_1} \wedge p_{ij_2})$$

Proof. By Theorem 11.2.4, $I\Delta_0(R)$ proves

$$a^{|a|} \leq b \rightarrow \mathrm{PHP}(R)_a^{2a}$$

Hence by Theorem 9.1.3 the propositional formula

$$\left\langle a^{|a|} \leq b \;\rightarrow\; \mathrm{PHP}(R)_a^{2a} \right\rangle_{a=n, b=n^{|n|}}$$

has $n^{O(\log(n))}$ size, constant-depth Frege proofs.

But this formula is clearly (shortly) provably equivalent to the preceding propositional version.   Q.E.D.

Analogously as before, using Theorem 11.2.3 in place of Theorem 11.2.4, we get the following theorem.

Theorem 12.1.2. *For every k there are constant-depth Frege proofs of the weak pigeonhole principle* $\mathrm{PHP}_n^{n^2}$

$$\bigvee_{i<n^2} \bigwedge_{j<n} \neg p_{ij} \;\vee\; \bigvee_{i_1<i_2<n^2} \bigvee_{j<n} (p_{i_1 j} \wedge p_{i_2 j})$$

$$\vee \bigvee_{j<n} \bigwedge_{i<n^2} \neg p_{ij} \;\vee\; \bigvee_{j_1<j_2<n} \bigvee_{i<n^2} (p_{ij_1} \wedge p_{ij_2})$$

of size $n^{O(\log^{(k)}(n))}$.

We shall mention a few more combinatorial principles: the Ramsey theorem and the Tournament principle. We shall confine the analysis to the simplest cases of these principles.

Ramsey theorem. *Every graph with n vertices contains either a clique or an independent set of size* $\geq (\log(n)/2)$

Tournament principle. *In a digraph $G = (V, E)$ with n vertices satisfying*

$$\forall v \neq w \in V, (v, w) \in E \equiv (w, v) \notin E$$

there exists a dominating set $W \subseteq V$

$$\forall v \in V \setminus W \; \exists w \in W, (w, v) \in E$$

of size $O(\log(n))$.

Note that as the sets whose existence is claimed in these principles are of a logarithmic size, they are coded by a number of size $n^{O(\log(n))}$, and hence, in fact, these principles are expressible as $\Sigma_\infty^b(R)$-formulas with the parameter n.

Theorem 12.1.3. *The theory $T_2^5(R)$ proves the* Ramsey theorem:

$$\exists X \subseteq \{0, \ldots, n-1\}, |X| \geq \left\lfloor \frac{\log(n)}{2} \right\rfloor$$
$$\wedge \left((\forall x, y \in X, R(x, y)) \vee (\forall x, y \in X, \neg R(x, y)) \right)$$

Proof. The idea of the proof is to reduce the Ramsey theorem to the weak pigeon-hole principle PHP$(S)_a^{2a}$ for S definable by a bounded formula from R.

Claim 1. *For $x, y < n$ define the function $F(x, y)$ by*

$$F(x, y) = \begin{cases} 1, & \text{if } R(x, y) \\ 0, & \text{if } \neg R(x, y) \end{cases}$$

Then there are $x_1 < x_2 < \cdots < x_s < n$ and $\epsilon_0, \ldots, \epsilon_{s-1} \in \{0, 1\}$, where $s = |n|$, such that

$$\forall i < j < s, \, F(x_i, x_j) = \epsilon_i$$

We shall prove the claim in a sequence of steps. First we define the relation $E \subseteq \{0, \ldots, n-1\}^{\leq s} \times \{0, 1\}^{\leq s}$ by

$$E((x_1, \ldots, x_j), (\epsilon_1, \ldots, \epsilon_k)) := x_1 = 0 \wedge x_1 < x_2 < \cdots < x_j$$
$$\wedge \, k + 1 = j \wedge \forall u < v \leq j, \, F(x_u, x_v) = \epsilon_u$$
$$\wedge \, \forall u < j \, \forall y < x_{u+1} \, \exists v \leq u, \, F(x_v, y) \neq \epsilon_v$$

The last conjunct means that x_{u+1} is the minimal x_{u+1} such that

$$\forall v \leq u, \, F(x_v, x_{u+1}) = \epsilon_v$$

Note that $E \in \Pi_1^b(R)$ as u, v are sharply bounded.

We want to show that for some $\overline{x} = (x_1, \ldots, x_s)$ and $\overline{\epsilon} = (\epsilon_1, \ldots, \epsilon_{s-1})$ the relation $E(\overline{x}, \overline{\epsilon})$ holds. For the sake of the construction assume that there is no such $\overline{x}$ and $\overline{\epsilon}$. We want to define a function

$$H : n \to \{0, 1\}^{\leq s-1} \quad (\leq n/2)$$

which will be 1–1 and hence will contradict the weak pigeonhole principle.

Define H by

    1. $H(0) := 0$
    2. for $0 < x < n$ put

$$H(x) := y$$

where $y = (\epsilon_1, \ldots, \epsilon_j) \in \{0, 1\}^j$, $j < s$, is the lexicographically minimal y (with minimal j) such that

$$\exists x_1 < \ldots x_j < x, \, E((x_1, \ldots, x_j, x), y)$$

Note that $H \in \mathcal{B}(\Sigma_1^b(E))$: That is, $H \in \mathcal{B}(\Sigma_2^b(R))$.

The range of H is included in $\{0, 1\}^{s-1}$ (that we identify via the dyadic representation of numbers with $n/2$) so it remains to prove that H is defined on the whole n and that it is 1–1.

Let $H(x) = H(x') = y$, where $y = (\epsilon_1, \ldots, \epsilon_j)$ and let it hold for $\overline{x} = (x_1, \ldots, x_j, x)$ and $\overline{x}' = (x_1', \ldots, x_j', x')$

$$E(\overline{x}, y) \wedge E(\overline{x}', y)$$

Then we show $x_i = x_i'$ by induction on i: It is true for $i = 1$ (as $x_1 = x_1' = 0$) and if $x_1 = x_1', \ldots, x_i = x_i'$ then necessarily $x_{i+1} = x_{i+1}'$ as x_{i+1} is determined by $x_1, \ldots, x_i$ and $\epsilon_1, \ldots, \epsilon_i$. This shows that H is 1–1.

To show that H is defined everywhere on n let $x < n$. If $x = 0$ then $H(x)$ is the empty word, that is, 0. Let $0 < x < n$. We define the sequence $x_1, x_2, \ldots$ and $\epsilon_1, \epsilon_2, \ldots$ by

1. $x_1 := 0$, $\epsilon_1 := F(x_1, x)$
2. x_{i+1} is the minimal $y > x_i$ such that for $j = 1, \ldots, i$

$$F(x_j, y) = \epsilon_j$$

3. $\epsilon_{i+1} := F(x_{i+1}, x)$

Clearly for all $i : E((x_1, \ldots, x_{i+1}), (\epsilon_1, \ldots, \epsilon_i))$; hence if $x_{i+1} = x$ it holds

$$H(x) = (\epsilon_1, \ldots, \epsilon_i)$$

Assume that $x_i \neq x$ for $i = 1, \ldots, s - 1$. That is,

$$x_1 < \cdots < x_{s-1} < x$$

Take $x_s \leq x$ to be the minimal $z \leq x$ such that

$$x_1 < \cdots < x_{s-1} < z$$

and

$$F(x_i, z) = \epsilon_i, \qquad i = 1, \ldots, s - 1$$

But then $E((x_1, \ldots, x_s), (\epsilon_1, \ldots, \epsilon_{s-1}))$ holds, thus contradicting our assumption that there are no such $\overline{x}, \overline{\epsilon}$.

This means that in the construction $x_i = x$, some $i < s$. That is,

$$H(x) = (\epsilon_1, \ldots, \epsilon_i), \text{ some } i < s$$

Hence H is defined everywhere on n.

By Theorem 11.2.4, $T_2^2(H)$ and so $T_2^5(R)$ proves that such H cannot exist. This proves the claim.

To prove the theorem take $x_1, \ldots, x_s$ and $\epsilon_1, \ldots, \epsilon_{s-1}$ such that

$$E(\overline{x}, \overline{\epsilon})$$

and let

$$A := \{x_i \mid \epsilon_i = 0\}, \qquad B := \{x_i \mid \epsilon_i = 1\}$$

Then either $|A| \geq (s/2)$ or $|B| \geq (s/2)$, and A is an independent set while B is a clique in the graph $(\{0, \ldots, n-1\}, R)$.   Q.E.D.

The propositional version of the Ramsey theorem is a formula built from $\binom{n}{2}$ atoms $p_{i,j}$, $\{i, j\} \subseteq \{0, \ldots, n-1\}$, $i \neq j$ and has the form RAM_n:

$$\bigvee_{i_1 < \cdots < i_s < n} \left(\bigwedge_{u < v \leq s} p_{i_u, i_v} \vee \bigwedge_{u < v \leq s} \neg p_{i_u, i_v} \right)$$

where $s = \lfloor \frac{\log(n)}{2} \rfloor$. The length of RAM_n is $O(n^{\log(n)} \cdot \log(n)^2)$.

Corollary 12.1.4. *The formula RAM_n has constant-depth F-proofs of size*

$$n^{O(\log(n))} = |\mathrm{RAM}_n|^{O(1)}$$

Proof. From the previous theorem follows, via Corollary 9.1.4, the bound of the form $n^{\log(n)^{O(1)}}$. The bound $n^{O(\log(n))}$ comes from the inspection of the proof of the Theorem 11.2.4 showing that in the $T_2^2(R)$ proof of $\mathrm{PHP}(R)_a^{2a}$ we use only numbers of size $\leq O(a\#a) = a^{O(\log(a))}$.   Q.E.D.

Despite the fact that proofs of the Tournament principle are similar to proofs of the Ramsey theorem, it is open whether it is provable in bounded arithmetic. This would be interesting to know in relation to the proof of Lemma 10.2.2.

12.2. Depth d versus depth $d + 1$

In this section we shall prove the exponential lower bound to the size of constant-depth LK-proofs. We shall also show a superpolynomial speed-up of the depth $d + 1$ system over the depth d system, which demonstrates that the depth d system does not polynomially simulate the depth $d + 1$ system (w.r.t. to the proofs of the depth d formulas). Note that by Theorem 4.4.15 constant-depth LK-proofs and constant-depth F-proofs are equally efficient, though the depth in LK (resp. in F) proofs might differ by a constant.

We shall begin with a slightly refined version of Theorem 12.1.2. In this section we shall denote the propositional version of PHP by $\mathrm{PHP}(A, B)$ to stress explicitly the domain A and the range B of the map; this is because these will not always have the form of an initial interval of numbers.

For technical reasons we shall work with set $\neg \mathrm{PHP}(A, B)$ of the formulas

1. $\bigvee_{j \in B} p_{ij}$, for each $i \in A$
2. $\neg p_{ik} \vee \neg p_{jk}$, for each $i \neq j$, $i, j \in A$ and $k \in B$
3. $\bigvee_{i \in A} p_{ij}$, for each $j \in B$
4. $\neg p_{ij} \vee \neg p_{ik}$, for each $j \neq k$, $j, k \in B$ and $i \in A$

and we shall study the LK-refutations of $\neg PHP(A, B)$, that is, the LK-proofs of the empty sequent from axioms of $\neg PHP(A, B)$.

Recall that we identify n with $\{0, \dots, n - 1\}$.

Lemma 12.2.1. *Let $0 < \epsilon < 1$, and let A and B be two sets of sizes $|A| = \lceil n^{1+\epsilon} \rceil$ and $|B| = n$.*

Then there is an LK-refutation of $\neg PHP(A, B)$ of depth 1 and length at most $n^{O(\log(n))}$. Moreover, every formula in the LK-proof is either a conjunction or a disjunction of size at most $O(\log(n))$.

Proof. By Theorem 11.2.4 the theory $T_2^2(R)$ proves $PHP(R)_n^{n^2}$. Assume that S is a graph of bijection between $n^{1+\epsilon}$ and n. Identify $n^{1+\epsilon}$ with the set $n \times n^\epsilon$ and $n^{1+2\epsilon}$ with $n^{1+\epsilon} \times n^\epsilon$, and for $(i, j) \in n^{1+\epsilon} \times n^\epsilon$ define the function S^1 by

$$S^1(i, j) := S(S(i), j)$$

Then S^1 is a bijection between $n^{1+2\epsilon}$ and n.

Similarly define the maps S^k between $n^{1+2^k\epsilon}$ and n. For $k := \lceil \log(\epsilon^{-1}) \rceil$, S^k is a bijection between n^2 and n, and it is clearly $\Delta_1^b(S)$-definable.

By the preceding $T_2^2(S^k)$, and so $T_2^2(S)$, proves $PHP(S)_n^{n^{1+\epsilon}}$.

Moreover, in the proof of Theorem 11.2.4 only numbers of size at most $n^{O(\log(n))}$ are used; in particular, all quantifiers in the $T_2^2(S)$-proof are bounded by $n^{O(\log(n))}$.

By Theorem 9.1.3 and Corollary 9.1.4 there is a constant-depth LK-refutation of $\neg PHP(n^{1+\epsilon}, n)$ of size $n^{O(\log(n))}$. Moreover, the proof of that theorem gives a depth 3, treelike refutation in which every depth 3 formula is a disjunction, and the number of formulas in any sequent in the refutation is bounded by constant c independent of n.

Claim. *Assume that a sequent of the form*

$$\bigvee_{i_1} \phi_{i_1}^1, \dots, \bigvee_{i_k} \phi_{i_k}^k, \Gamma \longrightarrow \Delta, \bigvee_{j_1} \psi_{j_1}^1, \dots, \bigvee_{j_\ell} \psi_{j_\ell}^\ell$$

occurs in the refutation, where Γ, Δ are cedents of depth ≤ 2 formulas, and other formulas are depth 3.

Then for any choice of $i_1, \dots, i_k$ the sequent

$$\phi_{i_1}^1, \dots, \phi_{i_k}^k, \Gamma \longrightarrow \Delta, \psi_1^1, \dots, \psi_1^2, \dots, \psi_1^\ell, \dots$$

has a depth 2, treelike LK-proof $\pi_{i_1,\dots,i_k}$ from the axioms $\neg PHP(n^{1+\epsilon}, n)$, of total size at most

$$\sum_{i_1,\dots,i_k} |\pi_{i_1,\dots,i_k}| \leq n^{O(\log(n))}$$

The claim is readily established by induction on the number of inferences in the original refutation above the sequent, utilizing the existence of the constant c.

Analogously with Lemma 4.6.4 any depth $d + 1$ treelike proof (of a depth d sequent) can be transformed with only a polynomial increase in size into a depth d proof of that sequent (not necessarily treelike).

By the claim applied to the last sequent of the refutation there is a treelike, depth 2 refutation of $\neg \text{PHP}(n^{1+\epsilon}, n)$ in which every depth 1 subformula has size $O(\log(n))$. Hence there is a depth 1 refutation (not necessarily treelike) of size $n^{O(\log(n))}$, in which every formula is of size at most $O(\log(n))$: That is, it is a conjunction or a disjunction of size $O(\log(n))$.   Q.E.D.

As in Sections 10.4 and 11.4, a formula ξ is $\Sigma_d^{S,t}$, $d \geq 0$, if it is *equivalent* to a formula ψ having the properties

1. $dp(\psi) \leq d + 1$,
2. if $dp(\psi) = d + 1$ then the outmost connective of ψ is $\bigvee$
3. ψ has $\leq S$ subformulas of depths $2, 3, \ldots, d + 1$
4. all depth 1 subformulas are (conjunctions or disjunctions) of arity at most t.

A formula is $\Pi_d^{S,t}$ iff its negation is $\Sigma_d^{S,t}$. A formula ξ is Δ_1^t if both ξ and $\neg\xi$ are expressible as disjunctions of conjunctions of arity at most t, that is, if they are both $\Sigma_1^{S,t}$.

For $d, n \geq 1$ arbitrary and any atom p_{ij}($i \in A$, $j \in B$) let $S_{i,j}^{d,n}$ be the depth d Sipser function built from the atoms $p_{ij}^{i_1, \ldots, i_d}$

$$S_{i,j}^{d,n} := \bigwedge_{i_1 < n} \bigvee_{i_2 < n} \cdots Q_{i_d < n} \; p_{ij}^{i_1, \ldots, i_d}$$

where Q is $\bigwedge$ iff d is odd. For $d = 0$, let $S_{i,j}^{d,n}$ be just p_{ij}. Let $\text{Var}(S_{i,j}^{d,n})$ denote the set of the atoms of $S_{i,j}^{d,n}$.

Recall also that for a formula ϕ and a restriction ρ, ϕ^ρ denotes the restricted formula.

We will make an essential use of the following technical proposition.

Lemma 12.2.2. *Let $c, d, n \geq 1$ and $|A| \leq n^2$, $|B| \leq n$. Let $S \leq 2^{n^{(1/3)}}$ and $t = \log(S)$.*

Assume that $\psi_1, \ldots \psi_u$ are formulas from $\Sigma_d^{S,t} \cup \Pi_d^{S,t}$ with the atoms from $V := \bigcup_{ij} \text{Var}(S_{ij}^{d,n})$ such that

$$\sum_{i \leq u} |\psi_i| \leq S$$

Assume moreover that $U_1, \ldots, U_r \subseteq A \times B$ are any sets, $r \leq n^c$.
Then there is a map ρ

$$\rho : V \to \{0, 1\} \cup \{p_{ij} \mid i \in A, \; j \in B\}$$

satisfying the following conditions:

1. *all formulas* ψ_i^ρ *are* Δ_1^t, $i \le u$
2. *for all* $i \in A$, $j \in B$, *the formula* $(S_{ij}^{d,n})^\rho$ *is either* 0 *or* p_{ij}
3. *for every* U_s, $s \le r$, *for at least* $|U_s| \cdot n^{-(1/2)}$ *pairs* $(i, j) \in U_s$ *it holds that*

$$(S_{ij}^{d,n})^\rho = p_{ij}$$

Proof. The proof of this lemma is almost identical to the combinatorial part of the proof of Theorem 11.4.3 where sets U_s are in place of cylinders. It utilizes Hastad's switching Lemma 10.4.9 analogously to the proof of Lemma 10.4.10.

Here is a brief sketch; the details are left to the reader.

For each $i \in A$, $j \in B$, $(B_v)_v$ partition the set $\mathrm{Var}(S_{ij}^{d,n})$ into n^{d-1} classes of the form

$$\{p_{ij}^{i_1,\ldots,i_{d-1},t} \mid t < n\}$$

one for each $i_1, \ldots, i_{d-1} < n$.

For each $i \in A$, $j \in B$, the probability spaces of random restrictions $R_{ij}^+(q)$, $R_{ij}^-(q)$ are defined for restrictions of atoms from $\mathrm{Var}(S_{ij}^{d,n})$. Hence for different i, j these restrictions are independent.

The random restrictions that are applied to $\psi_1, \ldots \psi_u$ are disjoint unions of ρ_{ij} independently drawn from $R_{ij}^{+/-}(q)$.

The rest of the argument is similar to that of Theorem 11.4.3. Q.E.D.

Definition 12.2.3. *The* Σ-depth *of an LK-proof* π *of length* $S = |\pi|$ *is the minimal* d *such that every formula in* π *is from* $\Sigma_d^{S,t} \cup \Pi_d^{S,t}$, *for* $t = \log(S)$.

Theorem 12.2.4 (Krajíček 1994). *Let* $d \ge 0$, $n \ge 1$, $1 > \epsilon > 0$, *and let* A, B *be two sets of size* $|A| = n^{1+\epsilon}$ *and* $|B| = n$. *For sufficiently large* n *the following two statements are valid:*

1. *Every* Σ-depth d, *treelike refutation of the set* $\neg PHP(A, B)(p_{ij}/S_{ij}^{d,n})$ *must have size at least*

$$\ge 2^{n^{1/5}}$$

2. *There are* Σ-depth d, *sequencelike, and* Σ-depth $d + 1$, *treelike refutations of the set* $\neg PHP(A, B)(p_{ij}/S_{ij}^{d,n})$ *of size at most* $n^{O(\log(n))}$.

Proof. Fix d, n, ϵ and A, B satisfying the hypothesis. We shall need the following modification of the pigeonhole principle.

Let

$$r : A \to \exp(B)$$

be a function satisfying

(a) $|r(i)| \ge n^{1/2}$, for all $i \in A$
(b) $|r^{(-1)}(j)| \ge n^{\frac{1}{2}+\epsilon}$, for all $j \in B$, where $R^{(-1)}(j) := \{i \in A \mid j \in r(i)\}$.

Then we *define* **modified** PHP, $\neg$ MPHP(A, B, r) to be the set of sequents

$$\longrightarrow \neg p_{ik} \vee \neg p_{jk}, \quad \text{for } i \neq j \in A, \ k \in r(i) \cap r(j)$$

$$\longrightarrow p_{ij_1}, \ldots, p_{ij_r}, \quad \text{for } i \in A, \ r(i) = \{j_1, \ldots, j_r\}$$

$$\longrightarrow \neg p_{ij} \vee \neg p_{ik}, \quad \text{for } j \neq k \in r(i), \ i \in A$$

$$\longrightarrow p_{i_1 j}, \ldots, p_{i_s j}, \quad \text{for } j \in B, \ r^{(-1)}(j) = \{i_1, \ldots, i_s\}$$

We shall prove the first part of the theorem now. Assume that π is an LK-refutation of $\neg$ PHP(A, B)($S_{ij}^{d,n}$) that is treelike and has size $S = |\pi| \leq 2^{n^{1/3}}$.

Let $\psi_1, \ldots, \psi_u$ be all formulas occurring in π (hence $\sum_{i \leq u} |\psi_i| \leq S$) and let $U_1, \ldots, U_r$ list all $n^{1+\epsilon} + n$ 1-dimensional cylinders in $A \times B$, that is, sets of the form $\{i\} \times B$ or $A \times \{j\}$.

By Lemma 12.2.2 there is a map ρ

$$\rho : \text{Var}(S_{ij}^{d,n}) \rightarrow \{0, 1, p_{ij}\} \text{ all } i, j$$

such that

(i) all ψ_i^ρ are $\Delta_1^t, t = \log(S)$

(ii) $(S_{ij}^{d,n})^\rho$ is 0 or p_{ij}

(iii) $|\{(i, j) \in U_s \mid (S_{ij}^{d,n})^\rho = p_{ij}\}| \geq n^{-\frac{1}{2}} \cdot |U_s|$.

Define a map $r : A \rightarrow \exp(B)$ by

$$r(i) := \{j \mid (S_{ij}^{d,n})^\rho = p_{ij}\}$$

Note that the map r satisfies (a) and (b).

Proof π^ρ, with every ψ_i replaced by ψ_i^ρ, is a refutation of $\neg$ PHP(A, B)(($S_{ij}^{d,n})^\rho$). This implies that there is a treelike LK-refutation π' of $\neg$ MPHP(A, B, r) such that:

1. $|\pi'| \leq S + n^{O(d)}$

2. every formula in π' is $\Delta_1^t, t = \log(S)$

This is because all formulas from $\neg$ PHP(A, B)(($S_{ij}^{d,n})^\rho$) follow from $\neg$ MPHP(A, B, r) by a treelike LK-proof of size $n^{O(d)}$ in which all formulas are Δ_1^t (just note that every $(S_{ij}^{d,n})^\rho$ is proved to be either 0 or p_{ij}, depending on whether $j \notin r(i)$ or $j \in r(i)$).

To finish the proof of the first part of the theorem it suffices to show that $\neg$ MPHP(A, B, r) cannot be refuted by an LK-proof satisfying 1 and 2.

Claim. *For n sufficiently large, $r : A \rightarrow \exp(B)$ a map satisfying (a) and (b), and π' a treelike LK-refutation of $\neg$ MPHP(A, B, r) satisfying 1 and 2, it must hold that*

$$S \geq 2^{n^{1/4} \cdot (1/2)}$$

For a binary tree T labeled by formulas or sequents (proof trees are examples of such trees) and a the node, denote by T^a the subtree with the root a and with the label of a changed to 0, and denote by T_a tree $T \setminus T^a$, with label of a (now a leaf) changed to 1.

We shall use a simple combinatorial fact that in a binary tree T there is node a such that

$$|T^a| \leq \frac{2}{3}|T| \quad \text{and} \quad |T_a| \leq \frac{2}{3}|T|$$

where $|T|$ denotes the number of nodes in the tree (cf. Lemma 4.3.10 or the claim in the proof of Lemma 9.3.2 taken from Spira theorem 3.1.15).

To prove the claim we shall construct a sequence of binary trees $T(1), T(2), \ldots$ obtained from the original LK-refutation π' satisfying the hypothesis of the claim, and a sequence $(R_0^+, R_0^-), (R_1^+, R_1^-), \ldots$ of pairs of subsets of $A \times B$ satisfying

1. $R_u^+ \cap R_u^- = \emptyset$
2. $|R_u^+ \cup R_u^-| \leq u \cdot t$
3. $(i, j) \in R_u^+ \rightarrow j \in r(i)$

Put $T(0) := \pi'$. Assume that we have $T(0) \ldots, T(i)$. To construct $T(i + 1)$ find in $T(i)$ a sequent $a = \Gamma \longrightarrow \Delta$ such that

$$|T(i)^a| \leq \frac{2}{3}|T(i)| \quad \text{and} \quad |T(i)_a| \leq \frac{2}{3}|T(i)|$$

Now distinguish two cases:

(i) there is $S \subseteq A \times B$ such that

    (a) $S \supseteq R_i^+$ and $S \cap R_i^- = \emptyset$

    (b) S is a graph of a 1–1 function from A to B

    (c) $(i, j) \in S \rightarrow j \in r(i)$

    (d) the evaluation

$$p_{ij} = \begin{cases} 1, & \text{if } (i, j) \in S \\ 0, & \text{if } (i, j) \notin S \end{cases}$$

        makes the sequent $\Gamma \longrightarrow \Delta$ true.

(ii) there is no such S.

In case (i), as $\Gamma \longrightarrow \Delta$ is equivalent to a $\Sigma_1^{S,t}$-formula (the disjunction of the negations of the formulas from Γ and of the formula from Δ), pick $R_{i+1}^+ \subseteq S$, $R_{i+1}^+ \supseteq R_i^+$ and $R_{i+1}^- \cap S = \emptyset$, $R_{i+1}^- \supseteq R_i^-$ satisfying conditions 1–3 (in particular, $|(R_{i+1}^+ \cup R_{i+1}^-) \setminus (R_i^+ \cup R_i^-)| \leq t$) forcing one of the disjuncts of size $\leq t$ of the $\Sigma_1^{S,t}$-formula to be true. Also set

$$T(i + 1) := T(i)_a$$

and label a by 1.

In case (ii) define $R_{i+1}^+ := R_i^+$, $R_{i+1}^- := R_i^-$ and put

$$T(i + 1) := T(i)^a$$

and label a by 0.

We obviously have the following two properties:

4. the nodes of $T(i)$ are labeled by the sequents of π' or by 0, 1
5. $|T(i)| \leq (\frac{2}{3})^i \cdot S$

Let $T(\ell)$ be the last tree in the sequence: That is, $T(\ell)$ is one node. We have

6. $\ell \leq \log_{3/2}(S) \leq t \cdot \log_{3/2}(2) \leq 2t$
7. the node $T(\ell)$ is labeled by 0 and was an initial sequent $\Pi \longrightarrow \Sigma$ of π'
8. for every $S \subseteq A \times B$ satisfying (a)–(d) with $i = \ell$, the evaluation

$$p_{ij} = \begin{cases} 1, & \text{if } (i, j) \in S \\ 0, & \text{if } (i, j) \notin S \end{cases}$$

makes $\Pi \longrightarrow \Sigma$ false.

Condition 6 is obvious (from 5); 7 and 8 are proved by induction on i (the root of every $T(i)$ is labeled by 0 and any S satisfying (a)–(d) makes the sequent in that node of π' false).

It follows that $\Pi \longrightarrow \Sigma$ must be a sequent of the form

$$\longrightarrow p_{ij_1}, \ldots, p_{ij_r}$$

for some $i \in A, r(i) = \{j_1, \ldots, j_r\}$ or of the form

$$\longrightarrow p_{i_1 j}, \ldots, p_{i_s j}$$

for some $j \in B, r^{(-1)}(j) = \{i_1, \ldots, i_s\}$.

But

$$|R_\ell^-| \leq \ell \cdot t$$

So for

$$\ell \cdot t < n^{1/2} \leq r$$

if $i \notin \text{dom}(R_\ell^+)$ we can find $j \in B$ such that $(i, j) \notin R_\ell^-$. This shows (as $\ell \leq 2t$) that if

$$2 \cdot t^2 < n^{1/2}$$

no R_ℓ^+, R_ℓ^- can force that all S considered make the first sequent false. Similarly no S can force the second sequent to fail. Hence we get

$$t \geq n^{1/4} \cdot \frac{1}{2}$$

and

$$S \geq 2^{n^{1/4} \cdot (1/2)}$$

This proves the first part of the theorem.

To prove part 2 note that substituting $S_{ij}^{d,n}$ for p_{ij} in the refutation guaranteed by Lemma 12.2.1 yields a Σ-depth d refutation of $\neg \mathrm{PHP}(A, B)(S_{ij}^{d,n})$ of size $n^{O(\log(n))}$.   Q.E.D.

Corollary 12.2.5. *For any $d \geq 0$ there is a superpolynomial speed-up (m versus $\exp(\exp(\Omega(\log^{1/2} m))))$ between the sequencelike and the treelike Σ-depth d refutations of sets of depth d sequents.*

We conclude this section by a problem.

Open problem. *Is there a constant $c \geq 0$ such that for every $d \geq c$ there is a sequence of sets $(T_n^d)_n$ of depth $\leq c$ sequents of the total length $n^{O(1)}$ satisfying*
   (i) any Σ-depth d, treelike refutation of T_n^d must have the size $\exp(n^{\Omega(1)})$,
   (ii) T_n^d has more than polynomially shorter Σ-depth $d+1$, treelike refutations?

12.3. Complete systems

This section is devoted to the main technical notion in the lower bound proofs for constant-depth Frege systems, the notion of *complete systems*. The complete systems generalize the truth tables or, from a different point of view, the disjunctive normal forms or decision trees. We defer the discussion of these connections until the beginning of section 12.4.

We shall consider two combinatorial situations, related to the pigeonhole principle and to the modular counting principle.

Let n be a natural number and D and R two sets of cardinalities $n + 1$ and n, respectively. $\mathcal{M}^{\mathrm{PHP}}$ denotes the set of partial 1–1 maps from D into R. For $\alpha \in \mathcal{M}^{\mathrm{PHP}}$, $|\alpha|$ denotes the number of the pairs forming α: That is, $|\alpha| = |\mathrm{dom}(\alpha)| = |\mathrm{rng}(\alpha)|$.

Let A be a set of the cardinality $a \cdot n + 1$. A *partial a-partition* of A is a set α of disjoint a-element subsets of A. Let $\mathcal{M}^{\mathrm{MOD}_a}$ be the set of all partial a-partitions of A. For $\alpha \in \mathcal{M}^{\mathrm{MOD}_a}$, $|\alpha|$ denotes the number of the equivalence classes of α.

Further $\mathcal{M}$ denotes either $\mathcal{M}^{\mathrm{PHP}}$ or $\mathcal{M}^{\mathrm{MOD}_a}$.

Definition 12.3.1.
   (a) For $H \subseteq \mathcal{M}$, the norm $||H||$ of H is

$$||H|| := \max_{\alpha \in H} |\alpha|$$

(b) The set $S \subseteq \mathcal{M}$ is a k-complete system *iff the following conditions are fulfilled:*

    1. $\forall \alpha, \beta \in S; \ \alpha \neq \beta \rightarrow \alpha \cup \beta \notin \mathcal{M}$

    2. $\forall \gamma \in \mathcal{M}; \ |\gamma| \leq n - k \rightarrow (\exists \alpha \in S; \ \gamma \cup \alpha \in \mathcal{M})$

    3. $\forall \alpha \in S; \ |\alpha| \leq k.$

    (Note that $S \neq \emptyset$ by 2 for $\alpha = \emptyset$.)

Lemma 12.3.2. *The following are examples of k-complete systems:*

1. *Let $i_0 \in D$ and let $S \subseteq \mathcal{M}^{PHP}$ be*

$$S := \{(i_0, j) \mid j \in R\}$$

Then S is 1-complete.

2. *Let $i_0 \in D$, $j_0 \in R$, and let $S \subseteq \mathcal{M}^{PHP}$ be*

$$S := \{(i_0, j_0)\} \cup \{\{(i_0, j), (i, j_0)\} \mid i_0 \neq i \in D, \ j_0 \neq j \in R\}$$

Then S is 2-complete.

3. *Let $X \subseteq A$ be a set with at most k elements and let S_X be the set of all partial a-partitions α of A such that*

    (a) *X is contained in the support of α $(= \bigcup \alpha)$*

    (b) *every equivalence class of α intersects X*

Then S is k-complete.

Definition 12.3.3.

1. *The set $S \subseteq \mathcal{M}$ refines the set $H \subseteq \mathcal{M}$, $H \lhd S$ in symbols, iff it holds for all $\alpha \in S$*

$$\forall \gamma \in H, \alpha \cup \gamma \notin \mathcal{M} \quad or \quad \exists \gamma \in H, \gamma \subseteq \alpha.$$

2. *For $S, T \subseteq \mathcal{M}$, a common refinement of S and T, $S \times T$ in symbols, is the set*

$$S \times T := \{\alpha \cup \beta \in \mathcal{M} \mid \alpha \in S, \beta \in T\}$$

3. *A projection $S(H)$ of $H \subseteq \mathcal{M}$ on $S \subseteq \mathcal{M}$ is the set*

$$S(H) := \{\alpha \mid \exists \gamma \in H, \gamma \subseteq \alpha\}.$$

Lemma 12.3.4. *Let $H \lhd S \lhd T$ for $H, S, T \subseteq \mathcal{M}$ and assume that S is k-complete and that $\|T\| + k \leq n$.*

    Then

$$H \lhd T$$

Proof. By the k-completeness of S and by $||T|| + k \leq n$

$$\forall \beta \in T \exists \alpha' \in S, \alpha' \cup \beta \in \mathcal{M}$$

hence by $S \lhd T$ it holds that

$$\forall \beta \in T \exists \alpha \in S, \alpha \subseteq \beta \qquad (*)$$

Assume now that $\gamma \in H$, $\beta \in T$ are such that $\gamma \cup \beta \in \mathcal{M}$. By $(*)$ there is $\alpha \in S$ s.t. $\alpha \subseteq \beta$. Thus $\gamma \cup \alpha \in \mathcal{M}$ too, and hence $\gamma' \subseteq \alpha$ for some $h' \in H$, by $H \lhd S$. We proved $\gamma' \subseteq \beta$ as it was required. Q.E.D.

Lemma 12.3.5. *Let $S, T \subseteq \mathcal{M}$ be such that S is k-complete, T is ℓ-complete, $||S|| + \ell \leq n$, $||T|| + k \leq n$, and $k + \ell \leq n$.*

Then the following two conditions hold:

1. *$S \lhd S \times T$, $T \lhd S \times T$.*
2. *$S \times T$ is $(k + \ell)$-complete.*

Proof. To prove the first part let $\alpha \in S$ and $\alpha \cup (\beta \cup \gamma) \in \mathcal{M}$ hold for some element $\beta \cup \gamma \in S \times T$. By k-completeness of S then $\alpha = \beta$ so $\alpha \subseteq \beta \cup \gamma$. This shows $S \lhd S \times T$. $T \lhd S \times T$ follows identically.

To prove the second part assume $\beta \cup \gamma$, $\beta' \cup \gamma' \in \mathcal{M}$ for $\beta \cup \gamma$, $\beta' \cup \gamma'$ two elements of $S \times T$. Then either $\beta \neq \beta'$ in which case $\beta \cup \beta' \notin \mathcal{M}$ by the completeness of S or $\gamma \neq \gamma'$ (i.e., $\gamma \cup \gamma' \notin \mathcal{M}$ by the completeness of T resp.). In both cases thus $(\beta \cup \gamma) \cup (\beta' \cup \gamma') \notin \mathcal{M}$. This proves condition (i) of Definition 12.3.1

To verify (ii) let $|\alpha| + k + \ell \leq n$. Then, as $||S|| \leq k$ and $||T|| \leq \ell$, by the completeness of S, T there are $\beta \in S$ and $\gamma \in T$ for which $\alpha \cup (\beta \cup \gamma) \in \mathcal{M}$.

Condition (iii) holds too as clearly $||S \times T|| \leq ||S|| + ||T|| \leq k + \ell$. Q.E.D.

Lemma 12.3.6. *Let $H, S, T \subseteq \mathcal{M}$, let S be k-complete and T be ℓ-complete. Assume that $||S|| + \ell \leq n$, $||T|| + k \leq n$ and that $H \lhd S \lhd T$.*

Then

$$T(S(H)) \doteq T(H) \quad \text{and} \quad T(S) = T$$

and

$$S(H) = S \text{ iff } T(H) = T$$

Proof. $T(S(H)) \subseteq T(H)$ is obvious from the definition of a projection.

To see $T(H) \subseteq T(S(H))$ let $\beta \in T(H)$ and $\gamma \subseteq \beta$ for some $\gamma \in H$. Then from $(*)$ in the proof of Lemma 12.3.4 $\alpha \subseteq \beta$ for some $\alpha \in S$. Hence $\gamma \cup \alpha \in \mathcal{M}$ and, as $H \lhd S$, we have $\gamma' \subseteq \alpha$ for some $\gamma' \in H$. Thus also $\gamma' \subseteq \alpha \subseteq \beta$, that is, $\beta \in T(S(H))$.

$T(S) = T$ follows by taking $H := \{\emptyset\}$.

Finally, to establish equivalence first assume $S(H) = S$. By the first part $T(S) = T$ and $T(S(H)) = T(H)$, and so $T(H) = T$.

For the opposite implication assume $T(H) = T$, and take $\alpha \in S$. By the ℓ-completeness of T (and by $|\alpha| + \ell \leq ||S|| + \ell \leq n$) $\alpha \cup \beta \in \mathcal{M}$ holds for some $\beta \in T$. By the assumption $\gamma \subseteq \beta$ holds for some $\gamma \in H$. Hence $\alpha \cup \gamma \in \mathcal{M}$ and $H \lhd S$ implies that $\gamma' \subseteq \alpha$ some $\gamma' \in H$. Thus $\alpha \in S(H)$.   Q.E.D.

Lemma 12.3.7.

1. *Let $S, H_i \in \mathcal{M}$, $i \in I$ be arbitrary sets.*
   Then:

$$S\left(\bigcup_I H_i\right) = \bigcup_I S(H_i)$$

2. *Let $S \subseteq \mathcal{M}$ be a k-complete set and $S_0, S_1 \subseteq S$ its two disjoint subsets, and let $T \subseteq \mathcal{M}$.*
   Then it holds that

$$T(S_0) \cap T(S_1) = \emptyset$$

3. *Let $S, T \subseteq \mathcal{M}$ be two sets such that $S \lhd T$, S is k-complete and $||T|| + k \leq n$. Let $S_0 \subseteq S$ be any set.*
   Then

$$T(S \setminus S_0) = T \setminus T(S_0)$$

Proof. The first two parts follow straightforwardly from the definition of a projection. In the third part Lemma 12.3.6 implies

$$T(S) = T$$

By parts 1 and 2

$$T(S \setminus S_0) = T \setminus T(S_0)$$

                                                              Q.E.D.

In the applications of the complete systems in the next section we shall move from a situation with n, D, R, $\mathcal{M}^{PHP}$ (or n, A, $\mathcal{M}^{MOD_a}$) and some $H, S, T, \ldots \subseteq \mathcal{M}$ to a situation where some information about a partial map (resp. a partial partition) will be available. The next definition and lemma formalize this.

Definition 12.3.8. *Let $\alpha, \rho \in \mathcal{M}$. Define the restriction α^ρ of α by ρ*

$$\alpha^\rho := \begin{cases} \alpha \setminus \rho, & \text{if } \alpha \cup \rho \in \mathcal{M} \\ \text{undefined}, & \text{if } \alpha \cup \rho \notin \mathcal{M} \end{cases}$$

Further put $H^\rho := \{\alpha^\rho \mid \alpha \in H\}$ *for* $H \subseteq \mathcal{M}$, $D^\rho := D \setminus \mathrm{dom}(\rho)$ *and* $R^\rho := R \setminus \mathrm{rng}(\rho)$ *in case of the PHP (and* $A^\rho := A \setminus \mathrm{supp}(\rho)$ *in case of the* MOD_a), *and* $n^\rho := n - |\rho|$.

Note that α^ρ *undefined* is not the same as $\alpha^\rho = \emptyset$. The *support* $\mathrm{supp}(\rho)$ is the set $\bigcup_{e \in \rho} e$. For $\rho \in \mathcal{M}^{PHP}$ this is $\mathrm{dom}(\rho) \cup \mathrm{rng}(\rho)$.

Lemma 12.3.9. *Let* $H, S, K \subseteq \mathcal{M}$ *and let* $\rho \in \mathcal{M}$ *be arbitrary.*
Then the following three conditions hold:

1. *if* $H \lhd S$ *then* $H^\rho \lhd S^\rho$,
2. *if* S *is* k-complete and $|\rho| + k \leq n$ *then* S^ρ *is also* k-complete,
3. *if* $K = S(H)$ *and* $H \lhd S$ *then* $K^\rho = S^\rho(H^\rho)$.

Proof. For the first part let $\gamma \in H$, $\alpha \in S$ satisfy $\gamma^\rho \cup \alpha^\rho \in \mathcal{M}^\rho$. Then $\gamma \cup \alpha \in \mathcal{M}$ and thus $\gamma' \subseteq \alpha$ for some $\gamma' \in H$. Clearly $(\gamma')^\rho \subseteq \alpha^\rho$.

For the second part assume that $\alpha_1^\rho \cup \alpha_2^\rho \in \mathcal{M}^\rho$ for $\alpha_1, \alpha_2 \in S$. Then also $\alpha_1 \cup \alpha_2 \subseteq \mathcal{M}$ and hence $\alpha_1 = \alpha_2$, that is, also $\alpha_1^\rho = \alpha_2^\rho$.

Now let $|\gamma| + k \leq (n)^\rho$ for some $\gamma \in \mathcal{M}^\rho \subseteq \mathcal{M}$. Since $|\rho| = n - (n)^\rho$ we get $|\gamma \cup \rho| + k \leq n$. By the k-completeness of S then for some $\alpha \in S$: $(\gamma \cup \rho) \cup \alpha \in \mathcal{M}$. But as $\gamma = \gamma^\rho$ clearly then $\gamma \cup \alpha^\rho \in \mathcal{M}^\rho$.

The inequality $||S^\rho|| \leq ||S|| \leq k$ is trivial.

For the last part of the lemma let $\kappa^\rho \in K^\rho$, for some $\kappa \in K$. Then we have $\kappa \in S$ and $\gamma \subseteq \kappa$ for some $\gamma \in H$, which implies $\kappa^\rho \in S^\rho$ and also $\gamma^\rho \subseteq \kappa^\rho$, and thus $\kappa^\rho \in S^\rho(H^\rho)$.

Finally let $\alpha^\rho \in S^\rho$, $\gamma^\rho \in H^\rho$ for which $\gamma^\rho \subseteq \alpha^\rho$. Then $\gamma \cup \alpha \in \mathcal{M}$ and $\gamma' \subseteq \alpha$ for some $\gamma' \in H$, as $H \lhd S$. Therefore, $\alpha \in S(H)$, which entails $\alpha \in K$, and so $\alpha^\rho \in K^\rho$. Q.E.D.

We will want an object $\rho \in \mathcal{M}$ to have certain particular properties. It is difficult to construct suitable ρ explicitly, and instead we shall show that such ρ exists by a counting argument.

The following lemma is a crucial technical result, a statement analogous to the switching Lemma 3.1.11 for this situation. See the discussion at the beginning of Section 12.4 and also Lemma 15.2.2.

Lemma 12.3.10. *Let* $H_1, \dots, H_N \subseteq \mathcal{M} = \mathcal{M}^{PHP}$, *such that for all* $i \leq N$, $||H_i|| \leq t \leq s$. *Let* $w \leq n$. *Assume that*

$$\frac{w^s}{(n+1-w)^{4s}t^{3s}} > N$$

Then there is $\rho \in \mathcal{M}$, $|\rho| = w$, *such that for every* $i \leq N$ *there is* $S_i \subseteq \mathcal{M}^\rho$ *such that*

1. S_i *is* $2s$-complete (w.r.t. $\mathcal{M}^\rho$)
2. $H_i^\rho \lhd S_i$, *all* $i \leq N$.

In particular, for $0 < \delta < \epsilon < (1/7)$, $w = n - n^\epsilon$ and $t = s = n^\delta$, the preceding inequality is satisfied for any $N < 2^{n^\delta}$. If $w = n - n^\epsilon$ and $N = n^{O(1)}$ then the inequality can be satisfied with $t = s = O(1)$ sufficiently large.

Proof. First we shall consider the case when $N = 1$. Denote $H := H_1$. We shall divide the proof into several steps.

1. Enumerate by $h^1, h^2, h^3, \ldots$ the elements of H, where $H \subseteq \mathcal{M}$, $\|H\| \le t \le s$.

2. We shall define a game played by two players. They construct a sequence of maps $\delta_0 \subseteq \delta_1 \subseteq \ldots$, $\delta_0 = \emptyset$, $\delta_i \in \mathcal{M}$.

   Assume that during a play the sequence $\delta_0 \subseteq \delta_1 \subseteq \ldots \subseteq \delta_\ell$ is already constructed.

   (a) Player I plays $h_{\ell+1}$: the first $h^{i_{\ell+1}}$ in the enumeration $h^1, h^2, \ldots$ such that $h_{\ell+1} \cup \delta_\ell \in \mathcal{M}$. Hence its move is completely determined by δ_ℓ and by the enumeration.

   (b) Assume that $h_{\ell+1} = p_{i_1 j_1} \ldots p_{i_t j_t}$. Player II chooses an arbitrary $\subseteq$-minimal $\delta_{\ell+1} \supseteq \delta_\ell$ such that $\{i_1, \ldots, i_t\} \subseteq \mathrm{dom}(\delta_{\ell+1})$ and $\{j_1, \ldots, j_t\} \subseteq \mathrm{rng}(\delta_{\ell+1})$.

   The play is finished iff either player I cannot make a move or player II constructed $\delta_{\ell+1} \supseteq h_{\ell+1}$.

3. Define the set S by

$$S := \{\delta_{k+1} \mid \text{some } \delta_0 \subseteq \delta_1 \subseteq \ldots \subseteq \delta_{k+1} \text{ is a sequence}$$

$$\text{constructed in a finished play}\}$$

Claim.

   (a) S is $\|S\|$-complete.

   (b) $H \lhd S$.

Assume that $\delta \ne \delta' \in S$. Let $\delta_\ell \ne \delta'_\ell$ be the first move of II that was different in the two plays. Clearly (as $\delta_\ell, \delta'_\ell$ are $\subseteq$-minimal) $\delta_\ell \cup \delta'_\ell \notin \mathcal{M}$, so $\delta \cup \delta' \notin \mathcal{M}$.

Take $\alpha \in \mathcal{M}$ such that $|\alpha| + \|S\| \le n$. Player II will follow the following strategy: On $\mathrm{supp}(\alpha)$ answer according to α; otherwise play arbitrarily but consistently with α. The inequality $|\alpha| + \|S\| \le n$ implies that he can always move. Clearly $\alpha \cup \delta \in \mathcal{M}$ for any output δ of such a game.

To see the second part of the claim notice that S is a refinement of H by definition when the play terminates.

In general it is not true that $\|S\| \le 2s$. We shall use map $\rho \in \mathcal{M}$ and show such an inequality for H^ρ. By Lemma 12.3.9 the claim will remain valid after the restriction by ρ.

4. Let $h_1, h_2, \ldots, h_{k+1}$ be the elements of H played by player I in a play $\delta_0 \subseteq \delta_1 \subseteq \ldots \subseteq \delta_{k+1}$. We shall call h_i the ith *critical map*.

   Now we consider the game played with H^ρ instead of H, but we shall use the same ordering on H^ρ induced by taking the first element of H among all h^i identified by restriction ρ.

5. Suppose that in a play we obtained the critical maps $h_1^\rho, \ldots, h_{k+1}^\rho$ and $\delta_0 \subseteq \delta_1 \subseteq \ldots \subseteq \delta_{k+1}$. The pairs of $h_i^\rho \setminus \delta_{i-1}$ will be called *critical pairs*.

6. We wish to show that for some $\rho \in \mathcal{M}$, $|\rho| = w$, in every play will occur less than s critical pairs. That would imply $\|S\| \leq 2s$.

   Assume, for the sake of contradiction, that for every such ρ there is a play in which at least s critical pairs occur.

   We assign a set of parameters for each such a play with $\geq s$ critical pairs. Denote by

$$\delta_0' := \emptyset$$

$$\delta_i' := \delta_i \setminus \delta_{i-1}$$

$$\delta_i^* = \big(\text{dom}(h_i) \times \text{rng}(h_i) \big) \cap \delta_i' = \big(\text{dom}(h_i^\rho) \times \text{rng}(h_i^\rho) \big) \cap \delta_i'$$

Let

$$s_i = |h_i^\rho \setminus \delta_{i-1}| \quad \text{for } 1 \leq i \leq k$$

be the number of critical pairs in h_i^ρ and put

$$s_{k+1} = s - s_1 \ldots - s_k$$

Also take $d_i = |\delta_i^*|$ and note that

$$|\delta_i'| = 2s_i - d_i$$

The sets $T_1, \ldots, T_{k+1} \subseteq \{1, \ldots, t\}$ are defined as follows. Let $i \leq k$ be fixed and let

$$h_i = \{e_1, \ldots, e_{t_i}\}$$

Then

$$h_i^\rho \setminus \delta_{i-1} = \{e_j\}_{j \in T_i}$$

The map h_i and the set T_i thus determine the critical pairs of h_i^ρ. We define T_{k+1} in a similar way, but we take only the initial part of $h_{k+1}^\rho \setminus \delta_k'$ consisting of the first s_{k+1} pairs.

7. Let $\gamma_1, \ldots, \gamma_k$ be the partial one-to-one maps with the support contained in $\{1, \ldots, t\}$ defined by the following. For $i \leq k$ fixed suppose

$$\text{dom}(h_i) = \{a_1, \ldots, a_{t_i}\} \quad \text{and} \quad a_1 < \cdots < a_{t_i}$$

$$\text{rng}(h_i) = \{b_1, \ldots, b_{t_i}\} \quad \text{and} \quad b_1 < \cdots < b_{t_i}$$

Then put

$$\delta_i^* = \big\{ (a_\ell, b_{\gamma_i(\ell)}) \mid \ell \in \mathrm{dom}(\gamma_i) \big\}.$$

Hence h_i and γ_i determine δ_i^* for all $i \leq k$.

8. Let τ be the union of ρ with the first s critical pairs. Hence $\tau \in \mathcal{M}$ and $|\tau| = w + s$.

9. We also define maps $\beta_1, \ldots, \beta_k$ from $\{1, \ldots, 2(s_i - d_i)\}$ into $\{1, \ldots, n + 1 - w - s\}$ such that (for $i = 1, \ldots, k$) β_i determines $\delta_i' - \delta_i^*$. Let

$$\{a_1', \ldots, a_{s_i - d_i}'\} = \big(\mathrm{dom}(\delta_i') - \mathrm{dom}(\delta_i^*) \big) \cap \mathrm{dom}(h_i),$$

with $a_1' < \cdots < a_{s_i - d_i}'$ and

$$\{b_1', \ldots, b_{s_i - d_i}'\} = \big(\mathrm{rng}(\delta_i') - \mathrm{rng}(\delta_i^*) \big) \cap \mathrm{rng}(h_i)$$

with $b_1' < \cdots < b_{s_i - d_i}'$.
Also let

$$D \backslash \mathrm{dom}(\tau) = \{u_1, \ldots, u_{n+1-w-s}\} \quad \text{and} \quad R \backslash \mathrm{rng}(\tau) = \{v_1, \ldots, v_{n-w-s}\}$$

Then

$$\delta_i(a_j') = v_{\beta_i(j)} \quad \text{for } j = 1, \ldots, s_i - d_i$$

and

$$\delta_i^{-1}(b_j') = u_{\beta_i(s_i - d_i + j)} \quad \text{for } j = 1, \ldots, s_i - d_i$$

10. The parameters π will be the pair (τ, π_0), where π_0 is the tuple consisting of the following objects

$$T_1, \ldots, T_{k+1}; \ \gamma_1, \ldots, \gamma_k; \ \beta_1, \ldots, \beta_k$$

11. The number of $T_1, \ldots, T_{k+1}$ is bounded by

$$\leq \binom{t_1}{s_1} \cdots \binom{t_{k+1}}{s_{k+1}} \leq \binom{t}{s_1} \cdots \binom{t}{s_{k+1}} \leq t^s$$

The number of $\gamma_1, \ldots, \gamma_k$ is clearly at most

$$\leq \binom{t_1}{d_1} \cdot t_1^{d_1} \cdot \cdots \cdot \binom{t_k}{d_k} \cdot t_k^{d_k} \leq t^{2s}$$

The number of $\beta_1, \ldots, \beta_k$ is also obviously bounded by the product

$$(n + 1 - w - s)^{2(s_1 - d_1)} \cdot \cdots \cdot (n + 1 - w - s)^{2(s_k - d_k)} \leq (n + 1 - w - s)^{2s}$$

Hence we get an upper bound to the number of tuples π_0

$$\leq t^{3s} \cdot (n + 1 - w - s)^{2s}$$

12. Take the parameters π. We shall show that they, in fact, determine the map ρ. This is done by the following process:

   (a) Put $\tau_1 := \tau$; τ_1 determines h_1 as the first h^{i_1} consistent with it. Knowing h_1 and T_1 yields the set κ_1 of the critical pairs in h_1. From κ_1 and the maps γ_1 and β_1 we reconstruct the first move δ_1 of player II.

   (b) Put $\tau_2 := \delta_1 \cup (\tau_1 \setminus \kappa_1)$. The map τ_2 determines h_2 as the first h^{i_2} consistent with it. Hence, as earlier, with the help of T_2 we get the set κ_2 of the critical pairs in h_2 and further from γ_2, β_2 also the map δ_2.

   (c) Repeating the process $k + 1$-times allows us to reconstruct the set of the first s critical pairs $\bigcup_{i \le k+1} \kappa_i$. But then

   $$\rho = \tau \setminus \bigcup_{i \le k+1} \kappa_i$$

13. As π defines ρ uniquely, the number of such parameters π cannot be smaller than the number of possible ρ. Define

   $$A := |\{\rho \in \mathcal{M} \mid |\rho| = w\}|$$

   and

   $$B := |\{\tau \in \mathcal{M} \mid |\tau| = w + s\}|$$

   and compute the number

   $$\frac{A}{B \cdot t^{3s}(n + 1 - w - s)^{2s}}$$

   which is equal to

   $$\frac{\binom{w+s}{s}}{\binom{n-w}{s}\binom{n+1-w}{s}s! \cdot t^{3s}(n + 1 - w - s)^{2s}}$$

   as $\binom{w+s}{s}$ is the number of ρ contained in one τ while $\binom{n-w}{s}\binom{n+1-w}{s}s!$ is the number of τ extending a given ρ.
   By a simple calculation we get that this fraction is at least

   $$\ge \frac{w^s}{(n + 1 - w)^{4s}t^{3s}}$$

   which is

   $$> N \ge 1$$

by the hypothesis of the lemma. Hence the number of possible maps ρ is bigger than the number of possible parameters π. Thus there must be ρ for which the assumption that there is at least one play with $\ge s$ critical pairs fails.

Now return to the case $N \geq 1$. Assuming that for each ρ there is at least one $i \leq N$ such that there is a play on h_i^ρ with $\geq s$ critical pairs, we may reconstruct ρ knowing i and the parameters π for such a play.

This shows that, in fact, the same argument works as long as

$$\frac{w^s}{(n+1-w)^{4s} t^{3s} N} > 1$$

Q.E.D.

We conclude this section by stating an analogous lemma for the case $\mathcal{M} = \mathcal{M}^{MOD_a}$. Its proof follows identical lines.

Lemma 12.3.11. *Let $a \geq 2$. Let $H_1, \ldots, H_N \subseteq \mathcal{M} = \mathcal{M}^{MOD_a}$, such that for all $i \leq N$, $\|H_i\| \leq t \leq s$. Let $w \leq n$. Assume that*

$$\frac{w^s}{[t \cdot (n-w)]^{c \cdot s}} > N$$

where $c > 1$ is a constant depending on a only.

Then there is $\rho \in \mathcal{M}$, $|\rho| = w$, such that for every $i \leq N$ there is $S_i \subseteq \mathcal{M}^\rho$ such that

    1. S_i is $(a \cdot s)$-complete (w.r.t. $\mathcal{M}^\rho$).

    2. $H_i^\rho \lhd S_i$, all $i \leq N$.

In particular, for $0 < \delta < \epsilon$ sufficiently small (depending on a) and $w = n - n^\epsilon$ and $t = s = n^\delta$, the preceding inequality is satisfied for any $N < 2^{n^\delta}$. If $w = n - n^\epsilon$ and $N = n^{O(1)}$ then the inequality can be satisfied with $t = s = O(1)$ sufficiently large.

12.4. k-evaluations

This section introduces the crucial notion of k-evaluation. First, however, we shall discuss the notion of a complete system as promised at the beginning of the previous section. We shall confine discussion to the situation with the Boolean variables $x_1, \ldots, x_n$ and the notion of a complete system w.r.t. the set $\mathcal{M}^{\text{trivi}}$ of all total truth assignments to the variables. That is a simpler situation than those considered in Section 12.3 but well explains the intuition behind the definitions and the statements of that section. See Section 15.2 also.

A complete system w.r.t. $\mathcal{M}^{\text{trivi}}$ is a set S of partial truth assignments to $x_1, \ldots, x_n$ such that

    1. any two $\alpha \neq \beta$ from S are incompatible

    2. every total truth assignment $\alpha \in \mathcal{M}^{\text{trivi}}$ contains some $\beta \in S$.

We identify the elements $\alpha \in S$ with the blocks

$$[\alpha] := \left\{ \beta \in \mathcal{M}^{\text{trivi}} \mid \alpha \subseteq \beta \right\}$$

of total assignments extending α. Then S is a complete system iff the set of the blocks $[\alpha]$, $\alpha \in S$, is a partition of $\mathcal{M}^{\text{trivi}}$.

Let $\phi(x_1, \ldots, x_n)$ be a propositional formula. The truth table of ϕ is simply a map assigning to every $\alpha \in \mathcal{M}^{\text{trivi}}$ the truth value $\phi(\alpha)$. If S is a complete system such that

$$\phi^{(-1)}(0) = \bigcup_{\alpha \in S_0} [\alpha] \quad \text{and} \quad \phi^{(-1)}(1) = \bigcup_{\alpha \in S_1} [\alpha]$$

for some partition $S_0 \cup S_1 = S$ of S then the truth table for ϕ can be abbreviated by a map from S into $\{0, 1\}$: map $\alpha \in S$ to 0 iff $\alpha \in S_0$.

Moreover, the formulas ϕ and $\neg\phi$ are expressible in particular disjunctive normal forms:

$$\phi \equiv \bigvee_{\alpha \in S_1} \alpha \subseteq \bar{x} \quad \text{and} \quad \neg\phi \equiv \bigvee_{\alpha \in S_0} \alpha \subseteq \bar{x}$$

where $\alpha \subseteq \bar{x}$ abbreviates the maximal conjunction of literals made true by α. Note that a complete system is just a particular disjunctive normal form of the formula 1, namely such a form in which the disjuncts are mutually incompatible.

Disjunctive normal forms for ϕ and $\neg\phi$ with this property are obtained from any decision tree for ϕ (see Section 3.1), and conversely, a complete system S allowing the expression of ϕ and $\neg\phi$ as earlier also yields a decision tree for ϕ (see Lemma 3.1.13). The depth of that decision tree will be $\leq \|S\|^2$. Hence it is of interest to have complete systems of small norm.

If $\phi = x_i$ then the complete system $S_{x_i} = \{x_i \mapsto 0, x_i \mapsto 1\}$ of the norm 1 allows the expression of ϕ and $\neg\phi$ as previously. If we have such a system S_ϕ for ϕ then it obviously works for $\neg\phi$ too. Hence the only difficult case in the construction of a complete system S_ϕ of small norm by induction on the depth of ϕ is the case when $\phi = \bigvee_i \phi_i$. A system S allows an expression of ϕ and $\neg\phi$ as earlier iff the sets $\phi^{(-1)}(0)$ and $\phi^{(-1)}(1)$ are unions of some blocks determined by S. However, having systems S_{ϕ_i} for ϕ_i allows an expression

$$\phi^{(-1)}(1) = \bigcup_{\alpha \in H} [\alpha]$$

of ϕ as a union of blocks, where

$$H = \bigcup_i \left\{ \alpha \in S_{\phi_i} \mid \alpha \in \phi_i^{(-1)}(1) \right\}$$

and it has the norm $\|H\| = \max_i \|S_{\phi_i}\|$ small. The problem is that H is not necessarily a subset of a complete system. But if S is a system *refining* H

$$\forall \alpha \in S, \ (\exists \beta \in H, \ \beta \subseteq \alpha) \vee (\forall \beta \in H, \ \beta \perp \alpha)$$

then

$$\phi \equiv \bigvee_{\alpha \in S_1} \alpha \subseteq \bar{x} \quad \text{and} \quad \neg\phi \equiv \bigvee_{\alpha \in S_0} \alpha \subseteq \bar{x}$$

where

$$S_1 = \{\alpha \in S \mid \exists \beta \in H,\ \beta \subseteq \alpha\}$$

and

$$S_0 = \{\alpha \in S \mid \forall \beta \in H,\ \beta \perp \alpha\}$$

A lemma stating that there is a partial truth assignment ρ such that there is a small norm S refining H^ρ thus replaces the switching Lemma 3.1.11 in this situation (see Lemma 15.2.2 and Lemmas 12.3.10 and 12.3.11).

Assume that we have for formulas ϕ complete systems S_ϕ allowing the expression of ϕ and $\neg\phi$ as earlier. This allows us to assign to ϕ the Boolean algebra $\exp(S_\phi)$ of the subsets of S_ϕ with a value H_ϕ

$$H_\phi := \left\{\alpha \in S_\phi \mid [\alpha] \subseteq \phi^{(-1)}(1)\right\}$$

We may think of H_ϕ as of those $\alpha \in S$ forcing ϕ true (see Section 12.7). Hence ϕ is a tautology iff $H_\phi = S_\phi$ iff all $\alpha \in S_\phi$ force ϕ true.

Now, in the particular definition of k-complete systems w.r.t. $\mathcal{M}^{PHP}$ no element α intuitively forces PHP true. Hence one expects to get an "evaluation" of formulas in which PHP will not be true. The notion of k-evaluation formalizes these ideas.

Definition 12.4.1. *Let Γ be a set of formulas closed under subformulas. The atoms of the formulas from Γ are p_{ij} if $\mathcal{M} = \mathcal{M}^{PHP}$ and p_X if $\mathcal{M} = \mathcal{M}^{MOD_a}$. A k-evaluation of Γ is a pair of mappings (H, S)*

$$H : \Gamma \to \exp(\mathcal{M}) \quad S : \Gamma \to \exp(\mathcal{M})$$

satisfying the conditions
1. *$\forall \varphi \in \Gamma,\ H_\varphi \subseteq S_\varphi \subseteq \mathcal{M}$*
2. *$\forall \varphi \in \Gamma,\ S_\varphi$ is k-complete*
3. *$H_0 = \emptyset,\ H_1 = \{\emptyset\},\ S_0 = S_1 = \{\emptyset\}$*
4. *For $\mathcal{M} = \mathcal{M}^{PHP}$:*

$$H_{p_{ij}} = \{\{(i, j)\}\}$$

$$S_{p_{ij}} = \left\{\{(i, j'), (i', j)\} \mid i' \neq i,\ j' \neq j\right\} \cup \{\{(i, j)\}\}$$

For $\mathcal{M} = \mathcal{M}^{MOD_a}$:

$$H_{p_X} = \{X\}$$

$$S_{p_X} = \left\{\alpha \in \mathcal{M} \mid X \subseteq \left(\bigcup \alpha\right) \wedge (\forall Y \in \alpha,\ Y \cap X \neq \emptyset)\right\}$$

5. *$\forall \neg\varphi \in \Gamma,\ H_{\neg\varphi} = S_\varphi \setminus H_\varphi$ and $S_{\neg\varphi} = S_\varphi$*

6.

$$\forall \varphi \in \Gamma, \ \varphi = \bigvee_{i \in I} \varphi_i \ \rightarrow \ \bigcup_{i \in I} H_{\varphi_i} \lhd S_\varphi \ \wedge \ H_\varphi = S_\varphi \left(\bigcup_{i \in I} H_{\varphi_i} \right)$$

Note that by Lemma 12.3.2 the system $S_{p_{ij}}$ (resp. S_{p_X}) from 4 is 2-complete (resp. *a*-complete).

In the sequel we denote by p_X the atoms of both forms p_{ij} and p_X to simplify the notation. In the former case $X = \{i, j\}$. The next lemma follows from Lemma 12.3.9 and from the observation that the set $\{\emptyset\}$ is refined by any other set.

Lemma 12.4.2. *Let $\rho \in \mathcal{M}$ and p_X be an atom. Define*

$$(p_X)^\rho = \begin{cases} 1 & \text{if } X \subseteq \rho \\ 0 & \text{if } X \cup \rho \notin \mathcal{M} \\ p_X & \text{otherwise} \end{cases}$$

For φ a formula define φ^ρ to be a formula obtained from φ by replacing all atoms p_X by $(p_X)^\rho$ (but performing no further simplifications). For Γ a set of formulas, put $\Gamma^\rho := \{\varphi^\rho \mid \varphi \in \Gamma\}$.

Then for any set of formulas Γ, any $\rho \in \mathcal{M}$ such that $|\rho| + k \leq n$, and any k-evaluation (H, S) of Γ, the pair (H^ρ, S^ρ) is a k-evaluation of Γ^ρ.

The following is a crucial theorem.

Theorem 12.4.3. *Let $d \geq 1$ and $0 < \epsilon$ be sufficiently small, and let $0 < \delta < \epsilon^{d-1}$. Let Γ be a set of formulas of depth d closed under the subformulas, and assume that $|\Gamma| < 2^{n^\delta}$.*

Then there exist $\rho \in \mathcal{M}$, $|\rho| \leq n - n^{\epsilon^{d-1}}$, and $k \leq O(n^\delta)$ such that there exists a k-evaluation of Γ^ρ.

Proof. For $d = 1$ the formulas in Γ are atoms and constants, and we have their evaluation by 2 and 3 of Definition 12.4.2. In this case $\rho = \emptyset$.

Assume now that the lemma is true for $d \geq 1$ and let Γ be a set of formulas of depth $d + 1$ that is closed under the subformulas. Let Γ' be the depth $\leq d$ formulas from Γ. Let $0 < \epsilon^d (= \epsilon^{d+1-1})$ be given.

By the assumption we have $\rho' \in \mathcal{M}$, $|\rho'| \leq n - n^{\epsilon^{d-1}}$, $k \leq O(n^\delta)$, and a k-evaluation (H', S') of $(\Gamma')^{\rho'}$.

Let $m = n - |\rho'|$, that is, $m \geq n^{\epsilon^{d-1}}$. We shall extend ρ' by a suitable ρ'' to form ρ. By Lemma 12.4.2 the restrictions of $\mathcal{M}^{\rho'}$ and S' by ρ'' will be $\leq k$-evaluations of $((\Gamma')^{\rho'})^{\rho''}$ again; thus we only need to find ρ'' so that we can extend this evaluation to the whole Γ.

Consider the case $\mathcal{M} = \mathcal{M}^{\text{PHP}}$. For a negation it is obvious for any ρ''. For a disjunction $\phi = \bigvee_i \phi_i$ apply Lemma 12.3.10 with n, D, R, and H replaced by

m, $D^{\rho'}$, $R^{\rho'}$, and $\bigcup_i H'_{\phi_i}$, and taking $t = s = n^\delta$, $w = m - m^\epsilon$. The following inequalities hold

$$n^\delta = n^{\epsilon^{d-1} \cdot (\delta/\epsilon^{d-1})} \leq m^{(\delta/\epsilon^{d-1})} \quad \text{and} \quad \frac{\delta}{\epsilon^{d-1}} < \epsilon$$

By Lemma 12.3.10 then there exist $\rho'' \in \mathcal{M}^{\rho'}$ and $S \subseteq \mathcal{M}^{\rho'\rho''}$ such that

$$\bigcup_i (H'_{\varphi_i})^{\rho''} \lhd S \quad \text{and} \quad |\rho''| = m - m^\epsilon$$

For such ρ'' extend the evaluation (H', S') to φ by defining

$$S_\varphi = S \quad \text{and} \quad H_\varphi = S\left(\bigcup_i (H'_{\varphi_i})^{\rho''}\right)$$

From the hypothesis that $|\Gamma| < 2^{n^\delta}$ it follows that there are $< 2^{n^\delta}$ such disjunctions and hence by Lemma 12.3.10 there is at least one ρ'' with these properties satisfied for all such $\varphi \in \Gamma$. This is enough as we also have

$$|\rho| = |\rho' \cup \rho''| \leq n - m + m - m^\epsilon = n - m^\epsilon \leq n - n^{\epsilon^d}.$$

The case of $\mathcal{M} = \mathcal{M}^{\text{MOD}_a}$ is treated analogously using Lemma 12.3.11.

Q.E.D.

Lemma 12.4.4. *Let*

$$R : \frac{\gamma_1, \ldots, \gamma_t}{\gamma_0}$$

be a Frege rule.

Then there is a constant r satisfying the following condition: Whenever

$$\frac{\gamma_1(\psi_1, \ldots, \psi_m), \ldots, \gamma_t(\psi_1, \ldots, \psi_m)}{\gamma_0(\psi_1, \ldots, \psi_m)}$$

is an instance of the rule and (H, S) is a k-evaluation of all subformulas occurring in the instance with $k \leq n/r$, and if $H_{\theta_i} = S_{\theta_i}$ for $i = 1, \ldots, t$ and $\theta_i = \gamma_i(\psi_1, \ldots, \psi_m)$, then also

$$H_{\theta_0} = S_{\theta_0}$$

Proof. Let R be given and take r to be greater than the number of subformulas in R.

Let (H, S) be a $k = (n/r)$-evaluation of the set Γ of formulas of the form $\gamma(\psi_1, \ldots, \psi_m)$, where γ is a subformula of some γ_i, $i = 0, \ldots, t$. Suppose that $H_{\theta_i} = S_{\theta_i}$ for $1 \leq i \leq t$.

Take T to be a common refinement of all S_γ, $\gamma \in \Gamma$. Such system exists by Lemma 12.3.5 and is $(n(r-1)/r)$-complete. Hence $||S_\gamma|| + ||T|| \leq n$ for every $\gamma \in \Gamma$.

Assume first that $\neg\gamma \in \Gamma$. Then

$$H_{\neg\gamma} = S_\gamma \setminus H_\gamma$$

and so by Lemma 12.3.7

$$T(H_{\neg\gamma}) = T \setminus T(H_\gamma)$$

Now suppose that α, β, $\alpha \vee \beta \in \Gamma$ with α and β having the forms $\bigvee_{i \in A} \xi_i$ and $\bigvee_{i \in B} \xi_i$, respectively. Hence, using Lemma 12.3.7 again,

$$H_{\alpha\vee\beta} = S_{\alpha\vee\beta}\left(\bigcup_{i\in A} H_{\xi_i}\right) \cup S_{\alpha\vee\beta}\left(\bigcup_{i\in\beta} H_{\xi_i}\right).$$

From Lemmas 12.3.7 and 12.3.6 we obtain

$$T(H_{\alpha\vee\beta}) = T\left(S_{\alpha\vee\beta}\left(\bigcup_{i\in A} H_{\xi_i}\right)\right) \cup T\left(S_{\alpha\vee\beta}\left(\bigcup_{i\in B} H_{\xi_i}\right)\right)$$

$$= T\left(\bigcup_{i\in A} H_{\xi_i}\right) \cup T\left(\bigcup_{i\in B} H_{\xi_i}\right)$$

$$= T\left(S_\alpha\left(\bigcup_{i\in A} H_{\xi_i}\right)\right) \cup T\left(S_\beta\left(\bigcup_{i\in b} H_{\xi_i}\right)\right) = T(H_\alpha) \cup T(H_\beta)$$

Moreover, by Lemma 12.3.6

$$T\left(H_{\theta_i}\right) = T\left(S_{\theta_i}\right) = T$$

for $i = 1, \ldots, r$, since $H_{\theta_i} = S_{\theta_i}$ and S_{θ_i} is complete.

The mapping

$$\gamma \to T(H_\gamma)$$

is thus a mapping of Γ into the Boolean algebra of subsets of T mapping $\neg$ on the operation of the complement, $\vee$ on the operation of the union and θ_i ($i = 1, \ldots, t$) on the whole set T.

Because any Frege rule is sound, we must have

$$T(H_{\theta_o}) = T$$

Hence by Lemma 12.3.6 also

$$H_{\theta_0} = S_{\theta_0}(H_{\theta_0}) = S_{\theta_0}$$

This concludes the proof of the lemma.   Q.E.D.

12.5. Lower bounds for the pigeonhole principle and for counting principles

In this section we obtain strong lower bounds to the size of constant-depth Frege proofs of the pigeonhole principle formulas PHP_n and the modular counting principles $Count_n^a$. First we shall define the formulas as we used some of them earlier in various forms. Recall that $[A]^a$ is the set of a-element subsets of A.

Definition 12.5.1.

1. For $|D| = n + 1$ and $|R| = n$ the formula PHP_n is built from atoms q_{uv}, $u \in D$ and $v \in R$:

$$\bigvee_{u_1 < u_2 < n+1} \bigvee_{v < n} (q_{u_1 v} \wedge q_{u_2 v}) \vee \bigvee_{u < n+1} \bigvee_{v_1 < v_2 < n} (q_{uv_1} \wedge q_{uv_2})$$

$$\vee \bigvee_{v < n} \bigwedge_{u < n+1} \neg q_{uv} \vee \bigvee_{u < n+1} \bigwedge_{v < n} \neg q_{uv}$$

2. For $|A| = a \cdot n + 1$ the formulas $Count_n^a$ are formed from atoms p_X, $X \in [A]^a$:

$$Count_n^a := \bigvee_{X \neq Y, X \cap Y \neq \emptyset} (p_X \wedge p_Y) \vee \bigvee_{i < n} \bigwedge_{i \in X} \neg p_X.$$

Note that PHP_n formalizes that there is no *bijection* between D and R and it is a weaker statement than the often used statement that there is no injection of D into R (and thus lower bounds will be stronger statements). The latter formula expresses that no *a-partition* of A can be total and it is a tautology as $|A| \equiv 1 \pmod{a}$.

We shall obtain exponential lower bounds to constant-depth proofs of PHP_n and $Count_n^a$. We shall also study the dependence among various instances of these principles.

It is easy to see that $PHP_m(\overline{q})$ follows from $Count_m^2(\phi_X)$ by a constant-depth size $O(m^2)$ *Frege* proof, where $A = \{0, \ldots, 2m\}$ and ϕ_X is q_{uv} for $X = \{u, v\}$, $u < m + 1, m + 1 \leq v < 2m + 1$, and ϕ_X is 0 otherwise. In the opposite direction Ajtai (1990) showed that there are no polynomial size constant-depth *Frege* proofs of $Count_n^2(p_X)$ from any instances $PHP_m(\phi_{uv})$ of the *pigeonhole principle*, ϕ_{uv} formulas in atoms p_X (implicitly of bounded depth). We shall also give a proof of a lower bound to the constant-depth Frege proofs of formulas $Count_m^a$ from the instances of formulas $Count_n^b$ for different primes a and b.

Lemma 12.5.2.

1. Let $\mathcal{M} = \mathcal{M}^{PHP}$ and let (H, S) be a k-evaluation of the set of subformulas of the formula PHP_n and let $k \leq (n/2) - 3$.
   Then $H_{PHP_n} = \emptyset$ and hence

$$H_{PHP_n} \neq S_{PHP_n}$$

In particular, if $\rho \in \mathcal{M}$ and $k \leq ((n - |\rho|)/2) - 3$, then

$$H_{(PHP_n)^\rho} \neq S_{(PHP_n)^\rho}$$

2. *Let $\mathcal{M} = \mathcal{M}^{MOD_a}$ and let (H, S) be a k-evaluation of the set of subformulas of the formula $Count_n^a$ and let $k \leq (n/2) - 2a$.*
Then $H_{Count_n^a} = \emptyset$ and hence

$$H_{Count_n^a} \neq S_{Count_n^a}$$

In particular, if $\rho \in \mathcal{M}$ and $k \leq ((n - |\rho|)/2) - 2a$, then

$$H_{(Count_n^a)^\rho} \neq S_{(Count_n^a)^\rho}$$

Proof. We shall prove part 2 of the lemma; part 1 is completely analogous.
$Count_n^a$ is a disjunction of the formulas of the form

$$p_X \wedge p_Y$$

where $X \cap Y \neq \emptyset$, $X \neq Y$, and

$$\bigwedge_{i \in X} \neg p_X$$

We shall show that for any such formula $\eta : H_\eta = \emptyset$. This will be enough as

$$H_{Count_n^a} = S_{Count_n^a} \left(\bigcup_\eta H_\eta \right) = S_{Count_n^a}(\emptyset) = \emptyset$$

Consider first the formula

$$\neg(\neg p_X \vee \neg p_Y)$$

for $X \cap Y \neq \emptyset$, $X \neq Y$. By Definition 12.4.1

$$H_{\neg p_X} = \{\alpha \in \mathcal{M} \mid X \subseteq supp(\alpha) \wedge \forall Z \in \alpha, Z \cap X \neq \emptyset\} \setminus \{X\}$$

and

$$H_{\neg p_Y} = \{\beta \in \mathcal{M} \mid Y \subseteq supp(\beta) \wedge \forall Z \in \beta, Z \cap Y \neq \emptyset\} \setminus \{Y\}$$

(recall $supp(\alpha) := \bigcup \alpha$). For $2a$-complete set T

$$T := \{\alpha \in \mathcal{M} \mid X \cup Y \subseteq supp(\alpha) \wedge \forall Z \in \alpha, Z \cap (X \cup Y) \neq \emptyset\}$$

it holds that

$$T(H_{\neg p_X} \cup H_{\neg p_Y}) = T$$

By Lemma 12.3.5 there is $k + 2a \leq n/2$-complete system W refining both T and $S_{\neg p_X \vee \neg p_Y}$. Hence Lemma 12.3.6 implies

$$W(H_{\neg p_X} \cup H_{\neg p_Y}) = W$$

and then also

$$H_{\neg p_X \vee \neg p_Y} = S_{\neg p_X \vee \neg p_Y}(H_{\neg p_X} \cup H_{\neg p_Y}) = S_{\neg p_X \vee \neg p_Y}$$

This gives

$$H_{p_X \wedge p_Y} = \emptyset$$

Now take the other formula

$$\eta := \vee_{i \in X} p_X$$

The set

$$T := \{X \in [A]^a \mid i \in X\}$$

is 1-complete and by Definition 12.4.1

$$H_\eta = S_\eta(T)$$

By Lemma 12.3.6

$$S_\eta(T) = S_\eta$$

hence

$$H_{\neg \eta} = \emptyset$$

The more general statement about the restricted formula $(\text{Count}\,{}^a_n)^\rho$ follows in exactly the same way observing that for any η of the preceding form if one of its atoms gets the value by ρ then all its atoms get a value and $(\eta)^\rho \equiv 0$, that is, $H_{(\eta)^\rho} = \emptyset$.   Q.E.D.

Now we are equipped to prove a strong lower bound.

Theorem 12.5.3 (Ajtai 1988, Beame et al. 1992). *Assume that F is a Frege proof system and d a constant, and let $n > 1$.*

Then in every depth d F-proof of the formula PHP$_n$ at least

$$2^{n^{(1/6)^d}}$$

different formulas must occur.

In particular, each depth d F-proof of PHP$_n$ must have size at least $2^{n^{(1/6)^d}}$ and must have at least $\Omega(2^{n^{(1/6)^d}})$ proof steps.

Proof. Let F be a Frege system and d a constant, let $n > 1$ be sufficiently large, and assume that $\pi = (\theta_1, \ldots, \theta_\ell)$ is a depth d F-proof of PHP$_n$. Take Γ to be the set of the formulas occurring in π as the subformulas.

Assume for the sake of contradiction that

$$|\Gamma| \le 2^{n^\delta}$$

for some $\delta < (1/5^{d-1})$.

Let f be a constant greater than the constant r assured by Lemma 12.4.4 for all rules R of F. Let $0 < \epsilon < 1/5$ such that $\delta < \epsilon^{d-1}$. By Theorem 12.4.3 there is $\rho \in \mathcal{M} = \mathcal{M}^{\text{PHP}}$ such that

$$|\rho| \leq n - n^{\epsilon^{d-1}}$$

and $k = 2n^{\delta}$-evaluation (H, S) of $(\Gamma)^{\rho}$. By Lemma 12.4.4, as $(n^{\rho}/f) \geq k$

$$H_{(\theta_i)^{\rho}} = S_{(\theta_i)^{\rho}}$$

for all steps θ_i in π. At the same time by Lemma 12.5.2, as $k \leq (n^{\rho}/2) - 3$

$$H_{(\text{PHP}_n)^{\rho}} \neq S_{(\text{PHP}_n)^{\rho}}$$

This is a contradiction; hence

$$|\Gamma| \geq 2^{n^{\delta}}$$

The lower bound to the size follows from trivial $|\pi| \geq |\Gamma|$, and the lower bound to the number of proof steps follows from Lemma 4.4.6.   Q.E.D.

The following theorem follows completely analogously.

Theorem 12.5.4. *Assume that F is a Frege proof system and d a constant, and let $a \geq 2$.*

Then there is $\epsilon > 0$ such that for sufficiently large n, in every depth d F-proof of the formula Count_n^a at least

$$2^{n^{\epsilon^d}}$$

different formulas must occur.

In particular, each depth d F-proof of Count_n^a must have size at least $2^{n^{\epsilon^d}}$ and must have at least $\Omega(2^{n^{\epsilon^d}})$ proof steps.

Theorem 9.1.3 and Corollary 9.1.4 then imply the following independence result for bounded arithmetic.

Corollary 12.5.5. *None of the $\Delta_0(R)$-formulas PHP(R) or $\text{Count}^a(R)$ is provable in $S_2(R)$. In fact, the formulas are not provable in any $S_k(R)$, $k \geq 1$.*

We now want to investigate the mutual relation of formulas PHP_n and Count_m^a. First note a simple lemma.

Lemma 12.5.6. *For $a \geq 2$, the theory*

$$I\Delta_0(R) + \forall x \; \text{Count}_x^a \, (\Delta_0(R))$$

proves the formula $\text{PHP}_n(R)$. In particular, there are constants $c, d \geq 1$ such that any propositional instance PHP_n of the pigeonhole principle has a depth d F-proof from some instances Count_m^a of size $\leq n^c$ (hence also $m \leq n^c$).

Proof. Assume $\neg\,\mathrm{PHP}_n(R)$ and define $A := \{0, \ldots, an\}$ and

$$\{x_1 < \cdots < x_a\} \in S \equiv \exists y < n + 1,$$

$$x_1 < n \;\wedge\; \bigwedge_{a > i \geq 2} x_i = x_{i-1} + (i-1)n \;\wedge\; x_a = (a-1)n + y \wedge R(y, x_1)$$

This clearly defines an a-partition of A.

The upper bounds to F-proofs follow from Theorem 9.1.3. Q.E.D.

Note that we used in an essential way that the pigeonhole principle speaks about *bijections* between $n + 1$ and n rather than just about an *injection* of $n + 1$ into n. For the latter version it is an open problem whether it can be proved from $\mathrm{Count}_n^a(\Delta_0(R))$.

The principle Count_n^a is, in fact stronger than PHP_n in the sense of the following theorem.

Theorem 12.5.7. *Let $a \geq 2$ and let d be fixed. Then there is $\epsilon > 0$ such that for sufficiently large n in any depth d Frege proof of $\mathrm{Count}_n^a(p_X)$ from the instances of the pigeonhole principle PHP_m at least*

$$2^{n^{\epsilon^d}}$$

distinct formulas must occur.

In particular, the size of all such proofs has to be at least $2^{n^{\epsilon^d}}$ and each such proof must have at least $\Omega(2^{n^{\epsilon^d}})$ steps.

Proof. Assume for the sake of contradiction that there is a depth d F-proof $\pi = (\theta_1, \ldots, \theta_r)$ of Count_n^a from some instances of PHP. Clearly we may assume w.l.o.g. (possibly increasing d by a constant and the proof polynomially) that only one instance

$$\mathrm{PHP}_m(\psi_{uv})$$

of PHP is used as an axiom in π (use the definition by cases to combine several instances into one). The formulas ψ_{uv} are built from the atoms p_X of Count_n^a.

Let Γ be the set of all subformulas occurring in π; in particular it contains formulas of the form

$$\neg \bigvee_{v < m} \psi_{uv} \quad \text{and} \quad \neg \bigvee_{u < m+1} \psi_{uv} \quad \text{and} \quad \psi_{uv_1} \wedge \psi_{uv_2} \quad \text{and} \quad \psi_{u_1v} \wedge \psi_{u_2v}$$

As in the proof of Theorems 12.5.3 and 12.5.4 let $\rho \in \mathcal{M}^{\mathrm{MOD}_a}$, $|\rho| \leq n - n^{\epsilon^{d-1}}$, be such that there is a $k \leq an^\delta$-evaluation (H, S) of Γ^ρ (constants ϵ, δ are the same as in Theorem 12.5.4).

Claim 1. *For $\xi = PHP_m(\psi_{uv})$*

$$H_\xi = S_\xi$$

To prove the claim assume $H_\xi \neq S_\xi$. We may assume then that, in fact, $H_\xi = \emptyset$ as otherwise we may restrict all formulas in $(\Gamma)^\rho$ further by some $\rho' \in S_\xi \setminus H_\xi$ that would collapse $(H_\xi)^\rho$ to $\emptyset$.

We first note two claims:

Claim 2. *Assume that* $H_\xi = \emptyset$, ξ *from Claim 1. Then for every* $u < m + 1$ *and every* $v < m$ *the sets*

$$\bigcup_{v < m} H_{\psi_{uv}} \quad and \quad \bigcup_{u < m+1} H_{\psi_{uv}}$$

are k-complete.

Claim 3. *We may assume w.l.o.g. that all* $2m + 1$ *complete systems from Claim 2 have the same cardinality, say* M.

Assuming Claims 2 and 3, Claim 1 follows as we may then count the sum

$$\sum_{u < m+1} |\bigcup_{v < m} H_{\psi_{uv}}| = (m + 1)M$$

also as

$$\sum_{v < m} |\bigcup_{u < m+1} H_{\psi_{uv}}| = m M$$

which is a contradiction.

So it remains to demonstrate Claims 2 and 3. Denote

$$S_u := \bigcup_{v < m} H_{\psi_{uv}}$$

and

$$S^v := \bigcup_{u < m+1} H_{\psi_{uv}}$$

As $H_\xi = \emptyset$ we have for $v_1 \neq v_2$

$$H_{\psi_{uv_1} \wedge \psi_{uv_2}} = \emptyset$$

which implies

$$\forall \alpha \in H_{\psi_{uv_1}} \forall \beta \in H_{\psi_{uv_2}}, \qquad \alpha \cup \beta \notin \mathcal{M}$$

On the other hand, $H_\xi = \emptyset$ also implies

$$H_{\neg \bigvee_v \psi_{uv}} = \emptyset$$

that is,

$$H_{\bigvee_v \psi_{uv}} = S_{\bigvee_v \psi_{uv}} \left(\bigcup_v H_{\psi_{uv}} \right) = S_{\bigvee_v \psi_{uv}}$$

Hence

$$\forall \alpha \in \mathcal{M}, |\alpha| \le n - k \rightarrow \exists \beta \in \bigcup_v H_{\psi_{uv}}, \qquad \alpha \cup \beta \in \mathcal{M}$$

This shows that S_u is k-complete. Similarly for S^v, and Claim 2 follows.

Claim 3 is demonstrated by a combinatorial construction of systems refining the systems from Claim 2 (using the idea behind Lemma 3.1.13 explained in Section 3.1). These refined systems resemble decision trees with all paths of equal length. Precisely, we say that a complete system is of the form of a depth i decision tree if (with $A = a \cdot n + 1$ and $[X]^k = \{Y \subseteq X \mid |Y| = k\}$)

1. for $i = 1$, S has the form

$$S = \big\{ \{Y\} \mid x \in Y, \ Y \in [A]^a \big\}$$

   for some $x \in A$
2. for $i > 1$, there is $x \in A$ such that for all $Y \in [A \setminus \{x\}]^{(a-1)}$ the set

$$\{\alpha \setminus (\{x\} \cup Y) \in S \mid \{x\} \cup Y \in \alpha\}$$

   is a complete system over $A \setminus (\{x\} \cup Y)$ of the form of a depth $(i - 1)$ decision tree.

Note a related example in 3 of Lemma 12.3.2.

Clearly all complete systems of the form of a depth i decision tree have the same cardinality.

Claim 1 implies with Lemma 12.4.4 that

$$H_{\theta_i} = S_{\theta_i}, \ i = 1, \dots, r$$

which contradicts Lemma 12.5.2. This proves the theorem.   Q.E.D.

The next corollary follows from the previous theorem by Corollary 9.1.4.

Corollary 12.5.8. *Let $k \ge 1$ and $a \ge 2$ be arbitrary. Then the formula $Count_n^a(R)$ is not provable in the theory*

$$S_k(R) + \forall x PHP_x(\Delta_0(R))$$

It remains to understand the relation of $Count_n^a$ and $Count_m^b$ principles for different $a, b \ge 2$. First a simple result in the spirit of Lemma 12.5.6. For it we generalize $Count_n^a$ to a conjunction of formulas saying that none of the sets $a \cdot n + k, k = 1, \dots, a - 1$, can be partitioned into a-element blocks. Denote this *generalized* principle $gCount_n^a$.

Lemma 12.5.9. *Let $a, b \ge 2$ and assume that b divides a. Then the theory*

$$I\Delta_0(R) + \forall x \ g Count_x^a(\Delta_0(R))$$

proves the formula $Count_n^b(R)$.

In particular, there are constants c, d such that for every n there is a depth d F-proof of size $\leq n^c$ of the formula $Count_n^b$ from some instances of the formulas $g\,Count_m^a$.

Proof. Work in $I\Delta_0(R)$ and assume that the R is a b-partition of the set B, $|B| = bn + 1$. Let $k := (a/b)$, put $A := an + k$, and let the partition S consist of the classes $X \in [A]^a$ such that

$$\exists Y \in R,\ X = \bigcup_{0 \leq i < k} \{y + (bn + 1)i \mid y \in Y\}$$

Then S is an a-partition of A, but as $1 \leq k < a$, $|A| \not\equiv 0 \pmod{a}$; that is, $g\,Count_n^a$ is violated.   Q.E.D.

The case when b does not divide a is also understood. In the following we confine discussion to the case when a, b are two distinct primes.

Theorem 12.5.10. *Assume $p, q \geq 2$ are two different primes. Then for every depth d and every $c > 1$ it holds that for sufficiently large n, in every depth d F-proof of the formula $Count_n^p$ from the instances of the $Count_m^q$ formula at least*

$$\geq n^c$$

different formulas must occur.
   In particular, the theory

$$I\Delta_0(R) + \forall x\ Count_x^q(\Delta_0(R))$$

does not prove the principle $Count_n^p(R)$.

Proof. Let $\mathcal{M} := \mathcal{M}^{\mathrm{MOD}_p}$ and let π be a depth d F- proof of $Count_n^p$ from one instance

$$Count_m^q(\psi_X)$$

$X \in [A]^q$, $|A| \equiv 1 \pmod{q}$ (as in the proof of Theorem 12.5.7 we assume w.l.o.g. that only one instance of $Count_m^q$ is used in π).
   Take Γ to be the set of all formulas occurring in π plus all formulas of the form $(i \in A)$

$$\bigvee_{i \in X} \psi_X \quad \text{and} \quad \bigvee_{i \in A} \bigvee_{i \in X} \psi_X$$

By the proof of Theorem 12.4.3 from Lemma 12.3.11 there are $\rho \in \mathcal{M}$, $|\rho| \leq n - n^{\epsilon^{d-1}}$, and a k-evaluation (H, S) of $(\Gamma)^\rho$ where k *is a constant* depending on c but independent of n (this is because if $|\Gamma| \leq n^c$ then Lemma 12.3.11 allows k to be a large enough constant).
   As in the proof of Theorem 12.5.7 it is then sufficient to establish the following claim.

Claim 1. *For* $\xi = Count_m^q(\psi_X)$ *it holds that*

$$H_\xi = S_\xi$$

In the proof of the claim we follow the strategy of the proof of Claim 1 in the proof of Theorem 12.5.7, but in this case it is complicated.

Claim 2. *Assume* $H_\xi = \emptyset$. *Then for every* $i \in A$ *the set* $\bigcup_{i \in X} H_{\psi_X}$ *is a k-complete system.*

Claim 2 is established similarly to Claim 2 in the proof of Theorem 12.5.7.
We defer the proof of Claim 1 from Claim 2 after Theorem 12.6.2.   Q.E.D.

12.6. Systems with counting gates

The formula $Count_n^q(R)$ formalizes a counting principle but it is natural to consider also a situation when the language allows direct counting. This was, in the context of bounded arithmetic, studied in Paris and Wilkie (1985). They extended language L_{PA} by a quantifier of the form $Q_a x < t$ with the meaning

$$Q_a x < t \phi(x) \text{ is true} \quad \text{iff} \quad |\{x < t \mid \phi(x) \text{ is true}\}| \equiv 0 (\bmod \, a)$$

Denote by $Q_a \Delta_0$ the set of the bounded formulas formed by also using the Q_a-quantifier. It is easy then to extend $I\Delta_0$ to the theory $IQ_a\Delta_0$ by adding a few rules for handling the new quantifier (and allowing all new bounded formulas in the induction scheme). We shall instead consider in greater detail the corresponding extension of a Frege system by new connectives.

Definition 12.6.1. *Let* $a \geq 2$ *be fixed and* $i = 0, \ldots, a - 1$. $MOD_{a,i}$ *is a propositional connective of unbounded arity such that*

$$MOD_{a,i}(\phi_1, \ldots, \phi_k) \text{ is true} \quad \text{iff} \quad |\{j \mid \phi_j \text{ true}\}| \equiv i(\bmod \, a)$$

MOD$_a$-axioms are the following axiom schemes
(a)

$$MOD_{a,0}(\emptyset)$$

(b)

$$\neg MOD_{a,i}(\emptyset), \quad for \, i = 1, \ldots, a - 1$$

(c)

$$MOD_{a,i}(\Gamma, \phi) \equiv \left[(MOD_{a,i}(\Gamma) \wedge \neg\phi) \vee (MOD_{a,i-1}(\Gamma) \wedge \phi) \right]$$

for $i = 0, \ldots, a - 1$, *where* $i - 1$ *means* $i - 1$ *modulo* a, *and where* Γ *stands for a sequence (possibly empty) of formulas.*

The system $F(MOD_a)$ or $LK(MOD_a)$ is the system F (resp. LK) whose language is extended by the connectives $MOD_{a,i}$, $i = 0, \ldots, a - 1$, and that is augmented by the preceding axioms.

The following theorem is straightforward, but, in fact, it is the only information about the constant-depth $F(MOD_a)$-proof systems. In particular, it is an open problem to apply the methods leading to the results about constant-depth circuits with MOD_p gates (see Theorem 3.1.14) to the lower bounds for constant-depth $F(MOD_p)$-proofs.

Theorem 12.6.2. *For $a \geq 2$, the formulas $Count_m^a$ have polynomial size, constant-depth $F(MOD_a)$-proofs.*

Proof. We shall denote by the symbol $\alpha \vdash_* \beta$ the fact that there are constant-depth size $n^{O(1)}$ $F(MOD_a)$-proofs of β from α. Thus we have

1.

$$\vdash_* MOD_{a,1}((1)_{i \in A})$$

as $1 \equiv |A| \pmod a$

2.

$$\neg Count_m^a \vdash_* MOD_{a,1}((p_X)_{i \in X})$$

for every $i \in A$ as $\neg Count_m^a$ implies that i is in exactly one X

3.

$$\vdash_* MOD_{a,0}((p_X)_{i \in X})$$

any fixed X as $|X| \equiv 0 \pmod a$

4.

$$\neg Count_m^a \vdash_* MOD_{a,1}((p_X)_{i \in A, i \in X})$$

from 1 and 2.

5.

$$\vdash_* MOD_{a,0}((p_X)_{i \in A, i \in X})$$

from 3.

6. From the last two steps we get

$$\neg Count_m^a \vdash_* MOD_{a,1}(\emptyset)$$

which contradicts an axiom.

Q.E.D.

The proof does not not really use propositional reasoning but rather only rudimentary counting modulo a. This suggests formulation of the counting principles in the language of rings and consideration of an equational logic as a propositional proof system.

Fix $a \geq 2$ and n, and $A := a \cdot n + 1$. Consider the equations

$$\sum_{i \in X} v_X = 1 \qquad\qquad (i)$$

$$v_X \cdot v_Y = 0 \qquad\qquad (X, Y)$$

for each $i \in A$ and each $X, Y \in [A]^a$ such that $X \perp Y$, which will be shorthand for $X \cap Y \neq \emptyset \wedge X \neq Y$.

The equations imply, in any field, that each v_X is 0 or 1. A particular solution determines the set

$$\alpha := \left\{ X \in [A]^a \mid v_X = 1 \right\}$$

and the equations force that α is a total a-partition of A, which is impossible. Hence the system has no solution in any field. In fact, the system consisting of equations (i) is not solvable in rings of characteristic a as the sum of the left sides is 0 while the sum of the right sides is 1.

Now we resume **the proof of Claim 1 from the proof of Theorem 12.5.10**. We continue using the notation from that proof.

For p and n take the system

$$\sum_{e \in E} u_E = 1 \qquad\qquad (e)$$

$$u_E \cdot u_F = 0 \qquad\qquad (E, F)$$

one equation (e) for each $e \in B = p \cdot n + 1$ and one for each $E, F \in [B]^p$, $E \perp F$. As earlier, the system expresses that $\{ E \mid u_E = 1 \}$ is a total p-partition of B and is contradictory in every field. We shall reach the contradiction, however, by making use of the alleged proof π of Count_n^p from $\mathrm{Count}_m^q (\psi_X)$.

Let $r(\bar{u}) = 0$ be an equation following from the (e) and (E, F) equations in a particular finite field $\mathbf{F}_q$. As the system is contradictory, every equation is its consequence, but for some such equations we will get a nontrivial information. Denote by w_e the polynomial

$$w_e := \sum_{e \in E} u_E - 1$$

and by $w_{E,F}$ the polynomial

$$w_{E,F} := u_E \cdot u_F$$

Hence the system (e)'s and (E, F)'s is equivalent to the system

$$w_e = 0 \quad \text{and} \quad w_{E,F} = 0$$

As it is contradictory, the polynomials generate in the ring $\mathbf{F}_q[\,\overline{u}\,]$ the trivial ideal, and hence for every $r(\overline{u})$ there are polynomials c_e and $c_{E,F}$ such that

$$\sum_e c_e \cdot w_e + \sum_{E,F} c_{E,F} \cdot w_{E,F} \equiv r(\overline{u})$$

is valid in $\mathbf{F}_q[\,\overline{u}\,]$ (this is a simple consequence of Hilbert's Nullstellensatz but has a simple direct proof too). In general we are able to say nothing useful about the polynomials c_e and $c_{E,F}$, but for some polynomials $r(\overline{u})$ there is extra information.

Lemma 12.6.3. *For $\alpha \in \mathcal{M} \setminus \{\emptyset\}$ define the monomials $\hat{\alpha}$ by*

$$\hat{\alpha} = u_{E_1} \cdot \,\cdots\, \cdot u_{E_t}$$

where $\alpha = \{E_1, \ldots, E_t\}$ and define $\hat{\emptyset} := 1$.
Let $S \subseteq \mathcal{M}$ be a k-complete system, $qk^2 < n$, and r_S the polynomial

$$r_S(\overline{u}) := \sum_{\alpha \in S} \hat{\alpha} \; - 1$$

Then r_S can be expressed in the ring $\mathbf{F}_q[\,\overline{u}\,]$ as a linear combination

$$r_S = \sum_e c_e \cdot w_e + \sum_{E,F} c_{E,F} \cdot w_{E,F}$$

in which the degree of all coefficients c_e, $c_{E,F}$ is at most $\leq qk^2$.

Proof. The lemma is proved by induction on $\|S\|$. Pick $\alpha \in S$ and let S_α be the kq-complete system

$$\left\{ \gamma \in \mathcal{M} \mid \bigcup \alpha \subseteq \bigcup \gamma \wedge \forall E \in \gamma,\, E \cap \bigcup \alpha \neq \emptyset \right\}$$

Any $\beta \in S \setminus \{\alpha\}$ is incompatible with α. So for every $\gamma \in S_\alpha$, either $|\beta^\gamma| \leq k-1$ or β^γ is undefined. Hence S_1

$$S_1 := \bigcup_{\gamma \in S_\alpha} \{\gamma\} \times S^\gamma$$

is a $kq + k - 1$-complete system. Each S^γ is k-complete (Lemma 12.3.9) and $|S^\gamma| \leq k - 1$; hence we may repeat the same process with each S^γ separately. In i steps this produces a $(k + (k - 1) + \cdots + (k - i + 1))q + (k - i)$-complete system S_i. That is, a S_k is $\binom{k}{2} \cdot q \leq qk^2$- complete system.

Call a complete system S *t-supported* iff there is $Z \subseteq B$ of size t such that

$$S = \left\{ \alpha \mid Z \subseteq \bigcup \alpha \wedge \forall E \in \alpha,\, E \cap Z \neq \emptyset \right\}$$

Define S to be *i-good* by induction on i
1. S is 0-good iff it is t-supported, some t

2. S is $(i + 1)$-good iff there is a supported complete system T such that S refines T and such that for every $\alpha \in T$ system S^α is i-good

Note that S_α is qk-supported and that S_k is k-good.

To simplify the notation we say that the *degree* of a linear combination L of w_e's and $w_{E,F}$'s is the maximum degree of its coefficients, and a "linear combination" automatically means a "linear combination of w_e's and $w_{E,F}$'s."

The lemma follows from three claims.

Claim 1. *If T is t-supported then there is a linear combination L_T of degree $\leq \|T\|$ such that*

$$\sum_{\alpha \in T} \hat{\alpha} = L_T + 1$$

Claim 2. *If T is i-good then there is a linear combination L_T of degree $\leq \|T\|$ such that*

$$\sum_{\alpha \in T} \hat{\alpha} = L_T + 1$$

Claim 3. *If T refines system U and T is i-good then there is a linear combination L_U of degree $\leq \|T\|$ such that*

$$\sum_{\alpha \in U} \hat{\alpha} = L_U + 1$$

The first claim is readily established by induction on t. To see the second claim write

$$\sum_{\alpha \in T} \hat{\alpha} = \sum_{\alpha \in T_0} \hat{\alpha} \left(\sum_{\beta \in T^\alpha} \hat{\beta} \right)$$

where T refines T_0 and T_0 is supported, and each T^α is $(i - 1)$-good. By the induction hypothesis there are linear combinations L_{T^α} such that

$$\sum_{\beta \in T^\alpha} \hat{\beta} = L_{T^\alpha} + 1$$

That is,

$$\sum_{\alpha \in T} \hat{\alpha} = \left(\sum_{\alpha \in T_0} \hat{\alpha} L_{T^\alpha} \right) + \sum_{\alpha \in T_0} \hat{\alpha}$$

and by Claim 1 the last sum is expressible as $L_{T_0} + 1$, L_{T_0} a linear combination. Clearly the total degree is $\leq \|T\|$.

The third claim is computed similarly:

$$\sum_{\beta \in T} \hat{\beta} = \sum_{\alpha \in U} \sum_{\beta \in T, \, \beta \supseteq \alpha} \hat{\beta} = \sum_{\alpha \in U} \hat{\alpha} \left(\sum_{\beta \in T^\alpha} \hat{\beta} \right) = \left(\sum_{\alpha \in U} \hat{\alpha} \cdot L_{T^\alpha} \right) + \sum_{\alpha \in U} \hat{\alpha}$$

where L_{T^α} are linear combinations of degree $\le \|T^\alpha\|$ such that

$$\sum_{\beta \in T^\alpha} \hat{\beta} = L_{T^\alpha} + 1$$

(their existence follows from Claim 2 as all T^α are clearly i-good if T is i-good). But the first sum is by Claim 2 expressible as $L_T + 1$. Hence

$$\sum_{\alpha \in U} \hat{\alpha} = \left(L_T - \sum_{\alpha \in U} \hat{\alpha} L_{T^\alpha} \right) + 1$$

The alert reader noticed that L_{T^α} are linear combinations over $B \setminus (\bigcup \alpha)$ and not over B. In particular, polynomials w_e miss unknowns u_F, for F intersecting $\bigcup \alpha$. However, if we add these missing terms, the difference it causes in the product $\hat{\alpha} \cdot L_{T^\alpha}$ is a linear combination of polynomials w_{F,E_i} and $u_{E_i} - u_{E_i}^2$, $E_i \in \alpha$. The latter are themselves degree 1 linear combinations over B. Hence each $\hat{\alpha} L_{T^\alpha}$ is a linear combination over B, even if L_{T^α} is not. This proves the claims.

The lemma then follows from Claim 3 as S_k is k-good refines S and $\|S_k\| \le qk^2$.

Q.E.D.

By Claim 2 in the proof of 12.5.10 each $\bigcup_{i \in X} H_{\psi X}$ is a k-complete system; denote it S_i. Then by the previous lemma there are linear combinations L_i of w_e's and $w_{E,F}$'s of degree $\le qk^2$ such that

$$\sum_{\alpha \in S_i} \hat{\alpha} = L_i + 1$$

and hence

$$\sum_{i \in A} \sum_{\alpha \in S_i} \hat{\alpha} = \sum_{i \in A} (L_i + 1) = \left(\sum_{i \in A} L_i \right) + 1$$

But the first sum can be also computed as

$$\sum_{i \in A} \sum_{\alpha \in S_i} \hat{\alpha} = \sum_{X \in [A]^q} \sum_{i \in X} \sum_{\alpha \in H_{\psi X}} \hat{\alpha} = 0$$

because $|X| \equiv 0 \pmod{q}$. Hence there is a linear combination $L = \sum_i L_i$ of degree $\le qk^2$ such that $L + 1 = 0$ in $\mathbf{F}_q[\overline{u}]$.

The proof of Claim 2 from the proof of Theorem 12.5.10, and thus the proof of the theorem itself, is concluded by the following lemma, taking $d = qk^2$.

Lemma 12.6.4. *Let p, q be different primes and d a constant. Then for sufficiently large n and for any linear combination L of w_e's and $w_{E,F}$'s of degree $\le d$ there is a partial p-partition α of $B = pn + 1$ such that*

$$L(\overline{u}_\alpha) = 0$$

where $\overline{u}_\alpha$ is defined by

$$u_E = \begin{cases} 1, & \text{if } E \in \alpha \\ 0, & \text{otherwise} \end{cases}$$

We shall not prove this lemma. The reader may consult Beame et al. (1994). Note that a lower bound d on the degree of a linear combination L for which $L + 1 = 0$ holds generally implies a lower bound in Theorem 12.5.10 of the form $n^{\Omega(d)}$. In particular, $d = n^{\Omega(1)}$ would yield an exponential lower bound in that theorem.

Open question. *Let $a, b \geq 2$ be two different numbers. When are there polynomial size, constant-depth $F(MOD_a)$-proofs of formulas $Count_n^b$?*

Theorem 12.5.10 does not yield any information about this problem.

12.7. Forcing in nonstandard models

In this section we reinterpret some of the previous material of this chapter as forcing constructions. We shall begin with a simple but illustrative result.

Recall that $\exists_1$-formulas are the existential formulas; we shall consider them in the form $\exists \overline{x} \phi(\overline{x}, \overline{u})$, ϕ open.

Theorem 12.7.1. *The fragment of Peano arithmetic axiomatized by Q and by the least number principle for existential formulas in the language $L_{PA}(R)$ does not prove the pigeonhole principle $PHP(R)$:*

$$L\exists_1(R) \nvdash PHP(R).$$

Proof. Let M be a countable, nonstandard model of true arithmetic $\text{Th}(N)$. Let $n \in M$ be a nonstandard number.

Let $\mathcal{P} = \{g \in M \mid g : \subseteq n + 1 \mapsto n, \ g \text{ injective}\}$. The symbol $g \in M$ means that g is coded in M.

Enumerate all elements of $n + 1$: $u_1, u_2, \ldots$ and also all $\exists_1(R)$-formulas $\theta_1(x)$, $\theta_2(x), \ldots$ with one free variable x and with parameters from M.

We shall build a sequence $g_0 \subseteq g_1 \subseteq \ldots \in \mathcal{P}$ such that

1. $\bigcup_i \text{dom}(g_i) = n + 1$
2. for $g := \bigcup_i g_i$,

    (a) $(M, g) \models L\exists_1(g)$

    (b) $(M, g) \models \text{"g is injective"}$

3. the cardinality of each g_i is standard

(Step 0) $g_0 := \emptyset$

(Step 2i) Choose any $g_{2i} \supseteq g_{2i-1}$ such that $u_i \in \text{dom}(g_{2i})$ and such that condition 3 is satisfied

(**Step** $2i+1$) Choose $g_{2i+1} \supseteq g_{2i}$ of standard size such that for any $F : n+1 \mapsto n$, $F \supseteq g_{2i+1}$, injective, if R is the graph of F, then

$$(M, F) \models \text{ least number principle for } \theta_i(x)$$

To perform the odd steps we need the following simple claim.

Claim. *Let* $\theta(x)$ *be a* $\exists_1(R)$*-formula. Then there is standard* k *such that for every* $g \in \mathcal{P}$ *and* $a \in \mathcal{M}$ *if*

$$\exists R \supseteq g, (M, R) \models \theta(a)$$

then there is $h \in \mathcal{P}, |h \setminus g| \leq k$ *such that*

$$\forall R \supseteq h, (M, R) \models \theta(a)$$

where R *ranges over the graphs of the total injective maps from* $n + 1$ *to* n *in the last two formulas.*

To see the claim observe that in the open kernel $\phi(x, \overline{u})$ of θ constantly many atomic subformulas of the form $R(s, t)$, where s and t are terms built from $x, \overline{u}$, occur.

If some $R \supseteq g$ makes $\theta(a)$ true in M we need to fix the truth value of only constantly many such subformulas. To make $R(i, j)$ true add the pair (i, j) to g; to make it false add (i, j') to g for some $j' \neq j$.

With the claim and $g = g_{2i}$ in an odd step, consider the formula

$$\eta(x) = \exists h \in \mathcal{P}, |h \setminus g| \leq k \wedge \text{``} R \supseteq h \text{ forces } \theta_i(x) \text{ true''}$$

where k is the number guaranteed for θ_i by the claim.

Take $a \in M$ to be the minimal element satisfying $\eta(x)$ in M, if there is any. Such an element exists as M satisfies the least number principle for all formulas. Let h be the map guaranteed for a by the claim. Put

$$g_{2i+1} := h$$

Q.E.D.

Note that the same proof works for those $L_{PA}(R)$-formulas in place of $\exists_1(R)$-formulas in which every universal quantifier is bounded by some m such that $m^c < n$ for all standard c: A statement analogous to the claim holds with the number k being replaced by a number of the form m^c, some c standard.

This is seen as follows. Let $\theta(x)$ be a formula of the form

$$\forall v_1 < m \exists u_1 \forall v_2 < m \ldots \exists u_\ell \phi(x, \overline{v}, \overline{u})$$

with ϕ open. Formula $\theta(x)$ is equivalent (in M) to the formula

$$\exists H_1, \ldots, H_\ell : m \to M \ \forall v_1, \ldots, v_\ell < m \ \phi(x, \overline{v}, u_i/H_i(x, v_1, \ldots, v_\ell))$$

where H_i are Skolem functions. The subformula

$$\forall v_1, \ldots, v_\ell < m \ \phi \ (x, \overline{v}, u_i / H_i(x, v_1, \ldots, v_\ell))$$

can be replaced by the conjunction

$$\bigwedge_{v_1, \ldots, v_\ell < m} \phi \ (x, \overline{v}, u_i / H_i(x, v_1, \ldots, v_\ell))$$

which contains $\leq c \cdot m^\ell$ distinct terms (in parameters $x, \overline{v}$ and also using the symbols H_i). Hence we may use this conjunction in place of the open kernel in the proof of the claim.

Note that this shows that the theory $\Sigma_1^b(R)$–MIN (which is equivalent to $T_2^1(R)$ by Lemma 5.2.7) does not prove PHP(R): That is, it is an alternative proof of Theorem 11.2.5. In fact, the proof of Corollary 11.3.2 from Riis (1993a) generalizes the proof of Theorem 12.7.1.

The rest of this section is devoted to an interpretation of the method of k-evaluations from Sections 12.3–5 as a forcing type of construction (we refer the reader to Takeuti and Zaring 1973 for basic information on forcing).

Let M be a countable nonstandard model of true arithmetic Th(N), $n \in M \setminus \omega$ its nonstandard element and $I \subseteq_e M$ the cut in which the elements $2^{|n|^\ell}$, $\ell < \omega$, are cofinal. Hence I is a model of S_2 (cf. Lemma 5.1.2). Our aim is to find a bijection $f : n + 1 \to n$ such that

$$(I, f) \models S_2(f) + \neg \text{PHP}(f)$$

The bijection f will be obtained by forcing. The set of the forcing conditions $\mathcal{P}$ is defined by

$$\mathcal{P} := \left\{ \alpha \in M \mid \alpha : \subseteq n + 1 \to n, \alpha \text{ is } 1 - 1, |\alpha| \leq n - n^{1/\ell} \quad \text{some } \ell < \omega \right\}$$

and partially ordered by inclusion.

Call subset $D \subseteq \mathcal{P}$ *dense* iff

$$\forall \alpha \in \mathcal{P} \exists \beta \in D, \alpha \subseteq \beta$$

and *definable* iff there is a formula $\phi(x)$ with parameters from M such that

$$D = \{\alpha \in \mathcal{P} \mid M \models \phi(\alpha)\}$$

A subset $G \subseteq \mathcal{P}$ is called *generic* if it satisfies four properties:
1. $G \neq \emptyset$
2. $\forall \alpha \in G \forall \beta \subseteq \alpha, \beta \in G$
3. $\forall \alpha, \beta \in G \exists \gamma \in G, \alpha \cup \beta \subseteq \gamma$
4. $D \cap G \neq \emptyset$, for every dense definable subset $D \subseteq \mathcal{P}$

The next lemma is standard and its proof is omitted.

Lemma 12.7.2.

1. Let $\alpha \in \mathcal{P}$ be arbitrary. Then there is generic G such that $\alpha \in G$.
2. Let G be generic and define the map f_G by

$$f_G := \bigcup G$$

That is,

$$f_G(x) = y \quad \text{iff} \quad \exists \alpha \in G, \alpha(x) = y$$

Then f_G is a bijection between $n + 1$ and n.

Let $\phi(f)$ be an $L(f)$-sentence with parameters from I. The notion of genericity allows us to define the notion of forcing.

For $\alpha \in \mathcal{P} : \alpha$ *forces* $\phi(f)$, $\alpha \Vdash \phi(f)$ in symbols, iff for every generic $G \subseteq \mathcal{P}, \alpha \in G$

$$(I, f_G) \models \phi(f)$$

This forcing notion satisfies the usual properties (cf. Takeuti and Zaring 1973).

We would like to find a generic G such that for every bounded $L(f)$-formula $\phi(x)$ with the parameters from I and with one free variable x some $\alpha \in G$ forces that $\phi(x)$ satisfies in (I, f_G) the least number principle. It is, however, not at all clear that such G exists and we will have to employ the restriction method from Sections 12.3–5 to achieve this.

Consider the set $\mathcal{F}$ of such bounded $L(f)$-formulas $\phi(x)$. Let $m \in M \setminus I$ be any fixed element and d a standard number. Define (in M) the set Γ_d of propositional formulas to be the set of formulas $\theta \in M$ that satisfy

1. $dp(\theta) < d$
2. $|\theta| < m$
3. the atoms of θ are among p_{ij}, $i < n + 1$ and $j < n$

Note that for any $\phi(x) \in \mathcal{F}$ there is d such that for all $u \in I$

$$\langle \phi \rangle_u \in \Gamma_d$$

where $\langle \phi \rangle$ is the translation from Definition 9.1.1.

As M is a model of true arithmetic, Theorem 12.4.3 holds in M. Choose constants $\epsilon = 1/6$ and $\delta = 1/\ell$, some fixed $\ell > \omega$. Hence in M it holds that for some $\rho \in \mathcal{P}, |\rho| \leq n - n^{(1/6^{d-1})}$, there is k-evaluation (H, S) of $(\Gamma)^\rho$ with $k \leq 2n^\delta$.

The following lemma is an important technical vehicle.

Lemma 12.7.3. *Let ϕ be a bounded $L(f)$-sentence with parameters from I and let $d < \omega$ be such that $\langle \phi \rangle \in \Gamma_d$. Let $\epsilon, \delta,$ and ρ be as earlier and let (H, S) be a $k = 2n^\delta$-evaluation of $(\Gamma_d)^\rho$.*

Then for a generic G containing ρ it holds that

$$G \Vdash \phi \quad \text{iff} \quad (I, f_G) \models \bigvee_{\alpha \in H_{\langle \phi \rangle^\rho}} \alpha$$

where $(I, f_G) \models \alpha$ iff $\alpha \subseteq f_G$.

Proof. The proof proceeds by induction on the logical complexity of ϕ. It is clear for the atomic ϕ and for the negation. We shall consider only the case when $\phi = \exists x < u \psi(x)$. For simplicity of notation we shall skip the reference to ρ.

By Definition 12.4.1 we have

$$H_{\langle \phi \rangle} = S_{\langle \phi \rangle} \left(\bigcup_{v < u} H_{\langle \psi \rangle_v} \right)$$

and

$$\bigvee_{\alpha \in H_{\langle \phi \rangle}} \alpha = \bigvee_{v < u} \bigvee_{\beta \in H_{\langle \psi \rangle_v}} \beta$$

The second equality follows from

$$\bigvee_{\alpha \in H} \alpha \equiv \bigvee_{\beta \in S(H)} \beta$$

whenever a complete system S refines H, which in turn follows from the identity

$$\bigvee_{\alpha \in S} \alpha \equiv 1$$

valid for every complete system S.

Hence $G \Vdash \phi$ iff $G \Vdash \exists v < u \psi(v)$ iff

$$(I, f_G) \models \bigvee_{v < u} \bigvee_{\beta \in H_{\langle \psi \rangle_v}} \beta$$

iff $(I, f_G) \models \bigvee_{\alpha \in H_{\langle \phi \rangle}} \alpha$. Q.E.D.

We are equipped now to describe a construction of a generic $G \subseteq \mathcal{P}$ for which

$$(I, f_G) \models S_2(f)$$

Let $\phi_1(x), \phi_2(x), \dots$ enumerate set $\mathcal{F}$ so that each ϕ_i appears twice. Put $\rho_0 := \emptyset$ and construct $\rho_0 \subseteq \rho_1 \subseteq \cdots \in \mathcal{P}$ in steps.

If we are in an even step, extend ρ_{2i} to some ρ_{2i+1} from the ith dense definable set. If we are in an odd step ρ_{2i+1} consider the formula ϕ_{i+1}. There are two possibilities:

1. for some $d < \omega$ s.t. $\langle \phi_{i+1} \rangle \in \Gamma_d$ there is k-evaluation (H, S) of $(\Gamma_d)^{\rho_{2i+1}}$ such that $k^c < n$, all $c < \omega$
2. otherwise.

In the first case pick in M the minimal $u < m$ such that

$$M \models \exists \rho \supseteq \rho_{2i+1}, |\rho \setminus \rho_{2i+1}| \leq k \wedge \rho \in \mathcal{P} \wedge \rho \Vdash \phi_{i+1}(u)$$

This is definable because of the bound $|\rho \setminus \rho_{2i+1}| \leq k$ and because $\rho \Vdash \phi_{i+1}(u)$ iff $\rho \models \bigvee_{\alpha \in H_{\langle \phi_{i+1} \rangle u}} \alpha$, which is clearly definable.

Pick $\rho_{2i+2} \supseteq \rho_{2i+1}$ any ρ witnessing the validity of the formula for the minimal u.

In the second case take any $d < \omega$ such that $\langle \phi_{i+1} \rangle \in \Gamma_d$ and any ρ_{2i+2} extending ρ_{2i+1} for which there is a k-evaluation of $(\Gamma_d)^{\rho_{2i+2}}$.

Take

$$G := \{\alpha \in \mathcal{P} \mid \rho_i \supseteq \alpha, \text{ some } i < \omega\}$$

By Lemma 12.7.2

$$(I, f_G) \models \neg \mathrm{PHP}_n(f_G)$$

and by Lemma 12.7.3

$$(I, f_G) \models \exists x \phi(x) \rightarrow \exists u \forall v < u, \phi(u) \wedge \neg \phi(v)$$

That is, also

$$(I, f_G) \models S_2(f_G)$$

12.8. Bibliographical and other remarks

Theorem 12.1.3 and Corollary 12.1.4 are due to Pudlák (1992a). Section 12.2 is after Krajíček (1994); Theorem 12.2.4 was the first exponential lower bound for the constant-depth systems. Earlier bounds (Ajtai 1988, Bellantoni, Pitassi, and Urquhart 1992) were slightly superpolynomial.

Sections 12.3 and 12.4 contain material from Krajíček et al. (1991). Lemma 12.3.10 was proved there using a probabilistic argument based on an unpublished work of Woods, who, in turn, built on Yao (1985), Cai (1989), and Hastad (1989). The present proof recasts the original as a counting argument; see Lemma 15.2.2 for a similar argument.

Ajtai (1988, 1990) investigated the lengths of proofs of PHP and Count_n^2 studied earlier in connection with the bounded arithmetic in Paris and Wilkie (1985) and Woods (1981). His arguments showed that there are no polynomial size constant-depth proofs of PHP, and of Count_n^2 from the instances of PHP. Bellantoni et al. (1992) extracted from Ajtai (1988) an explicit superpolynomial lower bound to PHP. The exponential lower bound was obtained independently in Krajíček et al. (1991) and Pitassi, Beame, and Impagliazzo (1993), and the exponential separation of PHP and Count_n^2 was obtained by Beame and Pitassi (1993) and Riis (1993b). The proof of Theorem 12.5.7 in part follows Riis (1993b). The separation of Count^p and Count^q for p, q primes (Theorem 12.5.10) was announced by Ajtai (1994a) and Riis (1994). The proof presented here is from Beame et al. (1994), where the dependence of the principles is completely characterized also for composite p, q.

Paris, Handley, and Wilkie (1984) and Paris and Wilkie (1985) characterize some of the classes $Q_a \Delta_0$ in terms of Bel'tyukov's machines (see Bel'tyukov 1979), and Gandy (unpublished, see Theorem 11 in Paris and Wilkie 1985) characterized class $Q_2 \Delta_0$ in terms of a limited primitive recursion.

Nothing is known about the proof systems with the counting gates. The construction of i-good refinement in the proof of Lemma 12.6.3 is similar to a construction of "canonical systems" in Riis (1993b), and it is based on the idea behind Lemma 3.1.13. The bounded arithmetic systems with the counting quantifiers are also not understood.

Theorem 12.7.1 is due to Paris and Wilkie (1985), and it is the first forcing argument in the context of weak arithmetic. Later forcing constructions of Ajtai (1988) or Riis (1993a,1993b) are based on the same principle, although technically more complicated. Our presentation is slightly modified from Riis (1993b).

13

Bounds for Frege and extended Frege systems

In this chapter we shall discuss the complexity of *Frege* systems without any restrictions on the depth. There is some nontrivial information, in particular nontrivial upper bounds, but no nontrivial lower bounds are known at present (only bounds from Lemma 4.4.12).

13.1. Counting in Frege systems

Theorems 9.1.5 and 9.1.6 are useful sufficient conditions guaranteeing the existence of the polynomial size EF-proofs and of quasipolynomial $n^{(\log n)^{O(1)}}$ size F-proofs, respectively. For example, U_1^1 proves the pigeonhole principle PHP(R) and hence there are quasipolynomial size F-proofs of PHP$_n$. A subtheory of U_1^1 corresponding to the polynomial size F-proofs, based on a version of inductive definitions, was considered by Arai (1991); see Section 9.6. Its axiomatization however, stresses a logical construction, whereas we would like a theory based on a more combinatorial principle.

The most important property of a Frege system relevant for the upper bounds is that it can *count*. We shall make this precise by showing that F simulates an extension of $I\Delta_0(\alpha)$ by counting functions, and that F p-simulates a propositional proof system *cutting planes*.

Definition 13.1.1.

    (a) *Let L_0 be the language of the second order bounded arithmetic but without the symbol #. The language L_{i+1} is obtained from the language L_i by adding for every bounded L_i-formula*

$$\theta(x_1, \ldots, x_k, y_1, \ldots, y_\ell)$$

with $k + \ell$ free variables a new function symbol $F_{\theta,\overline{y}}^{\overline{x}}(\overline{x}; \overline{y})$. The language L_ω is the union $\bigcup_i L_i$.

(b) The theory $I\Delta_0(\alpha)^{count}$ is a theory in the language L_ω axiomatized by PA^- (see Definition 5.1.1), the induction axioms for all bounded L_ω-formulas, and by all axioms of the form

$$F_{\theta,\overline{y}}^{\overline{x}}(\overline{0}; b_1, \ldots, b_\ell) := \overline{0}$$

and for $j = 1, \ldots, k$

$$F_{\theta,\overline{y}}^{\overline{x}}(a_1, \ldots, a_{j-1}, a+1, a_{j+1}, \ldots, a_k; b_1, \ldots, b_\ell)$$
$$:= F_{\theta,\overline{y}}^{\overline{x}}(a_1, \ldots, a_{j-1}, a, a_{j+1}, \ldots, a_k; b_1, \ldots, b_\ell)$$
$$+F_{\theta,\overline{y},x_j}^{x_1,\ldots,x_{j-1},x_{j+1},\ldots,x_k}(a_1, \ldots, a_{j-1}, a_{j+1}, \ldots, a_k; b_1, \ldots, b_\ell, a)$$

The meaning of the function $F_{\theta,\overline{y}}^{\overline{x}}(\overline{x}; \overline{y})$ is

$$F_{\theta,\overline{y}}^{\overline{x}}(\overline{a}; \overline{b}) = \left| \{ \overline{x} \in a_1 \times \ldots \times a_k \mid \theta(\overline{x}, \overline{b}) \} \right|$$

We note a simple logical property of $I\Delta_0(\alpha)^{count}$.

Lemma 13.1.2. Let T be the theory $I\Sigma_0^{1,b}$ (Definition 5.5.2) in the language of $I\Delta_0(\alpha)^{count}$ (i.e., without the # symbol, in particular), plus the axiom

$$\forall\alpha\exists u, f, \ Enum(f, u, \alpha)$$

where Enum is the formula from Lemma 5.5.14.

Then the theory T is conservative over the theory $I\Delta_0(\alpha)^{count}$ w.r.t. all $\forall\Delta_0(\alpha)$-formulas.

Proof. Note that any model of $I\Delta_0(\alpha)^{count}$ can be expanded to a model of the theory T, leaving the first order part unchanged. Q.E.D.

The following lemma is not surprising.

Lemma 13.1.3. The theory $I\Delta_0(\alpha)^{count}$ proves the pigeonhole principle $\forall x PHP(R, x)$.

Proof. Assume $\neg PHP(R, a)$, that is, $R(x_1, x_2)$ is a graph of a 1–1 map from $a + 1$ into a.

By induction on y and z the theory $I\Delta_0(\alpha)^{count}$ proves for $y \leq a+1$ and $z \leq a$

$$F_R^{x_1,x_2}(y, a;) = y \quad \text{and} \quad F_R^{x_1,x_2}(a+1, z;) \leq z$$

and hence

$$a + 1 = F_R^{x_1,x_2}(a+1, a;) \leq a$$

which is a contradiction. Q.E.D.

Next we define the *cutting planes proof system*.

Definition 13.1.4. *A CP-inequality in* $x_1, \ldots, x_n$ *is an inequality of the form*

$$a_1 x_1 + \ldots a_n x_n \geq b$$

where $a_1, \ldots, a_n, b$ *are integers.*

CP-inference rules are the following four rules allowing us to infer a CP-inequality from other CP-inequalities:

1. *initial*

$$\frac{}{x_i \geq 0} \quad and \quad \frac{}{-x_i \geq -1}$$

2. *addition*

$$\frac{\sum_i a_i x_i \geq b \quad \sum_i c_i x_i \geq d}{\sum_i (a_i + c_i) x_i \geq b + d}$$

3. *multiplication*

$$\frac{\sum_i a_i x_i \geq b}{\sum_i (c a_i) x_i \geq c b}$$

where c *is a nonnegative integer*

4. *division*

$$\frac{\sum_i a_i x_i \geq b}{\sum_i (a_i / k) x_i \geq \lceil b / k \rceil}$$

whenever k *is a positive integer dividing all* a_i

A cutting plane *refutation (a CP-refutation, shortly) of CP-inequalities* $I_1, \ldots, I_k$ *is a sequence* $J_1, \ldots, J_\ell$ *such that each* J_i *is either one of* $I_1, \ldots, I_k$ *or derived from the previous inequalities* $J_1, \ldots, J_{i-1}$ *by one of the CP-rules, and such that the last inequality* J_ℓ *is* $0 \geq 1$.

The size *of a CP-inequality* $\sum_i a_i x_i \geq b$ *is* $|b| + \sum_i |a_i|$. *The size of a CP-proof is the sum of the sizes of its steps.*

The proof system CP is complete, meaning that any set of CP-inequalities without a 0–1 solution has a CP-refutation (cf. Cook, Coullard, and Turán 1987).

For the following lemma define that a clause $C = \{x_{i_1}, \ldots, x_{i_a}, \neg x_{j_1}, \ldots, \neg x_{j_b}\}$ is represented by a CP-inequality $I(C)$:

$$\sum_{i \in U \setminus V} x_i - \sum_{j \in V \setminus U} x_j \geq 1 - b$$

where $U = \{i_1, \ldots, i_a\}$ and $V = \{j_1, \ldots, j_b\}$.

Lemma 13.1.5. *The cutting planes proof system p-simulates the resolution system.*

Proof. Let $x_1, \ldots, x_n$ be all atoms. If a clause contains both x_k and $\neg x_k$ then the CP-inequality representing the clause can be inferred from $0 \geq 0$ (which follows, e.g., from $x_i \geq 0$ by multiplying both sides by 0) by adding to it some initial inequalities $x_i \geq 0$ and $-x_i \geq -1$.

Let C_1 and C_2 be two clauses such that $x_j \in C_1$ and $\neg x_j \in C_2$, and let C be the resolvent of C_1 and C_2 w.r.t. x_j

$$C = \left(C_1 \setminus \{x_j\}\right) \cup \left(C_2 \setminus \{\neg x_j\}\right)$$

By the preceding material we may assume that for any $k \neq j$, at most one of the literals $x_k, \neg x_k$ occurs in $C_1 \cup C_2$. Moreover, we may also assume that $\neg x_j \notin C_1$ and $x_j \notin C_2$ as otherwise $I(C)$ can be obtained from $I(C_2)$ if $\neg x_j \in C_1$ (resp. from $I(C_1)$) if $x_j \in C_2$, by adding to it suitable initial inequalities $x_i \geq 0, -x_i \geq -1$.

Let $I(C_1) : \sum a_i x_i \geq 1 - b$ and $I(C_2) : \sum c_i x_i \geq 1 - d$. Note that all a_i, c_i are 0, 1 or -1. Add to the inequality $I(C_1) + I(C_2)$

$$\sum (a_i + c_i) x_i \geq 2 - b - d$$

all initial inequalities $x_i \geq 0$ if $a_i + c_i = 1$ and all $-x_i \geq -1$ if $a_i + c_i = -1$ to get an inequality J

$$\sum u_i x_i \geq 2 - 2b - 2d + 1$$

where all u_i are 0, 2 or -2. Note that $u_j = 0$ as $a_j + c_j = 0$. Apply to J the division rule with $k = 2$

$$\sum \frac{u_i}{2} x_i \geq \left\lceil 1 - (b + d) + \frac{1}{2} \right\rceil = 1 - (b + d - 1)$$

It is easy to verify that J represents the resolvent of C_1, C_2 w.r.t. x_j. Hence one resolution inference has a size $O(n)$ CP-simulation. This readily implies the lemma.

Q.E.D.

Lemma 13.1.6. *Let PHP_n^{CP} be the set of CP-inequalities representing the clauses of $\neg PHP_n$. That is, PHP_n^{CP} consists of*

$$x_{i1} + \cdots + x_{in} \geq 1, \qquad for\ 1 \leq i \leq n + 1$$

$$-x_{ik} - x_{jk} \geq -1, \qquad for\ 1 \leq i < j \leq n + 1, \qquad and\ 1 \leq k \leq n.$$

Then there is a polynomial size CP-refutation of the set PHP_n^{CP}.

Proof. Summing up all first inequalities gives

$$\sum_{ij} x_{ij} \geq n + 1$$

For a fixed k derive the inequalities ($r = 2, \ldots, n + 1$)

$$-x_{1k} - x_{2k} - \cdots - x_{rk} \geq -1$$

For $r = 2$ this is just an inequality of PHP_n^{CP}. The inequality for $r + 1$ is obtained by summing $r - 1$ copies of

$$-x_{1k} - x_{2k} - \cdots - x_{rk} \geq -1$$

with all inequalities of PHP_n^{CP} of the form

$$-x_{ik} - x_{r+1\,k} \geq -1, \qquad 1 \leq i \leq r$$

obtaining the inequality

$$-rx_{1k} - \cdots - rx_{r+1\,k} \geq 1 - 2r$$

which after division by r entails the inequality for $r + 1$.

Sum up all inequalities

$$-x_{1k} - \cdots - x_{n+1\,k} \geq -1$$

for $k = 1, \ldots, n$ to get

$$\sum_{ij} -x_{ij} \geq -n$$

and adding this to the first inequality of the proof yields

$$0 \geq 1$$

The length of any inequality is $O(n \cdot \log n)$ and there are $O(n^3)$ steps: That is, the total size is $n^{O(1)}$. \hfill Q.E.D.

Note that the lemma together with Theorem 12.5.3 implies that the constant-depth systems do not p-simulate CP. The following theorem is the main result of this section. Recall Definition 9.1.1.

Theorem 13.1.7. *Assume that $\theta(x)$ is a bounded $\Delta_0(\alpha)$-formula in the language L_ω and that*

$$I\Delta_0(\alpha)^{count} \vdash \forall x \theta(x)$$

Then there is a polynomial $p(x)$ such that all formulas $\langle \theta \rangle_n$ have size $\leq p(n)$ F-proofs.

Proof. The simulation is constructed as in Theorem 9.1.3. We only need to show how to simulate the new axioms from Definition 13.1.1, that is, how to define the formulas $p_{\bar{a},\bar{b},j}$ of the translation of formulas $\langle F_{\theta,\bar{y}}^{\bar{x}}(\bar{x}; \bar{y}) = z \rangle_{\bar{a},\bar{b},j}$ and how to prove the translations of the axioms. To simplify the notation we shall consider the formulas $\theta(x)$ with one free variable only (parameters implicit) and we shall use the atoms p_i instead of the formulas $\langle \theta \rangle_i$.

For any fixed n we need to define the formulas q_{ij} representing $F_\theta^x(i) = j$. A naive approach would be to use the inductive property of F to define $q_{00} := 0$ and

$$q_{i+1\ j} := (q_{ij} \wedge \neg p_i) \vee (q_{i\ j-1} \wedge p_i)$$

This, however, gives q_{nj} of size $\Omega(2^n)$. A slightly less naive method would be to define F on subintervals of n of the form $[u \cdot (n/2^v),\ (u+1) \cdot (n/2^v)),\ u < 2^v$, and proceed by induction on $v = \log n, \log n - 1, \ldots, 0$; this yields formulas of size $n^{O(\log n)}$ only.

We resolve this problem by using the *carry–save addition*. That is a technique allowing us to compute the sum of n numbers with n bits each by a Boolean formula of size $n^{O(1)}$. Thinking about $p_1, \ldots, p_n$ as n numbers with one bit each we shall find size $n^{O(1)}$ formulas computing the $\log n$ bits $r_k^0, \ldots, r_k^{\log n - 1}$ of $\sum_{i<k} p_i$ in terms of $p_1, \ldots, p_n$ and then put

$$q_{kj} := \bigwedge_{\tilde{j}(i)=1} r_k^i \wedge \bigwedge_{\tilde{j}(i)=0} \neg r_k^i$$

where $\tilde{j}(0), \ldots, \tilde{j}(\log n - 1)$ are the bits of j. From the construction of r_k^i it will be apparent that the system F can verify the inductive property

$$q_{kj} \equiv (q_{k-1,j} \wedge \neg p_{k-1}) \vee (q_{k-1,j-1} \wedge p_{k-1})$$

We define the formulas (in $p_1, \ldots, p_n$) a_{uv}^w and b_{uv}^w, where $u = 0, \ldots, \ell - 1$, $v = 0, \ldots, (n/2^u) - 1$, $w = 0, \ldots, \ell - 1$, and $\ell = \lceil \log n \rceil$. Denote by a_{uv} the ℓ-tuple $(a_{uv}^0, \ldots, a_{uv}^{\ell-1})$, similarly b_{uv}.

Put

$$a_{0v}^w = \begin{cases} p_v & \text{if } w = 0 \\ 0 & \text{if } w > 0 \end{cases}$$

and $b_{0v} := \bar{0}$, and then define

$$a_{u+1,v} := A(a_{u,2v}, b_{u,2v}, a_{u,2v+1}, b_{u,2v+1})$$

and

$$b_{u+1,v} := B(a_{u,2v}, b_{u,2v}, a_{u,2v+1}, b_{u,2v+1})$$

where A and B are ℓ-tuples of propositional formulas in 4ℓ atoms defining the *carries* and the *sums* in the sum of the four ℓ-bit-long numbers $a_{u,2v}, b_{u,2v}, a_{u,2v+1}, b_{u,2v+1}$, and are defined by

$$A := A_0(A_0(a_{u,2v}, b_{u,2v}, a_{u,2v+1}), B_0(a_{u,2v}, b_{u,2v}, a_{u,2v+1}), b_{u,2v+1})$$

and

$$B := B_0(A_0(a_{u,2v}, b_{u,2v}, a_{u,2v+1}), B_0(a_{u,2v}, b_{u,2v}, a_{u,2v+1}), b_{u,2v+1})$$

where

$$A_0^w(\overline{x}, \overline{y}, \overline{z}) := x^w \oplus y^w \oplus z^w$$

and $B_0^0(\overline{x}, \overline{y}, \overline{z}) := 0$ and

$$B_0^w(\overline{x}, \overline{y}, \overline{z}) := (x^{w-1} \wedge y^{w-1}) \vee (x^{w-1} \wedge z^{w-1}) \vee (z^{w-1} \wedge y^{w-1})$$

if $0 < w < \ell$.

The formula $t \oplus s$ is a shorthand for $(t \wedge \neg s) \vee (\neg t \wedge s)$. Note that A_0^w is the sum modulo 2 of the wth bits of $\overline{x}$, $\overline{y}$, $\overline{z}$ and that B_0^w is the carry on the wth place. Hence the sum (modulo 2^ℓ, but all our numbers are smaller than 2^ℓ) of numbers $\overline{x}$, $\overline{y}$, $\overline{z}$ is equal to the sum of $A_0(\overline{x}, \overline{y}, \overline{z})$ and $B_0(\overline{x}, \overline{y}, \overline{z})$.

We shall use some fixed formula $C(\overline{e}, \overline{f})$ in 2ℓ atoms representing the sum of two ℓ-bit-long numbers. We take for C the disjunctive normal form of the associated Boolean function. Such a formula has size $2^{O(\ell)} = n^{O(1)}$ and F can prove all the usual properties of the addition by simply considering all assignments.

Define

$$c_{uv} := C(a_{uv}, b_{uv})$$

Hence c_{uv} are the bits of $\sum_{v \cdot 2^u \le i < (v+1) \cdot 2^u} p_i$, and $c_{\ell-1\, 0}$ are the bits of the wanted sum $\sum_i p_i$. This allows us to put

$$r_k := c_{\ell-1\, 0}(p_1, \dots, p_{k-1}, 0, \dots, 0)$$

The inductive property of q_{kj} then follows from

$$p_k \to C(r_k, 0 \dots 0, 1) = r_{k+1}$$

$$\neg p_k \to r_k = r_{k+1}$$

which is verified by induction on u in

$$p_k \equiv (C(c_{uv}(p_k/0), 0 \dots 0, 1) = c_{uv})$$

where v is chosen to be the unique number such that $k \in [v \cdot 2^u, (v+1) \cdot 2^u)$.

All these verifications rest only on the F-provability of the basic properties of the addition for formula C that are available by our choice of C. Q.E.D.

The theorem and Lemma 13.1.3 entail the following corollary.

Corollary 13.1.8 (Buss 1987). *There are polynomial size Frege proofs of the pigeonhole principle PHP_n.*

From Theorem 12.2.4 we know that constant-depth systems cannot p-simulate system F. Note that the last corollary and Theorem 12.5.3 are also proof of this fact.

A CP-inequality with variables $x_1, \ldots, x_n$ can be coded by an $(n+1)$-tuple of numbers, each of them represented by a set of its bits. So the whole inequality is represented by a set, as is any CP-refutation. Recall a similar situation with the coding of F- and EF-proofs in Section 9.3.

Theorem 13.1.9. *The theory* $I\Delta_0(\alpha)^{count}$ *proves the soundness of the cutting planes proof system CP.*

Proof. It is enough to show that the theory T from Lemma 13.1.2 can define the addition, multiplication, and division of numbers coded by sets of their bits and prove the basic properties, say PA^-. This is done similarly to the proof of Lemma 5.5.4; there we used $\Delta_1^{1,b}$-induction to verify the basic properties, but we may replace that by a $\Sigma_0^{1,b}$-induction with the parameter (the multiplication table, etc.) witnessing the $\Sigma_1^{1,b}$-definition of the operation. These witnesses are definable by a $\Sigma_0^{1,b}$-formula with the help of the new counting functions. We leave the details to the reader.   Q.E.D.

The following corollary is obtained from the previous theorem and Theorem 13.1.7 analogously with 9.3.17 followed from 9.2.5 and 9.3.13.

Corollary 13.1.10. *Frege system F p-simulates cutting planes proof system CP.*

13.2. An approach to lower bounds

This section points out an open problem rather than presenting results.

The only known lower bound to the size of F-proofs is provided by Lemma 4.4.12. Its general form, with an identical proof, is the following lemma.

Lemma 13.2.1. *Let τ be a tautology that is not a substitution instance of any tautology of a smaller size. Denote by $s(\tau)$ the sum of the sizes of all subformulas of τ.*

Then every F-proof π of τ must have size at least

$$|\pi| = \Omega(s(\tau))$$

The weakness of the lemma is not so much in the poor bounds it provides (as always $s(\tau) = O(|\tau|^2)$) as in the fact that the value of $s(\tau)$ depends on a syntactic form of τ rather than on the meaning of τ. For example, a disjunction σ of n literals has the value $s(\sigma) = \Omega(n^2)$ if the brackets are associated to the left but only the value $s(\sigma) = O(n \cdot \log n)$ if the disjunctions are arranged in a balanced binary tree.

We would therefore be interested in obtaining a lower bound of the form at least $n^{1+\epsilon}$ but robust w.r.t. to the distribution of brackets and similar syntactical questions. In particular, we would like to obtain such a bound to the F-refutations of an unsatisfiable formula represented by a set of small clauses. Such a formulation seems to depend least on the syntactic form.

Tautologies that might be useful in this respect were considered by Tseitin (1968) . Let G be an undirected graph with n vertices labeled by 0 and 1 such that an odd number of vertices get label 1. Let each edge of G have an associated atom. Consider n sets of clauses $\{C_v^i\}_i$, one set for each vertex v, representing in the conjunctive normal form $\bigwedge_i C_v^i$ the formula

$$e_1 \oplus \cdots \oplus e_t = \ell_v$$

where $e_1, \ldots, e_t$ are the atoms associated to the edges incident with v and ℓ_v is the label of v. Clearly $i \leq 2^t$.

Denote by $C(G)$ the set of all clauses C_v^i, for all v and i. The set is unsatisfiable (as the sum of the labels of the nodes has to be even but is odd by the hypothesis).

There is a relation of the formula $C(G)$ to the instances of the counting principle Count_n^2. Let $G = (V, E)$ be a graph with labeling α of its edges satisfying $C(G)$. We may assume that G has exactly one vertex v_1 labeled by 1 and that for that vertex $\deg(v_1) = 1$ (a graph satisfying this can be defined from G in a simple way, e.g., by a $\Delta_0(G)$-formula). Then the labeling α of the edges defines a complete pairing on the set

$$A = \{\langle\{u, v\}, u\rangle \mid u \in V, \{u, v\} \in E\} \setminus \{\langle e, v_1\rangle\}$$

where e is the unique edge incident with v_1: pair $\langle\{u, v\}, u\rangle$ with $\langle\{u, v\}, v\rangle$ if $\alpha(\{u, v\}) = 0$ and $\langle\{u, v\}, v\rangle$ with $\langle\{u', v\}, v\rangle$ if $\{u, v\}$ and $\{u', v\}$ are the $(2i+1)$st and the $(2i+2)$nd edge incident with v and labeled 1 by α. The set A has an odd cardinality, however. This shows that $C(G)$ can be refuted from an instance of Count_n^2. Note that this refutation is polynomial size and is bounded depth if the degree of G is bounded.

The opposite implication also works. From any equivalence relation R falsifying Count_n^2 one can obtain a $\Delta_0(R)$-definable labeling α of the edges of the complete graph K_{2n+1} whose exactly one vertex is labeled by 1, such that α satisfies $C(K_{2n+1})$.

Assume that G has the degree bounded by an independent constant, then $|C(G)| = O(n)$. Urquhart (1987) showed that bounded degree expanders yield sets of clauses with exponential size resolution proofs. On the other hand, the clauses have polynomial size F-refutation for any graph G, utilizing the counting in F from the previous section. We would like to find bounded degree graphs for which any F-refutation of $C(G)$ must have the size $n^{1+\Omega(1)}$.

Next we consider a possible approach to showing such a lower bound (for suitable graphs). Consider the following search problem. Given a labeling α of the edges of G find a vertex v for which the parity condition

$$e_1 \oplus \cdots \oplus e_t = \ell_v$$

is violated. We want to solve the problem by using a modification of decision trees (cf. Definition 3.1.12) where at a node the tree branches into 2^t subtrees depending on the labels of the edges incident with a vertex v.

Denote by $D(G)$ the minimal height of such a tree solving the search problem for all labelings of the edges of G (labels of vertices are fixed: a part of the specification of G).

For example, if G is a complete graph with just one vertex labeled by 1 then $D(G) = n - 1$. If G is a circle (again with one vertex labeled by 1) then $D(G) \leq \log n$, by the binary search.

Assume we have an F-refutation π of $C(G)$. Knowing the truth values of all formulas in π allows us to construct a path $\psi_0, \ldots, \psi_r$ through π (similarly to the proof of Lemma 9.5.1) and to solve the search problem in this way. Define $newsize(\psi)$ for the formula ψ to be the minimal number k such that there is a set W of k vertices such that the truth value of ψ can be determined knowing only the labels of the edges incident with a vertex from W. Hence

$$D(G) \leq \ newsize(\pi) := \sum_{\psi} \ newsize(\psi)$$

where ψ runs over steps of π.

We would like to find a type of graph G and a space of random partial evaluations ρ leaving $\sim p \cdot n$ edges unlabeled and having with a high probability two properties
1. $D(G^\rho) = \Omega(pn)$
2. $newsize(\pi^\rho) = O\left(p^{1+\epsilon} \cdot newsize(\pi)\right)$
where $\epsilon > 0$. Choosing $p \sim n^{-1+\delta}$ gives

$$\Omega(pn) = D(G^\rho) \leq \ newsize(\pi^\rho) = O\left(p^{1+\epsilon} \ newsize(\pi)\right)$$

and thus

$$\Omega(p^{-\epsilon}n) = \Omega\left(n^{1+[1-\delta]\epsilon}\right) = \ newsize(\pi) \leq |\pi|$$

This is motivated by the method of Subbotovskaja (1961), which showed that the expected shrinking of a formula ψ in atoms $x_1, \ldots, x_n$ under random restrictions from Lemma 3.1.11 is by a factor $O(p^{3/2})$. We would need therefore a type of graph G for which the decision complexity $D(G)$ is $\Omega(n)$ and such that the restricted graphs G^ρ are of the same type (which would yield condition 1) and for which the shrinking of the formulas w.r.t. the newsize occurs with factor $p^{1+\Omega(1)}$. Again, expander-type graphs might be good candidates.

13.3. Boolean valuations

In this section we define the concept of *Boolean valuations*, which was introduced in Krajíček (1995a). It aims at describing a framework for proving the lower bounds to ℓ_F (recall the measure ℓ_F from Section 4.4: the number of distinct formulas in a proof). In the rest of the section we consider a Frege system F in the language $\{0, 1, \neg, \vee, \wedge\}$. We have to start with three rather formal definitions.

Definition 13.3.1. *Let* Γ *be a set of formulas. We say that a formula* τ *can be proved within* Γ, $\vdash_\Gamma$ *in symbols, if there is an F-proof of* τ *in which only formulas from the set* Γ *occur as subformulas.*

Definition 13.3.2.

1. *A* partial Boolean algebra *is a structure*

$$\mathcal{B}(0, 1, \wedge, \vee, \neg)$$

in the language of Boolean algebras $\{0, 1, \wedge, \vee, \neg\}$ *but in which the operations* $\wedge, \vee, \neg$ *are only partial; in which the elements 0 and 1 are distinct; and that satisfies those axioms* $t = s$ *of the* theory of Boolean algebras *BA for which both* t *and* s *are defined. For definiteness we take as the axioms of BA the instances of the following identities:*

   (a) $\neg 0 = 1$ $\neg 1 = 0$, $0 \vee u = u$ *and* $1 \wedge u = u$

   (b) $u \vee \neg u = 1$, $\qquad u \wedge \neg u = 0$

   (c) $u \vee v = v \vee u$, $\qquad u \wedge v = v \wedge u$

   (d) $u \vee (v \vee w) = (u \vee v) \vee w$, $\qquad u \wedge (v \wedge w) = (u \wedge v) \wedge w$

   (e) $u \vee (v \wedge w) = (u \vee v) \wedge (u \vee w)$, $\qquad u \wedge (v \vee w) = (u \wedge v) \vee (u \wedge w)$

2. *A* homomorphism $h : \mathcal{B}_1 \to \mathcal{B}_2$ *of a partial Boolean algebra* $\mathcal{B}_1$ *into* $\mathcal{B}_2$ *is a map* h *of the universe of* $\mathcal{B}_1$ *into the universe of* $\mathcal{B}_2$ *such that* $h(0_{\mathcal{B}_1}) = 0_{\mathcal{B}_2}$, $h(1_{\mathcal{B}_1}) = 1_{\mathcal{B}_2}$, *and*

   (a) $\neg h(u)$ *is defined and equal to* $h(\neg u)$, *whenever* $\neg u$ *is defined in* $\mathcal{B}_1$,

   (b) $h(u) \circ h(v)$ *is defined and equal to* $h(u \circ v)$, *whenever* $u \circ v$ *is defined in* $\mathcal{B}_1$ *(*$\circ = \wedge, \vee$*).*

3. *A* congruence relation *on a partial Boolean algebra* $\mathcal{B}$ *is a partition* $\cong$ *of the universe of* $\mathcal{B}$ *such that*

   (a) $u \cong v$ *implies* $(\neg u) \cong (\neg v)$, *provided both* $\neg u, \neg v$ *are defined in* $\mathcal{B}$,

   (b) $u \cong v$ *and* $u' \cong v'$ *imply* $(u \circ u') \cong (v \circ v')$, *provided both* $u \circ u'$ *and* $v \circ v'$ *are defined in* $\mathcal{B}$ *(*$\circ = \wedge, \vee$*).*

See Gratzer (1979) for details on partial algebras. The following definition is crucial.

Definition 13.3.3. *Let* Γ *be a set of formulas. A* Boolean valuation *of* Γ *is a map*

$$v : \Gamma \to \mathcal{B}$$

of the formulas from Γ *into a partial Boolean algebra* $\mathcal{B}(0, 1, \wedge, \vee, \neg)$ *satisfying the following conditions:*

1. $v(0) = 0_\mathcal{B}$ *and* $v(1) = 1_\mathcal{B}$, *if* $0 \in \Gamma$ *(resp.* $1 \in \Gamma$)
2. $\neg v(\psi)$ *is defined and equal to* $v(\neg \psi)$, *whenever* $\psi, \neg \psi \in \Gamma$
3. $v(\psi) \circ v(\phi)$ *is defined and equal to* $v(\psi \circ \phi)$, *whenever* $\psi, \phi, \psi \circ \phi \in \Gamma$, *for* $\circ = \vee, \wedge$.

Lemma 13.3.4. *There are* $c > 0$ *and an assignment* $*$ *assigning to any finite set* Γ *of formulas and any F-proof* π *of* τ *within* Γ *a set of formulas* $\Gamma_\pi^* \supseteq \Gamma$ *such that*

$$|\Gamma_\pi^*| \le c|\Gamma| \quad and \quad \max \{dp(\phi)| \ \phi \in \Gamma_\pi^*\} \le c + \max \{dp(\phi)| \ \phi \in \Gamma\}$$

and such that there is no Boolean valuation $v : \Gamma_\pi^* \to \mathcal{B}$ *in which* $v(\tau) \ne 1_\mathcal{B}$.

Proof. For R (and $p = (p_1, \ldots, p_n)$)

$$\frac{\gamma_1(p), \ldots, \gamma_r(p)}{\gamma_0(p)}$$

a rule of the system F let $T_R(p)$ be the set of all Boolean terms occurring as the subterms in some fixed equational proof e_R of the equation

$$\gamma_0(p) = 1$$

from the equations

$$\gamma_i(p) = 1, \qquad i = 1, \ldots, r$$

in the theory BA. Such a derivation exists as R is sound.

Let $\pi = \theta_1, \ldots, \theta_k$ be an F-proof of τ within Γ. Define the set Γ_π^* to be the smallest set satisfying

1. $\Gamma_\pi^* \supseteq \Gamma$
2. whenever

$$\frac{\gamma_1(\psi_1, \ldots, \psi_n), \ldots, \gamma_r(\psi_1, \ldots, \psi_n)}{\gamma_0(\psi_1, \ldots, \psi_n)}$$

is an inference in π using the rule R then

$$T_R(\psi_1, \ldots, \psi_n) \subseteq \Gamma_\pi^*$$

It is clear that the depth of Γ_π^* increases by a constant over $dp(\Gamma)$ and the cardinality of Γ_π^* is proportional to Γ.

Construct an equational derivation of

$$\tau = 1$$

constructing consecutively the derivations of

$$\theta_i = 1$$

using the derivations e_R if θ_i was inferred by using rule R. As the valuation v preserves the connectives on Γ_π^*, clearly

$$v(\theta_i) = 1_B$$

all i, and so also

$$v(\tau) = 1_B$$

That is a contradiction.   Q.E.D.

The following construction is a converse to the previous lemma.

Lemma 13.3.5. *There are a constant $c > 0$ and an assignment $+$ assigning to any finite set Γ of formulas a set of formulas $\Gamma^+ \supseteq \Gamma$ such that*

$$|\Gamma^+| \leq |\Gamma|^c \quad and \quad \max\{dp(\phi)|\ \phi \in \Gamma^+\} \leq c + \max\{dp(\phi)|\ \phi \in \Gamma\}$$

and such that for each $\tau \in \Gamma$, if there is no F-proof of τ within Γ^+ then there exists a Boolean valuation $v : \Gamma \to B$ in which $v(\tau) \neq 1_B$.

Proof. For $t_1 = t_2$ an identity of the language of BA define the set of formulas F_{t_1,t_2} in the following way:
1. for every subterm s of t_i, $i = 1, 2$, introduce a new atom q_s^i
2. for $\circ = \vee, \wedge, i = 1, 2$, and $s_1 \circ s_2$ a subterm of t_i put the formula

$$q_{s_1 \circ s_2}^i \equiv (q_{s_1}^i \circ q_{s_2}^i)$$

   into the set A_{t_1,t_2}
3. for $i = 1, 2$ and $\neg s$ a subterm of t_i put the formula

$$q_{\neg s}^i \equiv (\neg q_s^i)$$

   into the set A_{t_1,t_2}
4. set F_{t_1,t_2} is the set of all formulas occurring in a fixed F-proof π_{t_1,t_2} of the formula

$$\bigwedge A_{t_1,t_2} \to q_{t_1}^1 \equiv q_{t_2}^2$$

($\bigwedge$ bracketed to the left, say). Recall that we use $\equiv$ as an abbreviation.

Define the set Γ^+ to be the smallest set $\Gamma^+ \supseteq \Gamma$ closed under subformulas and satisfying the following conditions:

1. for any $\alpha, \beta, \gamma \in \Gamma$, all formulas

$$\alpha \equiv \beta \to \beta \equiv \alpha, \qquad \alpha \equiv \beta \to (\beta \equiv \gamma \to \alpha \equiv \gamma)$$

$$0 \equiv 1 \to 0, \qquad \alpha \equiv 1 \to \alpha$$

are in Γ^+

2. whenever $t_1 = t_2$ is one of the identities $\neg a = \neg a$, $a \vee b = a \vee b$, $a \wedge b = a \wedge b$ then

$$F_{t_1,t_2}(a/\alpha, b/\beta) \subseteq \Gamma^+$$

for all $\alpha, \beta \in \Gamma$

3. whenever $t_1 = t_2$ is one of the axioms of BA and $\alpha_s^i \in \Gamma$ then

$$F_{t_1,t_2}(q_s^i/\alpha_s^i) \subseteq \Gamma^+$$

Clearly $dp(\Gamma^+) \leq c + dp(\Gamma)$ and $|\Gamma^+| \leq |\Gamma|^c$ for some constant $c \geq 1$.

Define the relation $\sim$ on Γ by

$$\phi \sim \psi \quad \text{iff} \quad \vdash_{\Gamma^+} \phi \equiv \psi$$

Condition 1 implies that $\sim$ is an equivalence relation and 2 implies that it is even a congruence relation. From 1 it also follows that τ is not $\sim$-equivalent to 1 as by the hypothesis of the lemma τ has no F-proof included in Γ^+.

From $\sim$ define a partial Boolean algebra $\mathcal{B}$ and a valuation $v : \Gamma \to \mathcal{B}$ by

(a) the elements of $\mathcal{B}$ are the congruence classes from $\Gamma / \sim$
(b) $0_{\mathcal{B}} = 0/\sim$, $1_{\mathcal{B}} = 1/\sim$
(c) $\neg a = \neg \alpha / \sim$, whenever $\alpha \in \Gamma$ and $\alpha \in a$
(d) $a \circ b = (\alpha \circ \beta)/\sim$, whenever $\alpha \circ \beta \in \Gamma$ and $\alpha \in a$, $\beta \in b$, $\circ = \vee, \wedge$
(e) $v(\alpha) := \alpha / \sim$

It is straightforward to verify using 2 and 3 that the operations $\neg, \vee, \wedge$ are correctly defined and that v is a Boolean valuation. As $v(\tau) \neq 1_{\mathcal{B}}$, we are done.   Q.E.D.

We shall state the main theorem on Boolean valuations.

Theorem 13.3.6 (Krajíček 1995a). *Let τ be a propositional formula and let n_τ be the maximal number n such that for every set Δ of $\leq n$ formulas containing τ, there is a Boolean valuation $v : \Delta \to \mathcal{B}$ in which $v(\tau) \neq 1_{\mathcal{B}}$.*

Then

$$n_\tau = O(\ell_F(\tau))$$

and

$$\ell_F(\tau) = n_\tau^{O(1)}$$

The same is valid if we restrict F to the depth d systems and consider for Δ only the sets of depth $\leq d + c$ formulas, c a constant depending on F only.

Proof. Let π be an F-proof of τ with $\ell_F(\tau)$ distinct formulas. Take Γ to be the set of the formulas occurring in π and put $\Delta := \Gamma_\pi^*$. By Lemma 13.3.4 there is no Boolean valuation $v : \Delta \to B$ in which $v(\tau) \neq 1_B$. Also $|\Delta| = O(|\Gamma|)$.

For the second inequality take Γ a set of $n_\tau + 1$ formulas containing τ for which there is no Boolean valuation v in which $v(\tau) \neq 1_B$. By Lemma 13.3.5 there must be an F-proof of τ within Γ^+, hence

$$\ell_F(\tau) \leq |\Gamma^+| = |\Gamma|^{O(1)} = (n_\tau)^{O(1)}$$

The generalization to the constant-depth systems holds because of the bounds to the depth of Γ_π^*, Γ^+ in Lemmas 13.3.4 and 13.3.5.   Q.E.D.

The reader should find modifying the concept of Boolean valuations to the extensions of F by a set $A \subseteq$ TAUT of extra axioms straightforward; require that $v(\alpha) = 1_B$ for all $\alpha \in A$. This gives nontrivial information, as we shall see in the next chapter that for any propositional proof system P in the sense of Definition 4.1.1 there is a polynomial time subset $A_P \subseteq$ TAUT such that the minimal size of P-proof of τ is bounded by $O(\ell_{F(A_P)}(\tau))$, where $F(A_P)$ is F augmented by A_P as extra axioms.

It is apparently quite difficult to construct a Boolean valuation v giving to τ a value different from 1_B. This is because a nontrivial lower bound to n_τ implies by the previous theorem and by Lemma 4.4.6 a nontrivial lower bound to the number of proof-steps (in F), and by Lemma 4.5.7 also to the size of EF-proofs of τ. No nontrivial lower bounds are known for these two measures.

However, in the case of constant-depth Frege systems there are nontrivial bounds (cf. Sections 12.3–5), and, in fact, there is a general construction allowing us to construct a Boolean valuation from a k-evaluation. It is a modification of the algebraic *direct limit construction* (see Gratzer 1979). We first describe a general setting and then the particular case arising from k-evaluations.

Let $(I, \leq)$ be a partial order that is decomposed into three levels I_1, I_2, I_3 such that for any eight $a_1, \ldots, a_8 \in I_1$ there is $b \in I_2$ such that $a_1, \ldots, a_8 \leq b$, and such that for any two $b_1, b_2 \in I_2$ there is $c \in I_3$ for which $b_1, b_2 \leq c$. Assume that for every $a \in I$ there is a *total* Boolean algebra B_a, and that for every $a \leq b$ there is an *embedding* h_{ab} of B_a into B_b such that the whole collection $(B_a)_a$, $(h_{ab})_{ab}$ forms a commutative diagram. That is

$$h_{ab} \circ h_{bc} = h_{ac}$$

whenever $a \leq b \leq c$.

Let the set X be the disjoint union of the universes of algebras B_a for $a \in I_1$. For $u \in B_a$ and $v \in B_b, a, b \in I_1$ define

$$u \cong v \quad \text{iff} \quad h_{ac}(u) = h_{bc}(v), \quad \text{some } c \in I_2$$

The following observation follows from the condition that for any two $b_1, b_2 \in I_2$ the algebras B_{b_1} and B_{b_2} can be jointly embedded into some $B_c, c \in I_3$.

Lemma 13.3.7. *Let $a_1, a_2 \in I_1$ and $b_1, b_2 \in I_2$ such that $a_1, a_2 \leq b_1, b_2$, and let $u_1 \in B_{a_1}, u_2 \in B_{a_2}$. Then*

$$h_{a_1 b_1}(u_1) = h_{a_2 b_1}(u_2) \quad \text{iff} \quad h_{a_1 b_2}(u_1) = h_{a_2 b_2}(u_2)$$

Lemma 13.3.8. *For any collection $(B_a)_a$, $(h_{ab})_{ab}$ satisfying the preceding conditions, the relation $\cong$ is a partition of the set X.*

Proof. It is clear that $\cong$ is reflexive and symmetric. The transitivity follows from Lemma 13.3.7 as any three $B_{a_1}, B_{a_2}, B_{a_3}, a_1, a_2, a_3 \in I_1$, can be simultaneously embedded in some $B_b, b \in I_2$. $\hspace{3cm}$ Q.E.D.

Definition 13.3.9. *Denote by $[u]$ the $\cong$-equivalence class of $u \in X$. A partial structure $\mathcal{B}$ is defined as follows ($u \in B_a, v \in B_b, w \in B_c$, and $a, b, c \in I_1$):*
  1. *the universe of $\mathcal{B}$ is the quotient $X/\cong$,*
  2. *$0_\mathcal{B} = [0_{B_i}]$ and $1_\mathcal{B} = [1_{B_i}]$,*
  3. *$\neg[u] = [v]$, provided $\neg h_{ad}(u) = h_{bd}(v)$ for some $d \in I_2$,*
  4. *$[u] \circ [v] = [w]$, provided $h_{ae}(u) \circ h_{be}(v) = h_{ce}(w)$ for some $e \in I_2$, $\circ = \wedge, \vee$.*

The partial structure $\mathcal{B}$ is called the limit of the collection $(B_a)_a$, $(h_{ab})_{ab}$.

Lemma 13.3.10. *The limit $\mathcal{B}$ of the collection $(B_a)_a$, $(h_{ab})_{ab}$ satisfying the conditions in Definition 13.3.9 is a partial Boolean algebra.*

Proof. Take an identity of BA, say the distributivity law

$$u_1 \wedge (u_2 \vee u_3) = (u_1 \wedge u_2) \vee (u_1 \wedge u_3)$$

Assume that in $\mathcal{B}$ it holds

$$[u_2] \vee [u_3] = [u_4], \qquad [u_1] \wedge [u_4] = [u_5]$$

and

$$[u_1] \wedge [u_2] = [u_6], \qquad [u_1] \wedge [u_3] = [u_7], \qquad [u_6] \vee [u_7] = [u_8]$$

for $u_i \in B_{a_i}, a_i \in I_1, i = 1, \ldots, 8$. We need to show that $[u_5] = [u_8]$. Assume otherwise and let $b \geq a_1, \ldots, a_8$ be such that all B_{a_i} can be simultaneously embedded in B_b. Then by Lemma 13.3.7 the preceding equalities hold for $h_{a_i b}(u_i)$'s in place of $[u_i]$'s too but also $h_{a_5 b}(u_5) \neq h_{a_8 b}(u_8)$. That is impossible as B_b satisfies all identities of BA.

An analogous argument applies to all other identities of BA as they all contain at most eight different subterms. $\hspace{3cm}$ Q.E.D.

The following theorem is then clear.

Theorem 13.3.11. *Let Γ be a set of formulas and assume that $(I, \leq)$ and the collection $(B_a)_a$, $(h_{ab})_{ab}$ satisfies the preceding conditions. Let $\mathcal{B}$ be the limit of the collection $(B_a)_a$, $(h_{ab})_{ab}$ as defined. Assume that $I_1 = \Gamma$ and that there is a map $f : \Gamma \to X$ such that $f(\phi) \in B_\phi$ for all $\phi \in \Gamma$.*

Moreover assume that for any collection $\{\phi_1, \phi_2, \phi_3\} \subseteq \Gamma$ there is $b \in I_2$ such that the map

$$h_{\phi,b} \circ f : \phi \mapsto h_{\phi,b}(f(\phi))$$

is a Boolean valuation of $\{\phi_1, \phi_2, \phi_3\}$ into B_b.
Then the map

$$v : \phi \in \Gamma \mapsto [f(\phi)] \in \mathcal{B}$$

is a Boolean valuation *of Γ.*

Theorem 13.3.11 can be used to obtain Boolean valuations from complete systems. As an example we shall show that a *k-evaluation* of a set of formulas constructed in Section 12.5 gives rise via the limit construction to a *Boolean valuation* of that set.

The next lemma follows from Lemmas 12.3.6 and 12.3.7. $\mathcal{M} = \mathcal{M}^{MOD_a}$ in the rest of the section.

Lemma 13.3.12. *Let S and T be a k-complete system (resp. an ℓ-complete system), and let $\mathcal{B}_S$ and $\mathcal{B}_T$ be the Boolean algebras of the subsets of S and T, respectively. Assume that T refines S and that $k + \ell \leq n$. Define map $h : \mathcal{B}_S \to \mathcal{B}_T$ by*

$$h(H) := T(H)$$

Then h is an embedding *of $\mathcal{B}_S$ into $\mathcal{B}_T$.*

Let a pair (H, S) be a *k-evaluation* of the set Γ of formulas built from the atoms of the form p_X, $X \in [A]^a$, and assume that $24k \leq n$. Let $[\Gamma]^{\leq t}$ denote the set of $\leq t$-element subsets of Γ. Set $I_1 := \Gamma$, $I_2 := [\Gamma]^{\leq 8}$, and $I_3 := [\Gamma]^{\leq 16}$.

For $a \in I_i$ and $b \in I_j$ define $a \leq b$ iff $a \subseteq b$. To every set $b = \{\psi_1, \ldots, \psi_\ell\} \in I_j$ assign the Boolean algebra of the subsets of the *k-complete* system S_b

$$S_{\psi_1} \times \cdots \times S_{\psi_\ell}$$

defined as the *minimal* common refinement of all systems S_{ψ_u}, $u = 1, \ldots, \ell$. Such a system exists by Lemma 12.3.5 and is ℓk-complete. Define further the map $h_{ab} : \mathcal{B}_a \to \mathcal{B}_b$ by

$$h_{ab}(H) := S_b(H)$$

By 13.3.12 these maps are embeddings and by Lemma 12.3.6 the collection $(\mathcal{B}_a)_a$, $(h_{ab})_{ab}$ forms a commutative diagram. The next lemma is obvious.

Lemma 13.3.13. *For any three formulas $\psi_1, \psi_2, \psi_3 \in \Gamma$ there is $a \in I_2$ such that the composite map*

$$\psi \mapsto H_\psi \mapsto h_{\{\psi\}a} (H_\psi)$$

is a Boolean valuation of the set $\{\psi_1, \psi_2, \psi_3\}$.

Let $\mathcal{B}$ be a partial Boolean algebra that is the limit of the collection $(\mathcal{B}_a)_a$, $(h_{ab})_{ab}$. Define the map ν of formulas from Γ into $\mathcal{B}$ by

$$\nu(\psi) := [H_\psi]$$

That is, the value of ψ is the congruence class of H_ψ. The following lemma now follows from Theorem 13.3.11.

Lemma 13.3.14. *The map $\nu : \Gamma \to \mathcal{B}$ is a Boolean valuation of Γ. If $H_\phi = \emptyset$ for some $\phi \in \Gamma$ then $\nu(\phi) = 0_\mathcal{B}$.*

Theorem 12.4.3 implies the following lemma.

Lemma 13.3.15. *For every d there are $\epsilon > \delta > 0$ such that for every sufficiently large n and every set Γ of at most 2^{n^δ} depth d formulas built from the atoms p_X, $X \in [A]^a$, there is a partial a-partition $\rho \in \mathcal{M}$ such that*

$$|\rho| \leq n - n^\epsilon$$

and there exists a Boolean valuation

$$\nu : \Gamma^\rho \to \mathcal{B}$$

in which $\nu([Count_n^a]^\rho) \neq 1_\mathcal{B}$.

Note that the proofs of lower bounds from Theorems 12.5.7 and 12.5.10 consist of showing that the instances of PHP_n (resp. of $Count_n^q$ ($a = p$ in this case)) receive by ν the value $1_\mathcal{B}$ (cf. the remark after Theorem 13.3.6).

Let us close this section by giving an explicit description of the Boolean value in $\mathcal{B}$ assigned to $\phi \in \Gamma$ by the limit construction applied to k-evaluations, and by reexamining the role of the k-complete systems.

A formula ϕ gets the value

$$\nu(\phi) = \big\{(H, S) \mid (S \times S_\phi)(H) = (S \times S_\phi)(H_\phi)\big\}$$

where (H, S) runs over all $H \subseteq S$ and S a k-complete system.

Lemma 13.3.16. *Let $\mathcal{B}_a$ for some $a \in I_1$ be the Boolean algebra of the subsets of the k-complete system $S_a = \{\alpha_1, \ldots, \alpha_r\} \subseteq \mathcal{M}$. Let $b_i := [\{\alpha_i\}] \in \mathcal{B}$ be the values of $\{\alpha_i\}$ in $\mathcal{B}$.*

Then $b_1, \ldots, b_r$ form a partition of unity in $\mathcal{B}$:

1. $b_i \wedge b_j = 0_\mathcal{B}$ *for* $i \neq j$
2. $\bigvee_{1 \leq i \leq r} b_i = 1_\mathcal{B}$.

The proof is omitted.

13.4. Bibliographical and other remarks

The proof of PHP from the counting (13.1.3) was noted in Woods (1981) and Paris and Wilkie (1985). The proof system CP was defined in Cook et al. (1987), where Lemmas 13.1.5 and 13.1.6 are observed. Counting in F was developed in Buss (1987), where Corollary 13.1.8 was proved directly. Theorem 13.1.7 was noted in Krajíček (1995a) and we have followed Buss (1987) in the use of the carry-save addition. Corollary 13.1.10 is from Goerdt (1990). When the size of an inequality is measured by the sum of the absolute values of the coefficients, an exponential lower bound for CP was proved in Bonet, Pitassi, and Raz (1995) (see Krajicek (1995b) for an alternative proof via interpolation). Pudlák (unpublished) extended this to ordinary size. Cook et al. (1987) suggest that Tseitin's graph formulas based on bounded degree expanders are candidates for not having short CP-proofs.

For topics related to Lemma 13.2.1 see Bonet (1993). A communication complexity approach to the lower bounds for F is briefly discussed in Krajíček (1995a) (as are other related search problems). The graph tautologies were introduced in Tseitin (1968) and studied in Urquhart (1987). Hastad (1993) improved the shrinking factor to $p^{2-o(1)}$.

The next chapter is devoted to the question of *hard tautologies*. No natural combinatorial principles (like PHP or Count^a of Chapter 12) are known that would form plausible candidates for having only long proofs in F (but see Krajíček and Pudlák 1995). Arai (unpublished) suggested a principle of linear algebra

$$A^{n+1}\overline{x} = \overline{0} \rightarrow A^n\overline{x} = \overline{0}$$

(A an $n \times n$ 0–1 matrix, $\overline{x}$ a 0–1 vector), and Karchmer (unpublished) suggested that the Graham and Pollak theorem (saying that no $n-2$ complete bipartite subgraphs of the complete graph K_n can partition the edges) may be hard for F as all its known proofs utilize the linear algebra operations not known to be definable by polynomial size formulas. Both these principles are, however, provable in U_1^1 and thus admit quasi-polynomial size F-proofs (Theorem 9.1.6). This author suggested earlier (see Clote and Krajíček 1993) that a theorem of Bondy (1972) might be a good candidate, but Bonet, Buss, and Pitassi (1995) constructed its polynomial size F-proofs.

Urquhart (unpublished) posed the question whether the set $C(G)$, for G bounded degree expanders, must have exponential size constant-depth F-proofs. An affirmative answer might allow one to prove for constant-depth systems a result in the spirit of Chvátal and Szemerédi (1988).

The content of Section 13.3 is from Krajíček (1995a); the limit construction and Theorem 13.3.11 are new.

14

Hard tautologies and optimal proof systems

We shall study in this chapter the topic of hard tautologies: tautologies that are candidates for not having short proofs in a particular proof system. The closely related question is whether there is an optimal propositional proof system, that is, a proof system P such that no other system has more than a polynomial speed-up over P. We shall obtain a statement analogous to the NP-completeness results characterizing any propositional proof system as an extension of EF by a set of axioms of particular form. Recall the notions of a proof system and p-simulation from Section 4.1, the definitions of translations of arithmetic formulas into propositional ones in Section 9.2, and the relation between reflection principles (consistency statements) and p-simulations established in Section 9.3. We shall also use the notation previously used in Chapter 9.

14.1. Finitistic consistency statements and optimal proof systems

We shall denote by Taut (x) the Π_1^b-formula Taut$_0(x)$ from Section 9.3 defining the set of the (quantifierfree) tautologies, denoted TAUT itself.

Recall from Section 9.2 the definition of the translation

$$\phi(x) \rightarrow ||\phi||_{q(n)}^n(p_1, \ldots, p_n)$$

producing from a Π_1^b-formula a sequence of propositional formulas (Definition 9.2.1, Lemma 9.2.2). Also recall that a number a (or a formula, ... coded by a number a) is represented by a tuple $\tilde{a}$ of 0, 1: the bits of a. We use the notation $\geq_p$ for p-simulation from Definition 4.1.3.

Definition 14.1.1.

(a) *Let P be a propositional proof system. Function* $c_P(\tau) : TAUT \to N$ *is defined by*

$$c_P(\tau) := \min\{|\pi| \mid \pi \text{ is a } P\text{-proof of } \tau\}$$

(b) *Let P, Q be two propositional proof systems. Then system P is better than Q, $P \geq Q$ in symbols, iff there is a polynomial $p(x)$ such that*

$$\forall \tau \in TAUT, c_P(\tau) \leq p\big(c_Q(\tau)\big)$$

(c) *Propositional proof system P is optimal iff it is the greatest element of the quasi-order $\geq$.*

Observe that P is better than Q iff Q has a polynomial speed-up over P and that $P \geq_p Q$ implies $P \geq Q$ but the converse does not necessarily apply.

Problem. *Does there exist an optimal propositional proof system?*

Any proof system P that proves all tautologies in polynomial size is optimal; thus NP $=$ coNP implies the affirmative answer to the problem (cf. Theorem 4.1.2). It is unknown, however, whether the converse implication is also true.

Nontrivial information about the problem is provided by Theorem 9.3.17 and Corollary 9.3.18: *Relative to any theory S_2^i or T_2^i there is an optimal proof system.* That is, there is a $\geq$-greatest proof system among those whose consistency is provable in the theory. We use the idea of the proof of these results to obtain a particular representation of a general proof system.

Recall from Definition 9.3.11 the formula 0-RFN$(P)(x)$

$$\forall y(|y| \leq |x|)\forall z(|z| \leq |x|), P(y, z) \land \text{Fla}(z) \to \text{Taut}(z)$$

where $P(u, v)$ and Fla(v) are Δ_1^b-formulas defining the relations "u is a P-proof of v" and "v is a propositional formula," respectively.

Denote by $\|0\text{-RFN}(P)\|$ the sequence of the propositional formulas

$$\|0\text{-RFN}(P)\| := \Big\{\|0\text{-RFN}(P)(x)\|_{q(n)}^n\| \mid n < \omega\Big\}$$

where $q(x)$ is a fixed bounding polynomial of formula 0-RFN$(P)(x)$.

Theorem 14.1.2. *Let P be a propositional proof system. Let*

$$EF + \|0\text{-RFN}(P)\|$$

be the proof system obtained from EF by adding tautologies from $\|0\text{-RFN}(P)\|$ as extra axioms.
Then:

$$EF + \|0\text{-RFN}(P)\| \geq_p P$$

In particular:

$$EF + ||0\text{-}RFN(P)|| \geq P$$

Proof. Let π be a P-proof of τ. By Lemmas 9.3.13 and 4.6.3 there is a polynomial size EF-proof η_1 of

$$||P(u, v)||^m(\tilde{\pi}, \tilde{\tau}) \wedge || \text{Fla}(v)||^m(\tilde{\tau})$$

where $m = \max(|\pi|, |\tau|)$ (we skip the explicit reference to particular bounding polynomials, similarly as in Section 9.3). From this formula (and η_1) and the new axiom

$$||0\text{-}RFN(P)||^m$$

we get by substitution a polynomial size (EF $+ ||0\text{-}RFN(P)||$)-proof η_2 of

$$|| \text{Taut}(v)||^m(\tilde{\tau})$$

There is a polynomial size EF-proof η_3 of the implication (analogously to Lemma 9.3.15)

$$|| \text{Taut}(v)||^m(\tilde{\tau}) \to \tau$$

From η_2 and η_3 one obtains by modus ponens a polynomial size (EF $+ ||0\text{-}RFN(P)||$)-proof η_4 of τ.

Note that η_1 and η_3 are actually constructible by a polynomial time algorithm, and so this gives a p-simulation of P by (EF $+ ||0\text{-}RFN(P)||$). Q.E.D.

For P sufficiently strong (simulating the modus ponens and the substitution of constants in polynomial size, e.g., $P \supseteq$ EF) we could replace the reflection principle $||0\text{-}RFN(P)||$ by a bit more elegant consistency statement $\text{Con}(P)$ (cf. Lemma 9.3.12).

Note that a *natural* P (like the systems SF, G_i, G_i^*, G, ...) is, in fact, p-equivalent to EF $+|| 0\text{-}RFN(P)||$. This is because such a P admits a polynomial time construction of proofs of the formulas $|| \text{Con}(P)||^n$ (cf. Theorem 9.3.24).

Now we link the problem of the existence of an optimal proof system to two questions, one from logic and one from structural complexity theory. The logical question deals with the lengths of first order proofs of the so-called finitistic consistency statements. Let T be a consistent theory extending S_2^1 and with a polynomial time set of axioms. Then there is a Δ_1^b-formula $\text{Prf}_T(y, z)$ expressing that "y is a T-proof of formula z." Consider a formula $\text{Con}_T(a)$ naturally expressing that no T-proof of length $\leq a$ is a proof of $0 = 1$:

$$\forall y, \ |y| \leq a \to \neg \text{Prf}_P(y, \lceil 0 = 1 \rceil)$$

Note that $\text{Con}_T(a)$ is *not* a bounded formula.

It is a fundamental problem to estimate the length of the shortest proof of the sentence $\text{Con}_T(\tilde{n})$ in a theory S; see Pudlák (1986, 1987) for more background. The length of the numeral $\tilde{n}$ is $O(\log n)$; hence the only a priori lower bound to such proofs is the length of the formula: $\Omega(\log n)$. The next theorem sharply estimates the length of the shortest S-proofs in the case when $S = T$.

Theorem 14.1.3 (Pudlák 1986, 1987). *Let $T \supseteq S_2^1$ be a consistent theory with a polynomial time set of axioms and let $\text{Con}_T(a)$ be the formula defined previously.*

Then there are constants $\epsilon > 0$ and $c \geq 1$ such that for all n the minimal size m_n of a T-proof of the sentence $\text{Con}_T(\tilde{n})$ satisfies

$$n^\epsilon \leq m_n \leq n^c$$

Note that $|\text{Con}_T(\tilde{n})| << n^\epsilon$ and hence the lower bound is nontrivial. The upper bound is also nontrivial. To see this take, for example, $S = S_2^1$ and $T = \text{ZFC}$. There does not seem to be another way to prove $\text{Con}_T(\tilde{n})$ in S than to list (in S) all T-proofs of length $\leq n$ and check that none of them is a proof of $0 = 1$. This gives, however, only the estimate $2^{O(n)}$.

The question whether there is S admitting size $n^{O(1)}$ proofs of $\text{Con}_T(\tilde{n})$ for all T can be linked to the problem posed earlier.

Theorem 14.1.4. *The following two propositions are equivalent:*
1. *there exists an optimal propositional proof system*
2. *there exists a consistent theory $S \supseteq S_2^1$ with a polynomial-time set of axioms such that for every consistent theory $T \supseteq S_2^1$ with a polynomial-time set of axioms there is polynomial $p(x)$ such that for each n the sentence $\text{Con}_T(\tilde{n})$ has S-proof of size $\leq p(n)$.*

Proof. Assume that P is an optimal proof system. Define the theory S_P by

$$S_P := S_2^1 + 0\text{-RFN}(P)$$

Now let $T \supseteq S_2^1$ be a consistent theory with a polynomial-time set of axioms. Define the formula $\phi(x)$

$$\phi(x) := \text{Con}_T(|x|)$$

Then $\phi(x)$ is a Π_1^b-formula. Consider a proof system Q

$$Q := P + \{||\phi||^m \mid m < \omega\}$$

The formulas $||\phi||^m$ are tautologies as T is consistent.

Because P is optimal, there is a polynomial $q(x)$ such that each $||\phi||^m$ has P-proof of size $\leq q(m)$. Hence the theory S_P admits proofs of

$$\text{Taut}(||\phi||^m)$$

of size $m^{O(1)}$ and, similarly with the proof of Theorem 14.1.2, also size $m^{O(1)}$ proofs of

$$|x| \leq \tilde{m} \rightarrow \phi(x)$$

Hence $\phi(\tilde{n})$ has S_P-proof of size $\log(n)^{O(1)}$, and consequently $\mathrm{Con}_T(\tilde{N})$, equivalent to $\phi(2^N)$, has S_P-proof of size $N^{O(1)}$. This proves that the first statement implies the second.

Now let S be a theory satisfying the second statement. Define the propositional proof system P_S by

$$P_S(\pi, \tau) \text{ iff } \mathrm{Prf}_S(\pi, \lceil \mathrm{Taut}(\tilde{\tau}) \rceil)$$

Let Q be an arbitrary propositional proof system. By Theorem 14.1.2 Q is p-simulated by the system $EF + \|0\text{-RFN}(Q)\|$. As $S \supseteq S_2^1$, P_S p-simulates EF (by Corollary 9.3.18). It is thus sufficient to construct polynomial size P_S-proofs for the tautologies

$$\|0\text{-RFN}(Q)\|^m$$

Consider the theory T_Q

$$T_Q := S_2^1 + 0\text{-RFN}(Q)$$

By the hypothesis there are polynomial size S-proofs of

$$\mathrm{Con}_{T_Q}(\tilde{n})$$

Assume that π is the size $m = |\pi|$ Q-proof of τ. As $\neg\, \mathrm{Taut} \in \Sigma_1^b$, there is $k < \omega$ such that the implication

$$\neg\, \mathrm{Taut}(\tilde{\tau}) \rightarrow \exists y(|y| \leq m^k \wedge \mathrm{Prf}_{S_2^1}(y, \lceil \neg\, \mathrm{Taut}[\tilde{\tau}] \rceil)$$

is provable in S_2^1 and hence also in S: This is an instance of a general statement proved by induction on the logical complexity of $\phi(x)$.

Claim. *Let $\phi(x)$ be a Σ_1^b-formula. Then there is $k < \omega$ such that for every $u < \omega$*

$$N \models \phi(u) \;\rightarrow\; S_2^1 \vdash_{|u|^k} \phi(\tilde{u})$$

where $\vdash_m$ is an abbreviation for "provable by a proof of size $\leq m$."
In fact, this itself is formalizable in S_2^1. That is, S_2^1 proves

$$\phi(u) \rightarrow (\exists y(|y| \leq |\tilde{u}|^k)\, \mathrm{Prf}_{S_2^1}(y, \lceil \phi(\tilde{u}) \rceil)$$

where $\mathrm{Prf}_{S_2^1}$ is a Δ_1^b-formalization of "y is an S_2^1-proof of $\phi(u)$."

For the same reason there is a size $\leq m^k$ T_Q-proof of $Q(\tilde{\pi}, \tilde{\tau})$, and by the axioms $||0\text{–RFN}(Q)||$ T_Q also admits size $\leq m^k$ proofs of

$$\text{Taut}(\tilde{\tau})$$

and hence there are size $m^{O(1)}$ S-proofs of

$$\exists z, |z| \leq |\pi|^k \wedge \text{Prf}_{T_Q}(z, \ulcorner\text{Taut}[\tilde{\tau}]\urcorner)$$

This formula and the last but one entails that there is a size $m^{O(1)}$ S-proof of

$$\neg\,\text{Taut}(\tilde{\tau}) \rightarrow \neg\,\text{Con}_{T_Q}(\tilde{m}^\ell)$$

some fixed $\ell < \omega$.

By the hypothesis there is a constant $t < \omega$ such that all $\text{Con}_{T_Q}(\tilde{n})$ have size $\leq n^t$ S-proofs; hence there are size $\leq m^{\ell t}$ S-proofs of $\text{Taut}(\tilde{\tau})$.

By the definition of P_S this proof is also a P_S-proof of τ of size $\leq m^{\ell t} \leq |\pi|^{\ell t}$.

$$\text{Q.E.D.}$$

One may speculate about a construction of a theory T for which given S does not admit size $n^{O(1)}$ proofs of $\text{Con}_T(\tilde{n})$. Possible candidates are $T := S + \text{Con}_S$ or a theory formed from S by adding to the language a truth predicate for formulas in the language of S, Tarski's conditions on this predicate and the statement (using the new predicate) that all axioms of S are true (such theory is called "jump" of S in Buss 1986). However, if for S one can find T without short S-proofs of $\text{Con}_T(\tilde{n})$ it follows that S does not prove that $\text{NP} = \text{coNP}$. This is because the formula ϕ considered in the preceding proof is Π_1^b and $\text{NP} = \text{coNP}$ would allow us to express it also as a Σ_1^b-formula and so its instances (and consequently the instances of $\text{Con}_T(x)$) would have polynomial size proofs by the claim in the previous proof.

The next theorem links the problem of the existence of an optimal proof system to a problem in structural complexity theory.

Theorem 14.1.5. *The following two propositions are equivalent:*
1. *there exists an optimal propositional proof system*
2. *for every coNP-set X there exists a nondeterministic Turing machine M accepting exactly X and such that for every polynomial-time, sparse $Y \subseteq X$ there is a polynomial $p(x)$ such that every $u \in Y$ is accepted by M in time $\leq p(|u|)$.*

We refer the reader to Krajíček and Pudlák (1989a) for the proof.

14.2. Hard tautologies

The first definition formalizes a notion of hard tautologies. Recall the definition of the function $c_P(\tau)$ in Definition 14.1.1.

Definition 14.2.1. *A sequence* $\{\tau_n\}_{n<\omega}$ *of tautologies is* hard *for a propositional proof system P iff the following three conditions are fulfilled:*

1. *there exists a polynomial time machine computing from* $1^{(n)}$ *the formula* τ_n
2. $n \leq |\tau_n|$, *for all n*
3. *there is no polynomial* $p(x)$ *for which*

$$c_P(\tau_n) \leq p(|\tau_n|)$$

    would hold for all n

Note that $|\tau_n| = n^{O(1)}$. Conditions 1 and 2 imply that set $\{\tau_n \mid n < \omega\}$ is polynomial-time recognizable and so we may add it to P as extra axioms to form a new proof system $Q := P + \{\tau_n \mid n < \omega\}$. Adding extra axioms to a general proof system precisely means that π is a Q-proof of τ iff it is a P-proof of $\sigma \to \tau$, where σ is a conjunction of *substitution instances* of new axioms. P is then not better than Q (in the sense of Definition 14.1.1). Hence the task to construct a hard sequence $\{\tau_n\}_{n<\omega}$ for P is the same as the task to find proof system Q such that $P \not\geq Q$ and its *axiomatization* over P by a polynomial-time set of tautologies.

Assume $P \geq$ EF. Having Q for which $P \not\geq Q$ we may take the proof system $Q' := \text{EF} + ||0\text{–RFN}(Q)||$. By Theorem 14.1.2 $Q' \geq Q$; hence $P \not\geq Q'$ and the sequence $\{||0\text{–RFN}(Q)||^n\}_{n<\omega}$ is hard for P. This gives the following simple but useful statement.

Theorem 14.2.2. *Let P be a proof system and assume that* $P \geq EF$. *The following three statements are equivalent:*

1. *there exists a sequence of tautologies* $\{\tau_n\}_{n<\omega}$ *hard for P*
2. *there exists a proof system Q such that P is not better than* $Q : P \not\geq Q$
3. *there exists a proof system Q such that the sequence* $\{||0\text{–RFN}(Q)||^n\}_{n<\omega}$ *is hard for P*

The quasi-ordering $\geq$ (Definition 14.1.1) of proof systems induces a reducibility among sequences $\{\tau_n\}_{n<\omega}$, $\{\sigma_n\}_{n<\omega}$ over a given system P:

$$\{\tau_n\}_{n<\omega} \geq_P \{\sigma_n\}_{n<\omega} \quad \text{iff} \quad P + \{\tau_n \mid n < \omega\} \geq_P P + \{\sigma_n \mid n < \omega\}$$

that is, formulas σ_n can be deduced by polynomial size P-proofs from *substitution instances* of some τ_m's.

For example, by Lemma 12.5.6 and Theorem 12.5.7

$$\{\text{Count}_n^2\}_n \geq_P \{\text{PHP}_n\}_n$$

but not the converse for P being the depth d Frege system ($d \geq 5$).

The question whether there exists a $\geq_P$-complete sequence (i.e., a maximal element of the quasi-order $\geq_P$) is obviously equivalent with the main problem

of Section 14.1 whether there is an optimal proof system. If one restricts to the
sequences with polynomial size proofs in some system Q, then among those there
will be a $\geq_P$-complete one, provided Q has a polynomial-time axiomatization
over P.

We are interested in natural sequences $\{\tau_n\}_{n<\omega}$ that would be hard for systems
from Chapter 4, such as F or EF. We would like to find a $\Delta_0(\alpha)$-formula $\phi(x, \alpha)$,
expressing a natural combinatorial principle for finite structures, such that the
sequence

$$\{\langle\phi\rangle_n\}_{n<\omega}$$

is hard for F or EF (recall Definition 9.1.1 for translation $\langle\phi\rangle_n$). For example,
the formulas ϕ representing the pigeonhole principle or the modular counting
principles of Section 12.5 give rise to tautologies hard for constant-depth Frege
systems but not for F (by Theorems 12.5.3 and 12.5.4 and Corollary 13.1.8).
Similarly the Ramsey theorem from Section 12.1 is not hard for F.

In fact, if $\forall x \forall \alpha, \phi(x, \alpha)$ represents a scaled down Π_1^1-form of a combinatorial
principle whose usual proof depends on the existence of large numbers (like various
modifications of Ramsey theorem), then it is actually likely to have polynomial size
F-proofs. This is because the Π_1^1-version takes the large number as a parameter;
hence the length of the tautologies is also large.

Take a coNP set defined by a combinatorial property, for example, the set of
graphs without a large clique where *large* means of size greater than n^ϵ, n the
number of vertices and $\epsilon > 0$ fixed. For fixed n let p_{ij} be $\binom{n}{2}$ atoms for the
possible edges in a graph with n vertices and q_{uv}, $1 \leq u \leq \lceil n^\epsilon \rceil$, $1 \leq v \leq n$
another $n^{1+\epsilon}$ atoms. Consider the formula $A_n(\overline{p}, \overline{q})$

$$\bigwedge_u \bigvee_v q_{uv} \wedge \bigwedge_{u_1 \neq u_2} \bigwedge_v (\neg q_{u_1 v} \vee \neg q_{u_2 v})$$
$$\rightarrow \bigvee_{v_1 \neq v_2} \bigvee_{u_1 \neq u_2} (q_{u_1 v_1} \wedge q_{u_2 v_2} \wedge \neg p_{v_1 v_2})$$

expressing that if $\{(u, v) \mid q_{uv} = 1\}$ defines a 1–1 map then its range is not a clique
in the graph $\{(i, j) \mid p_{ij} = 1\}$. Note that any coNP-property can be represented
in a similar fashion as it can be defined by a Π_1^b-formula (use the translations of
Sections 9.1–2).

For a particular graph H of size n let $A_H(\overline{q})$ denote the instance of $A_n(\overline{p}, \overline{q})$
with p_{ij} replaced by 1 iff the edge (i, j) is in H. Then A_H is a tautology iff H
has no large clique, and such tautologies cannot have polynomial size proofs in
any proof system unless NP = coNP (cf. Theorem 4.1.2). Our aim is, given a
particular proof system P, to find a sequence of graphs H_n without large cliques
such that the sequence $\{A_{H_n}\}_{n<\omega}$ is hard for P; in particular, the graphs H_n will
be constructible by a p-time function.

Theorem 14.2.3. *Let $P \geq EF$ be a proof system that is not optimal. Then there exist graphs H_n such that the formulas $\{A_{H_n}\}_{n < \omega}$ form a sequence of tautologies hard for P.*

In particular, there is a p-time function f constructing H_n from $1^{(n)}$ and the set $\{H_n \mid n < \omega\}$ is itself p-time.

Proof. Let X denote the set of graphs without a large clique and $\phi(x)$ be its Π_1^b-definition in S_2^1. The theory S_2^1 obviously defines the usual reduction of TAUT to X (cf. Garey and Johnson 1979); that is, there is a p-time function g (Δ_1^b-definable in S_2^1) such that S_2^1 proves

$$\text{Taut}(a) \equiv \phi(g(a))$$

Take a proof system Q such that P is not better than $EF + \|0\text{-RFN}(Q)\|$; we may assume the existence of such Q by the hypothesis of the theorem and by Theorem 14.2.2.

The function f defined by

$$f\left(1^{(n)}\right) := g\left(\|0\text{-RFN}(Q)\|^n\right)$$

has the required properties. Q.E.D.

We remark that if $P = EF$ and, for example, $EF \not\geq G_i$ then H_n can be found that, in addition, the formulas A_{H_n} have polynomial size G_i-proofs. This is because G_i admits polynomial size proofs of the instances of $\|0\text{-RFN}(G_i)\|$, by Theorems 9.3.16 and 9.2.5.

14.3. Bibliographical and other remarks

The question whether there is an optimal proof system was studied in Krajíček and Pudlák (1989a) and the statements of Section 14.1 are from there, except Theorem 14.1.3, which is due to Pudlák (1986, 1987).

Pitassi and Urquhart (1992) show how to translate non-three-colorable graphs into tautologies, and vice versa, such that the minimal size of an EF-proof of τ is polynomially related with the minimal number of steps needed in a construction of the associated graph G_τ using the rules of the Hajós calculus. A corollary to Theorem 14.1.2 is that any nondeterministic acceptor of non-three-colorable graphs can be p-simulated by the Hajós calculus augmented by a p-time set of extra initial graphs (associated to the formulas from $\|0\text{-RFN}(Q)\|$, for suitable Q).

15

Strength of bounded arithmetic

The previous chapters dealt mostly with the metamathematical properties of the systems of bounded arithmetic and of the propositional proof systems. We studied the provability and the definability in these systems and their various relations. The reader has by now perhaps some feeling for the strength of the systems. In this chapter we shall consider the provability of several combinatorial facts in bounded arithmetic.

In the first section we study the counting functions for predicates in PH, the *bounded* PHP, the approximate counting, and the provability of the infinitude of primes. In the second section we demonstrate that a lower bound on the size of constant-depth circuits can be meaningfully formalized and proved in bounded arithmetic. The last, third section studies some questions related to the main problem whether there is a model of S_2 in which the polynomial-time hierarchy does not collapse.

15.1. Counting

A crucial property that allows a theory to prove a lot of elementary combinatorial facts is *counting*. In the context of bounded arithmetic this would require having Σ_∞^b-definitions of the counting functions for Σ_∞^b-predicates.

The uniform counting is not available.

Theorem 15.1.1. *There is no $\Sigma_\infty^b(\alpha)$-formula $\theta(\alpha, a)$ that would define for each set α and each $n < \omega$ the parity of the set $\{x < n \mid \alpha(x)\}$.*

Proof. Assume $\theta(\alpha, a)$ defines the parity of the sets. Then the propositional formula $\langle\theta\rangle_n$, provided by Definition 9.1.1, is of size $\leq 2^{(\log n)^{O(1)}} \ll 2^{n^{\Omega(1)}}$ and of constant depth (Lemma 9.1.2) and computes the parity of $\{i < n \mid p_i = 1\}$ for $p_i := \langle\alpha\rangle_i$. This contradicts Theorem 3.1.10.   Q.E.D.

We may still hope for a nonuniform counting. That is, the definitions of the counting functions would not be instances of one $\Sigma^b_\infty(\alpha)$-formula, but each Σ^b_∞-definable set would require a special Σ^b_∞-definition of its counting function.

As mentioned at the end of Section 2.2, Toda (1989) puts a strong block to such a hope. Recall the definitions of the function $\#R$ and of the class $\#P$ from the end of Section 2.2 and define the function $\oplus R$ by

$$\oplus R(x) := \text{ the parity of the set } \{y \mid R(x, y)\}$$

and let $\oplus P$ be the class of the functions $\oplus R$ with $R \in P$ satisfying the condition $R(x, y) \rightarrow |y| \leq |x|^{O(1)}$. Note that the counting function for a predicate $A(x)$ is the function $\#R$ for $R(x, y) := (A(x) \wedge y < x)$.

We state Toda's result explicitly; recall that $\square^p_\infty$ denotes the class of the functions computable by a polynomial time machine with an oracle from PH.

Theorem 15.1.2 (Toda 1989). *Assume that PH does not collapse. Then*

$$\#P \not\subseteq \square^p_\infty \quad \text{and} \quad \text{even} \ \oplus P \not\subseteq \square^p_\infty$$

A form of the counting that is available in S_2 is provided by the next simple but useful lemma.

Lemma 15.1.3. *Let $g(x)$ be a function with a Σ^b_∞-graph and majorized by $2^{x^{O(1)}}$. Then there is a function $f(a, b)$, Σ^b_∞-definable in S_2, such that S_2 proves*

$$f(a, 0) = 0 \wedge \forall i < |a|, \ f(a, i + 1) = f(a, i) + g(i)$$

In particular, if A is a Σ^b_∞-definable set then there is a function $h(a, i)$, Σ^b_∞-definable in S_2, such that S_2 proves $h(a, 0) = 0$ and that for all $i < |a|$

$$h(a, i + 1) = \begin{cases} h(a, i) + 1 & \text{if } A(i) \\ h(a, i) & \text{if } \neg A(i) \end{cases}$$

Proof. For $i < |a|$ the length $|g(i)|$ is bounded by $i^{O(1)} = |a|^{O(1)}$. Hence the sequence

$$(\, 0, g(0), g(0) + g(1), \ldots, g(0) + \cdots + g(|a| - 1) \,)$$

consisting of the values of $f(a, i)$ for $i \leq |a|$ is coded by a number $\leq 2^{|a|^{O(1)}}$ and bounded induction can be applied to prove the existence of such a sequence for all a.

The second part follows, applying the first part to a particular g, the characteristic function of A.   Q.E.D.

Note that the lemma implies that all Σ^b_∞-consequences of $S_2(\alpha)^{\text{count}}$ (cf. Section 13.1) are provable in the theory $S^1_2 + 1 - \text{Exp}$ of Section 5.5.

Although the counting is not available within PH, an *approximate counting* is available.

Theorem 15.1.4 (Sipser 1983b). *Let A be a Σ_∞^b-definable set and let $\epsilon > 0$ be fixed. Then there is a function $f(x)$ with a Σ_∞^b-graph such that for any a*

$$|A \cap a| \leq f(a) \leq (1 + \epsilon)|A \cap a|$$

Proof. Fix a of the form $a = 2^n$ and identify a with the n-dimensional vector space over $\mathbf{F}_2$. Denote $A_a := A \cap a$ and assume that $2^{t-1} \leq |A_a| < 2^t$. Take U to be the set of $(1 + t) \times n$, 0–1 matrices.

Claim 1. *For any $x \in A_a$ there are at least $\geq |U|/2 = 2^{t+n}$ matrices $M \in U$ for which*

$$\forall y \in A_a \setminus \{x\}, Mx \neq My$$

For any $y \neq x$, $mx = my$ holds for 2^{n-1} vectors m; hence $Mx = My$ holds for $|U|/2^{t+1} = 2^n$ matrices M (as they have $t + 1$ rows). Hence the number of matrices satisfying the formula from the claim is at least

$$\geq |U| - \frac{|U|}{2^{t+1}}(|A_a| - 1) \geq \frac{|U|}{2} = 2^{t+n}$$

using $2^{t-1} \leq |A_a| < 2^t$.

Claim 2. *There are $M \in U$ and $B \subseteq A_a$ such that $|B| \geq |A|/2$ and such that*

$$\forall x \in B \forall y \in A_a \setminus B, Mx \neq My$$

The claim follows as by Claim 1 there are at least $\geq (|U|/2)|A_a|$ pairs $(M, x) \in U \times A_a$ satisfying Claim 1, whereas the failure of Claim 2 would imply that there are $< |U||A_a|/2$ such pairs.

Iterating Claim 2 $\leq \log(|A_a|) = t$-times yields

Claim 3. *There are matrices $M_1, \ldots, M_t \in U$ and sets $B_0 = \emptyset \subseteq B_1 \subseteq \cdots \subseteq B_t = A_a$ such that*

$$B_{i+1} = B_i \cup \{x \in A_a \setminus B_i \mid \forall y \in A_a \setminus B_i, x \neq y \to M_{i+1}x \neq M_{i+1}y\}$$

Define a map $F : a = 2^n \to 2^{(1+t)}t < 2|A_a| \log(|A_a|)$ by

$$F(x) := (M_1 x, \ldots, M_t x)$$

The map F is 1–1 on the set A_a. To make F Σ_∞^b-definable take such F for the minimal $t \leq n = |a|$ and for the lexicographi⸱ $M_1, \ldots, M_t \in U$ for which it is 1–1 on A_a.

Take k minimal such that

$$(2kn) \leq (1 + \epsilon)^k$$

Clearly $k \leq O(n) = O(|a|)$ and so we may define a set A' of the sequences of elements of A_a of length k, identified with elements of a^k. Hence $|A'| = |A_a|^k$.

Applying Claim 3 to the set A' we get a map $F' : A' \rightarrow 2|A'| \log(|A'|)$ that is 1–1 on A'. By the choice of k we have

$$(2|A'| \log(|A'|))^{k^{-1}} = (2|A_a|^k k \log(|A_a|))^{k^{-1}} \leq |A_a|(2kn)^{k^{-1}} \leq (1 + \epsilon)|A_a|$$

This allows us to define a function $f(a)$ to be the kth-root of the supremum of the range of F'. Then

$$|A_a| \leq f(a) \leq (1 + \epsilon)|A_a|$$

as required, and f is Σ^b_∞-definable.   Q.E.D.

The inability to count in bounded arithmetic can sometimes be replaced by the bounded PHP, which was pioneered by Woods (1981). To ensure that the presentation is unambiguous we define the principle once again.

Definition 15.1.5. *The $\Sigma^b_\infty PHP^b_a$ is the set of axioms of the form*

$$\exists x < b \forall y < a, \neg\theta(x, y) \vee \exists x_1 < x_2 < b \exists y < a, \theta(x_1, y) \wedge \theta(x_2, y)$$

where $\theta \in \Sigma^b_\infty$ and may contain other parameters (different from a, b).
$\Delta_0 PHP^b_a$ is the same scheme but for Δ_0-formulas θ only.
$\Sigma^b_\infty PHP$ and $\Delta_0 PHP$ denote the scheme for $b = a + 1$.

Note that we formulate the pigeonhole principle as a statement about injective functions rather than about bijections as earlier.

We shall state explicitly an important open problem.

Problem. *Does the theory S_2 prove $\Sigma^b_\infty PHP$? Does the theory $I\Delta_0$ prove $\Delta_0 PHP$?*

By (the proof of the) Theorem 11.2.4 the theory S_2 proves $\Sigma^b_\infty PHP^b_a$ for $b = a^2, 2a$; this is open for the Δ_0-case. On the other hand, $S_2(\alpha)$ does not prove $PHP^{a+1}_a(\alpha)$ (Corollary 12.5.5).

Lemma 15.1.6. *Assume that S_2 is not finitely axiomatizable.*
Then no theory S^i_2 proves all formulas from $\Sigma^b_\infty PHP$.

Proof. Assume that $\theta(x)$ is a formula for which the induction fails.

$$\theta(0) \wedge (\forall x < a, \theta(x) \rightarrow \theta(x + 1)) \wedge \neg\theta(a)$$

Define a 1–1 function $f : a + 1 \to a$ by

$$f(x) = \begin{cases} x & \text{if } \theta(x) \\ x - 1 & \text{if } \neg\theta(x) \end{cases}$$

Hence $\Sigma^b_\infty \text{PHP}$ proves S_2. That is, $S^i_2 \vdash \Sigma^b_\infty \text{PHP}$ would imply $S^i_2 = S_2$ and the lemma follows by Theorem 10.1.1.   Q.E.D.

An important case when the counting (or the existence of exponentially large numbers) could be replaced by $\Sigma^b_\infty \text{PHP}$ is the proof of the existence of infinitely many primes. The usual proof of the existence of a prime bigger than a requires us to take $a! + 1$, which is not definable in S_2 (by Theorem 5.1.4).

Theorem 15.1.7 (Woods 1981, Paris et al. 1988). *The theory S_2 proves that there exist unboundedly many primes*

$$\forall x \exists p > x, \; Prime(p)$$

where $Prime(p)$ is a Π^b_1-definition of primes

$$\forall y, z < p, \, y \cdot z \neq p$$

In fact, the theory S_2 proves Sylvester's theorem

$$\forall x \geq 1 \forall y > x \exists z, \, y < z \leq y + x \; \wedge \; (\exists p \leq z, \, Prime(p) \wedge p > x \wedge p|z)$$

where $p|z$ denotes the divisibility

$$\exists u \leq z, \, p \cdot u = z$$

The reader may find the proof of the first part in Paris et al. (1988) and of the second part in Woods (1981).

It is an open problem whether the theory $I\Delta_0$ can prove the infinitude of primes. It is also unknown which of the basic elementary number-theoretic theorems, such as Fermat little theorem or Lagrange's four-square theorem, can be proved in $I\Delta_0$; see Berarducci-Intrigila (1991). Woods (1981) conjectured that the theory $I\Delta_0(\pi)$ can prove the infinitude of primes, where the formulas $\pi(0) := 0$ and

$$\pi(n + 1) = \begin{cases} \pi(n) + 1 & \text{if } n + 1 \text{ is a prime} \\ \pi(n) & \text{otherwise} \end{cases}$$

are added to PA^-.

15.2. A circuit lower bound

In this Section we show that a weaker form of Theorem 3.1.10 can be proved in a subtheory of S_2. Recall from Section 3.1 the relevant definitions.

We shall adopt the following framework. The input 0–1 vectors are coded by *numbers* $a, b, \ldots$, hence the number of the input bits is a *length* $n = |a|, \ldots$. The set of the input vectors of a given length is thus definable, but it is not coded by a number.

The circuits are also coded by numbers. This implicitly indicates that we cannot define circuits of a superpolynomial size as numbers of length $2^{|a|^\epsilon}$ are not definable (Theorem 5.1.4). Clearly the size and the depth of a circuit are Δ_1^b-definable, as is the function computed by a circuit (the circuit is a parameter in the definition).

Boolean functions are identified with subsets of $2^n = a$ and thus are not coded by numbers. But all interesting functions, in particular all characteristic functions of the sets from PH, are definable by a bounded formula (by Δ_1^b- resp. Σ_1^b- in the case of P-resp. NP-sets; cf. Theorem 3.2.12).

This framework does not allow us to prove (even to formulate) Theorem 3.1.7 or to prove an exponential lower bound. It allows us, however, to rephrase such bounds as the nonexistence of polynomial upper bounds. By Theorem 3.1.4 the existence of polynomial size circuits for NP-functions is the chief problem and that is perfectly formalizable in the framework.

We begin by proving (a weaker form of) Theorem 3.1.10. The notions from the following definition occurred in a slightly different form in Section 12.3 and in the discussion at the beginning of Section 12.4.

Definition 15.2.1. Let $x_1, \ldots, x_n$ be Boolean variables and let $\mathcal{M}$ be the set of partial truth assignments to $x_1, \ldots, x_n$.

1. A complete system *is a subset* $S \subseteq \mathcal{M}$ *such that every total truth assignment extends exactly one* $\alpha \in S$.
2. *For* $H, S \subseteq \mathcal{M}$, S refines H *if*

$$\forall \alpha \in S, (\forall h \in H, h \perp \alpha) \vee (\exists h \in H, h \subseteq \alpha)$$

where $h \perp \alpha$ *means that* h, α *are contradictory.*
3. *For* $H \subseteq \mathcal{M}$, *the norm of* H *is* $\|H\| := \max_{h \in H} |\operatorname{dom}(h)|$.

For $\alpha \in \mathcal{M}$ denote by $[\alpha]$ the set of the tuples a for which $a_i = \alpha(x_i)$, if $x_i \in \operatorname{dom}(\alpha)$. Then $\bigcup_{\alpha \in S}[\alpha]$ partitions 2^n if S is a complete system. If f is a Boolean function such that for some complete system S and a partition $S_0 \cup S_1 = S$ of S it holds that

$$f^{-1}(i) = \bigcup_{\alpha \in S_i}[\alpha], \qquad i = 0, 1$$

then both f and $\neg f$ are expressible in a disjunctive normal form whose conjunctions are of size $\leq \|S\|$. Assume $f = \bigvee_u f_u$ and

$$f_u^{-1}(1) = \bigcup_{\alpha \in S_1^u}[\alpha]$$

where $\|S^u\| \leq t$ for all u. To express f and $\neg f$ as a disjunction of conjunctions of size $< s$ it is sufficient to find a complete system S refining $H := \bigcup_u S_1^u$ and satisfying $\|S\| < s$ as then

$$f^{-1}(1) = \bigcup_{\beta \in S_1} [\beta]$$

where

$$S_1 := \{\beta \in S \mid \exists u \exists \alpha \in S_1^u, \; \alpha \subseteq \beta\}$$

Hence the following lemma substitutes the switching Lemma 3.1.11 in this approach to collapsing circuits.

For $X \subseteq \mathcal{M}$ and $\rho \in \mathcal{M}$

$$X^\rho := \{\alpha \setminus \rho \mid \alpha \cup \rho \in \mathcal{M}, \; \alpha \in X\}$$

In the next proof we follow the strategy of the proof of Lemma 12.3.10.

Lemma 15.2.2. *Let $H_\ell \subseteq \mathcal{M}$ be sets for which $\|H_\ell\| \leq t \leq s$, $\ell = 1, \ldots, m$. Assume $2m < (n^{1-\epsilon} - 1/8t)^s$, some fixed $\epsilon > 0$.*

Then for sufficiently large n there is $\rho \in \mathcal{M}$ for which $|dom(\rho)| = n - n^\epsilon$ and such that

$$\exists \text{ a complete } S_\ell \subseteq \mathcal{M}^\rho, \; \|S_\ell\| < s \quad \text{and} \quad S_\ell \text{ refines } H_\ell$$

holds for all ℓ.

Proof. First fix ℓ and denote H_ℓ simply by H. Also fix $\rho \in \mathcal{M}$.

Let $H = h^1, \ldots, h^r$ be a fixed listing of H. Construct the following two-player game.

> **Step** 1: Player I picks $h_1 :=$ the first h^j consistent with ρ; player II picks assignment σ_1 on $dom(h_1) \setminus dom(\rho)$.
>
> **Step** i: Player I picks $h_i :=$ the first h^j consistent with $\rho \cup \sigma_1 \cup \ldots \cup \sigma_{i-1}$; player II picks σ_i an assignment on $dom(h_i) \setminus (dom(\rho) \cup \bigcup_{j<i} dom(\sigma_j))$.

The game ends when either all h^j's are exhausted or $h_i \subseteq \rho \cup \bigcup_{j \leq i} \sigma_j$.

The following claim is analogous to the claim in the proof of Lemma 12.3.10.

Claim. *The set of all*

$$\sigma_1 \cup \cdots \cup \sigma_k$$

in a finished game forms a complete system refining H^ρ.

We would like to show that for some ρ, in no game can II assign $\geq s$ values.

A *marked value sequence* is $v \in \{0,1\}^s$ together with $j_0 = 0 < j_1 < \cdots < j_k = s$. We think of a marked value sequence as of a coding that II played $j_i - j_{i-1}$ values $v_{j_{i-1}+1}, \ldots, v_{j_i}$ when facing h_i.

The *history* of the game (specifying it uniquely) is the list of $h_1, \ldots, h_k$ that II faced plus the sequence $D = (D_1, \ldots, D_k)$ s.t. D_i is a $j_i - j_{i-1}$-element subset of t, specifying the positions of the atoms in h_i that II had to evaluate.

Notice that because by definition h_i is the first h^j not given 0 by $\rho \cup \bigcup_{j<i} \sigma_j$, instead of knowing the history of the game it is enough to know the (unique) assignment τ with the domain $\text{dom}(\rho) \cup \bigcup_{i \leq k} \text{dom}(\sigma_i)$ that extends ρ and s.t.

$$\rho \cup \sigma_1 \cup \cdots \cup \sigma_{i-1} \cup (\tau \downarrow \text{dom}(\sigma_i))$$

makes h_i true, all $i \leq k$. Then from τ, using the marked value sequence v and D, it is easy to reconstruct all $h_1, \sigma_1, \ldots, h_k, \sigma_k$, and ρ.

Thus the triple v, D, τ determines ρ and hence a map

$$F_H(\rho) := \langle v, D, \tau \rangle$$

where v, D are associated to a fixed game in which II assigns $\geq s$ values is a 1–1 map.

Assume that $|\text{dom}(\rho)| = k$. Then there are $\binom{n}{k} 2^k$ such ρ's and at most $\binom{n}{k+s} 2^{k+s}$ partial assignments τ. Clearly there are $\leq 4^s$ marked value sequences and $\leq t^s$ sequences D. Hence F_H is a 1–1 map from $A := \binom{n}{k} 2^k$ into $B := \binom{n}{k+s} 2^{k+s} (4t)^s$. An elementary computation shows

$$\frac{A}{B} \geq (8t)^{-s} \left(\frac{k+1}{n-k-s+1} \right)^s \geq (8t)^{-s} \left(\frac{k+1}{n-k+1} \right)^s \geq \left(\frac{k}{8t(n-k)} \right)^s$$

Choosing $k = n - n^\epsilon$ we get

$$\frac{A}{B} \geq \left(\frac{n^{1-\epsilon} - 1}{8t} \right)^s$$

that is $2^{n^{\Omega(1)}}$ for $t \leq s \leq n^\delta$ such that $\epsilon + \delta < 1$, and is $\geq n^{\Omega(s)}$, for t a fixed constant.

Thus the map F_H violates the weak PHP in both cases.

Now assume that for every ρ, $|\text{dom}(\rho)| = k$, there is H_ℓ for which player II assigns in a game on H_ℓ^ρ at least s values. Define

$$F(\rho) := \left(\ell, F_{H_\ell}(\rho) \right)$$

Then F is a 1–1 map from A into $B \cdot m$ and by the hypothesis of the lemma $(A/Bm) \geq 2$. Hence F violates the weak PHP and there must be ρ for which it is undefined, that is, for which player II assigns $< s$ values in any game on any H_ℓ^ρ.

Q.E.D.

Now we use the lemma to formalize a lower bound to the constant-depth circuits in the theory $PV_1 + WPHP(PV_1)$; recall PV_1 from Section 5.3 and $WPHP(PV_1)$ from Section 7.3.

Theorem 15.2.3. *Let d be a constant and $p(x)$ a polynomial. Then the theory $PV_1 + WPHP(PV_1)$ proves that for sufficiently large n there are no depth d size $\leq p(n)$ circuits computing the parity function $\oplus(x_1, \ldots, x_n)$.*

Proof. Work in a nonstandard model M of $PV_1 + WPHP(PV_1)$ and let $n \in$ $Log(M) \setminus \omega$. Let $C \in M$ be a depth d size $\leq p(n)$ circuit in $x_1, \ldots, x_n$.

As in the standard proof of Theorem 3.1.10 we show that for some ρ, $|\,dom(\rho)| \geq$ $n - n^{\epsilon^d}$, the function computed by the restricted circuit C^ρ has an associated complete system of the norm $< s \ll n^{\epsilon^d}$. Such a ρ is obtained by iterating Lemma 15.2.2 $\leq d$-times (as outlined before that lemma). Hence we only need to show that the lemma holds in M with $m = p(n)$.

We shall take $t = s$ a large (w.r.t. d and $p(x)$) but fixed constant. Then the Δ_1^b-definable subsets of $\mathcal{M}$ of norm $< s$ are coded by numbers $\leq 2^{(2n)^s} \leq t(C), t$ an L-term. Elements of $\mathcal{M}$ are also coded, and there are only finitely many $(< \omega)$ marked value sequences and tuples D; specifically, a quantification over them is sharply bounded. Taking in the definition of $F_H(\rho)$ the first v, D (τ is unique) for which it is defined makes F_H Δ_1^b-definable.

As we cannot count in M the domain and the range of F, we cannot follow the proof of Lemma 15.2.2 in demonstrating that F violates WPHP. Instead we shall certify such a violation by constructing a Δ_1^b-definable 1–1 map G of two disjoint copies of (a set containing) $Rng(F)$ into $Dom(F)$. $Dom(F)$ is supposed to be the set of all $\rho \in \mathcal{M}$, $|dom(\rho)| = k = n - n^\epsilon$. The composed function $F \circ G$ violates the WPHP.

Assume we have a Δ_1^b-definable 1–1 map G_0 of $2 \cdot (8t)^s \cdot m$ copies of $[n]^{k+s}$ into $[n]^k$. Construct the map G (we use the notation from the proof of Lemma 15.2.2) as follows

1. identify the elements of the form $\langle \ell, \langle v, D, \tau \rangle \rangle$ with $m \cdot (4t)^s$ copies of τ
2. identify the set of $\tau \in \mathcal{M}$, $|\tau| = k + s$, with 2^s copies of the set of pairs (X, η), where $X \in [n]^{k+s}$ and $\eta \in \mathcal{M}$ with $|\eta| = k$ (thinking of $X = dom(\tau)$ and η the restriction of τ to the first k elements of X, and s-tuples of the values of $\tau \setminus \eta$ determining one of the 2^s copies)
3. map the pair (X, η) to ρ, where $dom(\rho) = G_0(X)$ and such that the values of ρ are given by η

Map G is 1–1 and clearly Δ_1^b-definable. It remains to construct G_0, which is simple and left as an exercise.   Q.E.D.

15.3. Polynomial hierarchy in models of bounded arithmetic

In this section we study the complexity classes within the models of bounded arithmetic systems and we show some connections among various open problems mentioned earlier in the book.

We begin with a definition; see also Section 7.6.

Definition 15.3.1. *Let M be a model of PV.*

1. *$P(M)$, the class of the polynomial-time subsets of M, is the class of the subsets of M definable by an open PV-formula with parameters from M.*
   $\Sigma_i^p(M)$, the class of the Σ_i^p-subsets of M, is the class of the subsets of M definable by a Σ_i^b-formula with parameters from M.
   $\Pi_i^p(M)$, the class of the Π_i^p-subsets of M, is the class of the subsets of M definable by a Π_i^b-formula with parameters from M.
   The classes $\Sigma_1^p(M)$ and $\Pi_1^p(M)$ are also denoted $NP(M)$ and $coNP(M)$.
   $PH(M)$ is the class of the subsets of M definable by a Σ_∞^b-formula with parameters from M.
2. *The classes $P^{stan}(M)$, $(\Sigma_i^p)^{stan}(M)$, $(\Pi_i^p)^{stan}(M)$, and $PH^{stan}(M)$ are defined analogously to the classes in 1 but allowing no parameters.*
3. *The classes $(P/poly)(M)$, $(\Sigma_i^p/poly)(M)$, $(\Pi_i^p/poly)(M)$, and $(PH/poly)(M)$ are defined analogously to the classes in 1 but allowing different parameters for every length $n \in Log(M)$ of elements of M.*

For example, $X \in (P/\text{poly})(M)$ if there is an open PV-formula $\theta(x, y)$ such that for every $n \in Log(M)$ there is $c_n \in M$ such that for any $a \in M$, $|a| = n$

$$a \in X \quad \text{iff} \quad M \models \theta(a, c_n)$$

The following is one of the main open problems.

Fundamental problem. *Is there a model $M \models S_2$ such that $PH(M)$ does not collapse? That is: such that $PH(M) \neq \Sigma_i^p(M)$, all $i < \omega$?*

The problem is equivalent, by Corollary 10.2.6, to the fundamental problem whether S_2 is finitely axiomatizable. A little is known about the problem, but there are some surprising connections with the bounded PHP. We first prove two results about end-extensions. $K \supseteq M$ is an end-extension of M if M is a cut in K. K is a proper end-extension of M if, moreover, $K \setminus M \neq \emptyset$.

Lemma 15.3.2. *Let M, K be two models of $I\Delta_0$ and assume that K is a proper end-extension of M. Then $M \models B\Sigma_1^0$.*

Proof. Recall the collection scheme B from Definition 5.2.11. Let $\theta(x, y, z)$ be a Δ_0-formula with parameters from M and assume

$$M \models \forall x < a \exists y \exists z \theta(x, y, z)$$

As M is a cut in K, Lemma 5.1.3 implies that for all $b \in K \setminus M$

$$K \models \forall x < a \exists y, z < b \, \theta(x, y, z)$$

K satisfies the least number principle for Δ_0-formulas; hence there is the minimal b with this property and clearly $b \in M$. Then (by Lemma 5.1.3 again)

$$M \models \forall x < a \exists y < b \exists z \theta(x, y, z)$$

Q.E.D.

The following lemma constructs a particular end-extension.

Lemma 15.3.3. *Let M be a model of $I\Delta_0$ and $a, t \in M \setminus \omega$ such that there is $b \in M$, $M \models b = a^t$. Assume that for some $R(x, y) \in \Delta_0(M)$*

$$M \models \neg\, PHP(R)_a^{a^2}$$

That is, the function F with graph R violates the weak PHP.

Then there is an end-extension K of M, a model of $I\Delta_0$, such that for some $c \in K$, $K \models c = a^{t^2}$.

Proof. Assume $a^{t^2} \notin M$ as otherwise we could take $K := M$. Similarly, as in the proof of Theorem 11.2.3, we may find a function

$$H : a \times 2^t \to a$$

with a $\Delta_0(M)$-graph such that for any $G : 2^t \to a$, $G \in \Delta_0(M)$, there is $g < a$ such that

$$M \models \forall x < 2^t, H(g, x) = G(x)$$

Thinking about $g < a$ as about a "number"

$$\tilde{g} := \sum_{i < 2^t} H(g, i) \cdot a^i$$

we may Δ_0-define in M the structure M_0 with the universe $\{\tilde{g} \mid g < a\}$ and define the operations $+, \cdot$ and relations $=, \leq$ on it, simulating the construction from Lemma 5.5.4.

In M_0 there is a largest element. We take a cut $I \subseteq M$ such that $t^2 \in I$ but $2^t \notin I$, closed under addition, and then define K to be the substructure of M_0 with $\tilde{g} \in K$ iff $\max\{i \mid H(g, i) \neq 0\} \in I$.

The model M itself embeds in K as a cut, and $a^{t^2} \in K$ is the element $\tilde{g}$ with

$$H(g, i) = \begin{cases} 1 & \text{if } i = t^2 \\ 0 & \text{otherwise} \end{cases}$$

Q.E.D.

We are now ready to prove a theorem with an interesting proof.

Theorem 15.3.4 (Paris et al. 1988). *Assume that $I\Delta_0$ does not prove $\Delta_0 PHP_a^{a^2}$. Then $I\Delta_0$ is not finitely axiomatizable.*

Proof. Assume that $I\Delta_0$ does not prove $\Delta_0\mathrm{PHP}_a^{a^2}$. By a compactness argument it follows from Theorem 11.2.3 that there is a model $M \models I\Delta_0$ and $a, t \in M \setminus \omega$ such that

1. a^t exists in M
2. $a^{kt}, k = 1, 2, \ldots$, are cofinal in M
3. $M \models \neg\mathrm{PHP}(R)_a^{a^2}$, some $R \in \Delta_0(M)$.

By 2 a^{t^2} is not in M and so Lemma 15.3.3 implies that M has a proper end-extension and hence satisfies by Lemma 15.3.2 the collection scheme $B\Sigma_1^0$.

Now assume that $I\Delta_0$ is finitely axiomatizable. We use this to construct a model M satisfying 1–3 but not satisfying $B\Sigma_1^0$, which will be a contradiction.

From the finite axiomatizability assumption it follows that in any model of $I\Delta_0$ there are finitely many Skolem functions $f_1, \ldots, f_r$, Δ_0-definable in M, such that any substructure closed under them is a model of $I\Delta_0$ too.

Take some M_0 satisfying 1–3 and define a function $Z : N \to M_0$ by

  (i) $Z(0) := \langle a, t, a^t, b \rangle$,

    where b are all parameters in the definition of R

  (ii) $Z(\langle i, u \rangle) := f_i(Z(u))$, for $i = 1, \ldots, r$

  (iii) $Z(\langle r + 1, u \rangle) := Z(u_1) + Z(u_2)$, for $u = \langle u_1, u_2 \rangle$

  (iv) $Z(\langle r + 2, u \rangle) := Z(u_1) \cdot Z(u_2)$, for $u = \langle u_1, u_2 \rangle$

  (v) $Z(\langle s, u \rangle) := 0$, for $s > r + 2$

Let $M \subseteq M_0$ be the subset of M_0 defined by

$$c \in M \quad \text{iff} \quad c = Z(k), \text{ some } k \in N$$

Then M is, in fact, a substructure of M_0 (by (iii) and (iv)), is a model of $I\Delta_0$ (as it is by (ii) closed under the Skolem functions), contains a, t, a^t (by (i)), and hence R is Δ_0-definable in M too and therefore violates $\mathrm{PHP}(R)_a^{a^2}$. But also $a^{t^2} \notin M$.

To see that $B\Sigma_1^0$ fails in M note that the function Z has a Σ_1^0-definition (common for M_0 and M) and that

$$M \models \forall x < a \exists k, y, (Z(k) = x \wedge y = a^{kt})$$

But by 2 there is no bound in M one could put on y.   Q.E.D.

The idea behind the proof of Lemma 15.3.3 can be used to prove another interesting result. Recall Definition 3.2.1 of the classes E_i. The symbols E_i^{stan} and Δ_0^{stan} indicate that again parameters are not allowed.

Theorem 15.3.5 (Paris and Wilkie 1985). *Let M be a model of $I\Delta_0$, $a, t \in M \setminus \omega$ such that a^t exists in M and assume that $R \in \Delta_0(M)$ violates $\mathrm{PHP}(R)_a^{a^2}$.*

Then the hierarchy $\Delta_0^{stan}(M)$ does not collapse. That is, there is no i such that

$$\Delta_0^{stan}(M) = E_i^{stan}(M)$$

Proof. Denote by $E_i^{\overline{u}}$ and $\Delta_0^{\overline{u}}$ the classes of E_i- respectively of Δ_0-formulas with parameters among $\overline{u}$.

Without loss of generality we may assume that all parameters in the definition of R are smaller than a^t. As in the previous propositions let $G : a^t \rightarrow a$ be an injective Δ_0-definable map; we may assume that it is $\Delta_0^{a^t,a}$-definable as the minimal parameters in the definition of R that yield a 1–1 map G are definable from a^t, a.

Claim 1. *For any j there is a $\Delta_0^{a^t,a}$-formula $Tr_j(x, y)$ such that for any $E_j^{a^t,a}$-formula $\theta(x)$ there is $n < \omega$ such that*

$$\forall x < a^t, \theta(x) \equiv Tr_j(x, \underline{n})$$

Any $E_j^{a^t,a}$-formula θ is for $x < a^t$ equivalent to a formula of the form

$$\exists y_1^1, \ldots, y_1^k < a^t \forall y_2^1, \ldots, y_2^k < a^t \ldots Q_j \overline{y}_j < a^t \, \theta_0(x, \overline{y}, a^t, a)$$

where θ_0 is an open formula with atomic subformulas of the form $t_1 = t_2, t_1 + t_2 = t_3$, or $t_1 \cdot t_2 = t_3$ with $t_1, t_2, t_3 \in \{0, 1, a^t, a, x, \overline{y}\}$. We shall simply assume that θ is of this form.

Let $\eta(x, \overline{y}, u)$ be a $\Delta_0^{a^t,a}$-formula, a partial truth definition, such that for every formula θ_0 of the preceding form there is $\ell < \omega$ such that for $x, \overline{y} < a^t$

$$\theta_0 \left(x, \overline{y}, a^t, a\right) \equiv \eta \left(x, \overline{y}, \underline{\ell}\right)$$

Such a formula is constructed easily.

Using the map G we may rewrite $\theta(x)$ in the form

$$\exists \langle z_1^1, \ldots, z_1^k \rangle < a^t \wedge \overline{z}_1 \subseteq \mathrm{Rng}(G),$$

$$\forall \langle z_2^1, \ldots, z_2^k \rangle < a^t, \overline{z}_2 \subseteq \mathrm{Rng}(G) \rightarrow \cdots \psi(x, \overline{z}, a^t, a)$$

where ψ is a $\Delta_0^{a^t,a}$-formula obtained from $\eta(x, \overline{y}, \underline{\ell})$ by replacing variable y by (the $\Delta_0^{a^t,a}$-definition of) $G^{(-1)}(z)$.

In this way we constructed a $\Delta_0^{a^t,a}$-formula $Tr_j(x, v)$ such that every $E_j^{a^t,a}$-formula $\theta(x)$ is for $x < a^t$ equivalent to $Tr_j(x, \underline{n})$, suitable n encoding both k and ℓ.

Claim 2. *There is no j such that*

$$\Delta_0^{a^t,a}(M) \subseteq E_j^{a^t,a}(M)$$

The claim follows from the previous claim as, for example, the $\Delta_0^{a^t,a}$-formula $\neg \, Tr_j(x, x)$ cannot be expressed as an $E_j^{a^t,a}$-formula.

As any $\Delta_0^{\overline{u}}(M)$-set of the form $\{x \mid \phi(x, \overline{u})\}$ can be retrieved from the $\Delta_0^{\text{stan}}(M)$-set $\{\langle x, \overline{y}\rangle \mid \phi(x, \overline{y})\}$, the assumption that for some i

$$\Delta_0^{\text{stan}}(M) \subseteq E_i^{\text{stan}}(M)$$

would imply that for some j

$$\Delta_0^{\overline{u}}(M) \subseteq E_j^{\overline{u}}(M)$$

But this contradicts Claim 2, and the theorem is proved.   Q.E.D.

We conclude the first part of the Section by stating a proposition relating the PHP and consistency statements. It rests upon two facts: that $S_2 + \text{Exp}$ proves $\Sigma_\infty^b \text{PHP}$ and that $S_2 \vdash NP = coNP$ allows us to express each bounded formula as Σ_1^b and apply the provable Σ_1^b-completeness (cf. the claim in the proof of Theorem 14.1.4).

Theorem 15.3.6. *Assume that the theory $S_2 + Con(S_2)$ does not prove $\Sigma_\infty^b PHP$. Then S_2 does not prove that $NP = coNP$. That is, there is a Π_1^b-formula not equivalent in S_2 to any Σ_1^b-formula.*

We turn our attention to the simplest case of the fundamental problem.

Problem. *Does there exist a model M of PV in which $NP \neq coNP$, that is, $NP(M) \neq coNP(M)$?*

Further we say that a theory proves $NP = coNP$ if $NP(M) = coNP(M)$ holds for each model of the theory, and similarly for other classes. We shall also freely move between the languages of PV and S_2^1; note that the Δ_1^b-formulas in L are equivalent to the open PV-formulas (cf. Section 5.3).

Theorem 15.3.7. *Consider three propositions:*

   1. $NP = coNP$
   2. $P = NP$
   3. EF is a complete proof system

Then if S_2^1 proves one of the propositions, it proves all three, and then also PV proves all three propositions.

Proof. S_2^1 proves 1 iff it proves 2 by Corollary 7.2.5. If S_2^1 proves

$$\forall \tau, \neg \text{Taut}(\tau) \vee \text{EF} \vdash \tau$$

then there is, by Theorem 7.2.3, a polynomial-time function deciding whether $\neg \tau$ is satisfiable or τ a tautology (as by Theorem 9.3.16 and Lemma 4.6.3 S_2^1 proves that EF is sound). That is, $S_2^1 \vdash P = NP$.

Assume that S_2^1 proves $NP = coNP$. That is, $\text{Taut}(a) \equiv (\exists y \leq t(a), \theta(a, y))$, for some $\theta \in \Delta_1^b$. Then we may think of $b \leq t(a) \wedge \theta(a, b)$ as a proof system

P (b a proof of a) whose soundness (and completeness) is provable in S_2^1. Hence by Corollary 9.3.18 S_2^1 proves that EF polynomially simulates P and so it is also complete.

The provability of the propositions in PV follows from Corollary 7.2.4 as they are all $\forall\exists\Delta_1^b$.   Q.E.D.

A simple corollary to Corollary 7.2.5 is the following.

Corollary 15.3.8. *Assume that $P \neq NP$. Then there is a model of S_2^1 in which NP $\neq$ coNP.*

We link the completeness of EF in a model of PV to the extendability of the model. First we need a technical lemma. Recall Definition 9.2.1. To simplify the notation we shall not show the bounding polynomials explicitly.

Lemma 15.3.9. *Let M be a model of PV and let $\phi(a)$ be a Π_1^b-formula. Let $m \in M$. Then the following two conditions are equivalent:*
  1. *there is an extension N of M (necessarily Δ_1^b-elementary) such that*

$$N \models PV + \neg\phi(m)$$

  2. *$M \models EF \nvdash \|\phi\|^{|m|}(\tilde{m})$*

Proof. Assume that condition 1 holds while 2 fails and let π be an EF-proof of $\|\phi\|^{|m|}(\tilde{m})$ in M. Then $\pi \in N$ and it is an EF-proof in N too. As EF is sound in models of PV, $\|\phi\|^{|m|}(\tilde{m})$ is a tautology in N and hence by Lemma 9.3.12 $\phi(m)$ holds in N. That contradicts the assumption.

Now assume that 1 fails. This implies that the theory PV + Diag(M) in the language with a constant for every element of M, where Diag(M) is the diagram of M, proves the formula $\phi(m)$. That means that PV proves an implication

$$\sigma(m, u) \to \phi(m)$$

where u are some other constants and σ is an open formula. By Theorem 9.2.6 in M there is EF-proof of

$$\|\sigma\|^{|m|,|u|}(\tilde{m}, \tilde{u}) \to \|\phi\|^{|m|}(\tilde{m})$$

As σ is true in M, by Lemma 9.3.12 $\|\sigma\|^{|m|,|u|}(\tilde{m}, \tilde{u})$ has in M an EF-proof and thus also $\|\phi\|^{|m|}(\tilde{m})$ has (by modus ponens) an EF-proof in M. This shows that 2 fails.   Q.E.D.

We get the following corollary.

Corollary 15.3.10. *Let M be a model of PV. Then the following two conditions are equivalent:*

1. *The system EF is complete in M.*
2. *Any extension of M is Σ_1^b-elementary.*

Proof. Let $\phi(m)$ be a Π_1^b-sentence. Then by Lemma 9.3.12 it holds in M iff $\|\phi\|^{|m|}(\tilde{m})$ is a tautology in M. The rest follows by Lemma 15.3.9.   Q.E.D.

The last two statements demonstrate that a way to show that EF is not complete in a model M is to find a suitable tautology τ and an extension N of M containing a new truth assignment not satisfying τ. A forcing-type construction of such an extension is provided (in the context of the theory V_1^1 related to S_2^1 by the RSUV-isomorphism) by Theorem 9.4.2. Such a construction can be also viewed as a construction of a particular Boolean valuation ν (outside the model) of the set of the propositional formulas coded in M that gives τ a value $\nu(\tau) \neq 1_B$ (cf. Section 13.3). The problem is to produce such an evaluation without assuming that EF is not complete in M.

Another possibility how to prove that PV does not prove NP = coNP would be to demonstrate a superpolynomial lower bound to the size of EF-proofs. The following lemma is a direct corollary of Parikh's Theorem 5.1.4.

Lemma 15.3.11. *Assume that the theory S_2 proves that the system EF is complete. Then all tautologies have polynomial size EF-proofs.*

We shall show at least that PV itself does not prove such superpolynomial lower bounds. Let Bound(f) be the formula

$$\forall x \exists \tau > x, \text{Taut}(\tau) \wedge \text{EF} \not\vdash_{f(|\tau|)} \tau$$

where $\not\vdash_m$ means "not provable by a proof of size $\leq m$."

We say that f is *provably in PV superpolynomial* if for every polynomial $p(x)$ PV proves

$$\forall x \exists y > x \exists z < y, f(z) = y \wedge p(z) < y$$

Theorem 15.3.12. *Let f be a function definable in PV and provably in PV superpolynomial. Then*

$$PV \not\vdash \text{Bound}(f)$$

Proof. Let M be a countable nonstandard model of PV such that for some element $a \in M$ the numbers $|a|^k$, $k = 1, 2, \ldots$, are cofinal in $\text{Log}(M)$.

We shall construct a countable chain of cofinal extensions of M : $M_0 = M \subseteq M_1 \subseteq \ldots$. Fix in advance a countable language L' with names for all elements of all M_i and let $c_1, \ldots$ be an enumeration of L' with infinite repetitions. Put $M_0 := M$ and assume that we have M_i. Consider two cases:

1. $c_i \in M_i$ and $M_i \models \text{Taut}(c_i) \wedge \text{EF} \not\vdash c_i$
2. otherwise.

In the first case take an extension N_i of M_i in which $\neg c_i$ is satisfiable and put

$$M_{i+1} := \{a \in N_i \mid \exists b \in M_i, N_i \models a < b\}$$

hence M_{i+1} is a cofinal extension of M_i.

In the second case put $M_{i+1} := M_i$.

Note that if $c \in L'$ is a propositional formula in some M_i then in some M_j either it has an EF-proof or its negation is satisfiable. Hence $M' := \bigcup_i M_i$ is a cofinal extension of M in which every tautology has an EF-proof of size $\leq |a|^k$, some $k < \omega$. This implies that $M' \models \neg \text{Bound}(f)$ whenever f is PV-provably superpolynomial.   Q.E.D.

In the model M' in the previous proof the exponentiation is not a total function. It would be interesting to improve the construction to obtain a model of $S_2^1 + \neg \text{Exp}$ in which EF is complete. Such model would exist, in particular, if any countable model of S_2^1 would admit a Δ_1^b-elementary extension to a model of S_2^1 in which EF is complete. This is because we could take for the ground model a model satisfying the negation of some $\forall \Pi_1^b$-consequence of $S_2^1 + \text{Exp}$ (such formulas exist as $S_2^1 + \text{Exp}$ is not $\forall \Pi_1^b$-conservative over S_2^1; cf. Section 10.5).

Another direction in which it would be interesting to improve the construction is to find a model of $\text{PV} + \neg \text{Exp}$ in which EF is complete and that satisfies the collection scheme $B\Sigma_1^b$ (Definition 5.2.11). This would eliminate a possible pathological property of M' in which the shortest EF-proofs of tautologies $\leq a$, for some a, are cofinal in the model. It would also imply that the function $f(n) := \max_{|\tau| \leq n} c_{\text{EF}}(\tau)$ is total in M' and subexponential (see Definition 14.1.1 for the definition of $c_{\text{EF}}(\tau)$).

We conclude the Section recalling that a nonuniform version of the problem, a construction of a model of PV in which $\text{NP}/\text{poly} \neq \text{coNP}/\text{poly}$, would imply that $\text{PV} \neq S_2^1(\text{PV})$ (Theorem 10.2.4, second part). Hence even the very first step toward the fundamental problem has important consequences. Indeed, I believe that a construction of a model of PV in which EF is not complete would yield a hint of how to construct Boolean valuations of sets of polynomially many formulas and how to prove the nonexistence of a polynomial upper bound to the size of EF-proofs.

15.4. Bibliographical and other remarks

Theorem 15.1.1 follows from Yao (1985) and Hastad (1989); the superpolynomial lower bounds of Ajtai (1983) and Furst, Saxe, and Sipser (1984) imply only the nonexistence of a $\Delta_0(\alpha)$-counting.

Paris and Wilkie (1985) show that f in Lemma 15.1.2 can be Δ_0 if A is Δ_0. The proof of Theorem 15.1.4 follows an unpublished presentation of Sipser (1983b) by A. Wilkie; Paris and Wilkie (1985) show by another proof that for $A \in \Delta_0$ a function f satisfying a weaker inequality $|A_a| \leq f(a) < |A_a|^{1+\epsilon}$ can also be Δ_0.

Bounded PHP was studied in Woods (1981), where Sylvester's theorem is deduced by using $\Delta_0 PHP_a^{2a}$.

The proof of Lemma 15.2.2 follows to a large extent an unpublished proof of Woods of the switching lemma 3.1.11. He estimates the probability (over ρ) that in a game player II assigns $\geq s$ values and that that game follows a given marked value sequence v, splitting the computation further according to the history of the game. Prior to our presentation Razborov (1994a) found a similar presentation of Hastad's original proof of 3.1.11.

Razborov (1995) considers a framework in which Boolean complexity is formalized in second order bounded arithmetic systems. Input vectors are still coded by numbers, but Boolean functions and circuits are coded by sets. This allows us, in particular, to speak directly about exponentially large circuits. He shows that in such formalization all main known lower bounds can be proved in the theory V_1^1 (and often in a weaker theory). By Lemma 5.5.14 there is no problem with direct counting arguments. The only potential problem may arise in formalizing proofs where Boolean functions are quantified (e.g., in various conditional probabilities as in the original proof of 3.1.11 in Hastad 1989), which could increase the complexity of induction formulas. Razborov (1994a) overcomes this in the case of the switching lemma by a new proof; another option is to replace the original quantification over arbitrary functions by a quantification over a $\Delta_1^{1,b}$-definable sequence of functions.

Razborov (1994) complements this by a result in the opposite direction: namely that the theory $S_2^2(\alpha)$ does not prove a superpolynomial lower bound to the circuits computing the satisfiability predicate unless a standard cryptographic assumption fails; see also Krajíček (1995b) for an alternative proof. This assumption, saying that strong pseudo-random number generators exist, implies in particular that NP $\not\subseteq$ P/poly.

Lemma 15.3.2 is due to Paris and Kirby (1978). Lemma 15.3.3 and Theorem 15.3.4 are from Paris et al. (1988); Theorem 15.3.5 is from Paris and Wilkie (1985). Theorem 15.3.6 is from Paris and Wilkie (1987b) and relates to the interpretability questions; see also Hájek and Pudlák (1993) for details and related topics.

Lemma 15.3.9 is from Krajíček and Pudlák (1990b); A. Wilkie (unpublished) gave a model-theoretic proof. Theorem 15.3.12 is also from Krajíček and Pudlák (1990b); Cook and Urquhart (1993) proved a similar result for the intuitionistic version of S_2^1, and Buss (1990b) showed that the two results are actually equivalent.

A research program some reader may contemplate is to show that a statement like NP $\neq$ coNP is not provable in PV, for example. This is a task to construct a model of PV in which NP = coNP and it is somewhat interesting. Theorem 7.6.1 is in this spirit.

I shall close the book, however, recalling the Preface: In my opinion, instead of constructing a model in which NP = coNP and thus confirming what we already know, namely that the problem is not elementary and that the available methods are insufficient, it would be far more interesting to construct a model M for which one could prove NP(M) $\neq$ coNP(M) *by an argument using the underlying idea behind our belief that the conjecture* NP $\neq$ coNP *is true*. Results in the opposite direction, as well as many oracle-independence results, have less to do with the problem and rather exploit technical details of the definitions.

REFERENCES

Adleman, L., and Manders, K. (1977) Reducibility, randomness, and intractability, in: *Proceedings of the 9th Annual ACM Symposium on Theory of Computing*, pp. 151–63. ACM Press.

Ajtai, M. (1983) Σ_1^1-formulae on finite structures, *Annals of Pure and Applied Logic*, **24**: 1–48.

Ajtai, M. (1988) The complexity of the pigeonhole principle, in: *Proceedings of the IEEE 29th Annual Symposium on Foundations of Computer Science*, pp. 346–55.

Ajtai, M. (1990) Parity and the pigeonhole principle, in: *Feasible Mathematics*, eds. S.R.Buss and P.J.Scott, pp. 1–24. Birkhäuser.

Ajtai, M. (1994a) The independence of the modulo p counting principles, in: *Proceedings of the 26th Annual ACM Symposium on Theory of Computing*, pp. 402–11. ACM Press.

Ajtai, M. (1994b) Symmetric systems of linear equations modulo p, preprint.

Ajtai, M. (1995) On the existence of modulo p cardinality functions, in: *Feasible Mathematics II*, eds. P. Clote and J. Remmel, pp. 1–14. Birkhäuser.

Aleliunas, R., Karp, R. M., Lovász, L., Lipton, M. J., and Rackoff, C. (1979) Random walks, universal traversal sequences, and the maze problems, in: *Proceedings of the IEEE 20th Annual Symposium on Foundations of Computer Science*, pp. 218–23.

Alon, N., and Boppana, R. (1987) The monotone circuit complexity of boolean functions, *Combinatorica*, **7(1)**: 1–22.

Andreev, A. E. (1985) On a method for obtaining lower bounds for the complexity of individual monotone functions (in Russian), *Doklady AN SSSR*, **282(5)**: 1033–7.

Andreev, A. E. (1987) On a method for obtaining more than quadratic effective bounds for the complexity of π-schemes (in Russian), *Vestnik Moskovskove Univ. Matemat.*, **42(1)**: 70–3.

Arai, T. (1991) Frege system, *ALOGTIME* and bounded arithmetic, manuscript.

Baker, T., Gill, J., and Solovay, R. (1975) Relativizations of the $P = ? NP$ question, *SIAM J. Computing*, **4**: 431–42.

Balcazár, J. L., Díaz, J., and Gabarró, J. (1988) *Structural Complexity I*. Berlin, Springer.

Balcazár, J. L., Díaz, J., and Gabarró, J. (1990) *Structural Complexity II*. Berlin, Springer.

Barrington, D. A. (1989) Bounded-width polynomial-time branching programs recognize exactly those languages in NC^1, *Comp. Systems Sci.*, **38(1)**: 150–64.

Beame, P., Impagliazzo, R., Krajíček, J., Pitassi, T., Pudlák, P., and Woods, A. (1992) Exponential lower bounds for the pigeonhole principle, in: *Proceedings of the 24th Annual ACM Symposium on Theory of Computing*, pp. 200–21. ACM Press.

Beame, P., Impagliazzo, R., Krajíček, J., Pitassi, T., and Pudlák, P. (1994) Lower bounds on Hilbert's Nullstellensatz and propositional proofs, submitted.

Beame, P., and Pitassi, T. (1993) Exponential separation between the matching principles and the pigeonhole principle, submitted.

Bellantoni, S., Pitassi, T., and Urquhart, A. (1992) Approximation and small depth Frege proofs, *SIAM J. Computing*, **21(6)**: 1161–79.

327

Bel'tyukov, A. (1979) A computer description and a hierarchy of initial Grzegorczyk classes, *Zapiski Nauc. Sem. LOMI*, **88**, Steklov Institute.

Bennett, J. H. (1962) *On spectra*, Ph.D. Thesis, Princeton University.

Berarducci, A., and Intrigila, B. (1991) Combinatorial principles in elementary number theory, *Annals of Pure and Applied Logic*, **55**: 35–50.

Beth, E. W. (1959) The foundations of mathematics. Amsterdam, North-Holland.

Blake, A. (1937) Canonical expressions in boolean algebra, Ph.D. Thesis, University of Chicago.

Bondy, J.A. (1972) Induced subsets, *J. Combinatorial Theory (B)*, **12**: 201–2.

Bonet, M. (1993) Number of symbols in Frege proofs with and without the deduction rule, in: *Arithmetic, Proof Theory and Computational Complexity*, eds. P. Clote and J. Krajíček, pp. 61–95. Oxford, Oxford University Press.

Bonet, M., and Buss, S. (1993) The deduction rule and linear and near-linear proof simulations, *J. Symbolic Logic*, **58(2)**: 688–709.

Bonet, M., Buss, S. and Pitassi, T. (1995) Are there hard examples for Frege systems? in: *Feasible Mathematics II*, eds. P. Clote and J. Remmel, pp. 30–56. Birkhäuser.

Bonet, M., Pitassi, T., and Raz, R. (1995) Lower bounds for cutting planes proofs with small coefficients, preprint.

Boppana, R., and Sipser, M. (1990) Complexity of finite functions, in: *Handbook of Theoretical Computer Science*, ed. J. van Leeuwen, pp. 758–804. Amsterdam, Elsevier.

Buss, S. R. (1986) *Bounded Arithmetic*. Naples, Bibliopolis. (Revision of 1985 Princeton University Ph.D. thesis.)

Buss, S. R. (1987) The propositional pigeonhole principle has polynomial size Frege proofs, *J. Symbolic Logic*, **52**: 916–27.

Buss, S. R. (1988) Weak formal systems and connections to computational complexity, lecture notes, University of California, Berkeley.

Buss, S. R. (1990a) Axiomatizations and conservation results for fragments of bounded arithmetic, in: *Logic and Computation, Contemporary Mathematics*, **106**: 57–84. Providence, American Mathematical Society.

Buss, S. R. (1990b) On model theory for intuitionistic bounded arithmetic with applications to independence results, in: *Feasible Mathematics*, eds. S.R. Buss and P.J. Scott, pp. 27–47. Birkhäuser.

Buss, S. R. (1991) Propositional consistency proofs, *Annals of Pure and Applied Logic*, **52**: 3–29.

Buss, S. R. (1993a) Some remarks on lengths of propositional proofs, submitted.

Buss, S. R. (1993b) Relating the bounded arithmetic and polynomial time hierarchies, submitted.

Buss, S. R., and Hay, L. (1988) On truth-table reducibility to *SAT* and the difference hierarchy over *NP*, *Proceedings Structure in Complexity*, IEEE Computer Society Press, 224–33.

Buss, S. R., and Krajíček, J. (1994) An application of boolean complexity to separation problems in bounded arithmetic, *Proceedings of the London Mathematical Society*, **69(3)**: 1–21.

Buss, S. R., Krajíček, J., and Takeuti, G. (1993) On provably total functions in bounded arithmetic theories R_3^i, U_2^i and V_2^i, in: *Arithmetic, Proof Theory and Computational Complexity*, eds. P. Clote and J. Krajíček, pp. 116–61. Oxford, Oxford University Press.

Buss, S. R., and Turán, G. (1988) Resolution proofs of generalized pigeonhole principles, *Theoretical Computer Science* **62**: 311–17.

Cai, J. (1989) With probability one, a random oracle separates *PSPACE* from the polynomial-time hierarchy, *J. Comput. System Sci.*, **38**: 68–85.

Chiari, M., and Krajíček, J. (1994) Witnessing functions in bounded arithmetic and search problems, submitted.

Chrapchenko, V. M. (1971) A method of determining lower bounds for the complexity of π-schemes (in Russian), *Matemat. Zametki*, **10(1)**: 83–92.

Chvátal, V., and Szemerédi, E. (1988) Many hard examples for resolution, *J. of ACM*, **35(4)**: 759–68.

Clote, P. (1992) ALOGTIME and a conjecture of S.A. Cook, *Annals of Mathematics and Artificial Inteligence*, **6**: 57–106.

Clote, P., and Krajíček, J. (1993) Open problems, in: *Arithmetic, Proof Theory and Computational Complexity*, eds. P. Clote and J. Krajíček, pp. 1–19. Oxford, Oxford University Press.

Clote, P., and Takeuti, G. (1986) Exponential time and bounded arithmetic, in: *Proceedings Structure in Complexity*, LN Comp. Sci. **223**, Springer-Verlag.

Clote, P., and Takeuti, G. (1992) Bounded arithmetic for *NC, ALOGTIME, L* and *NL, Annals of Pure and Applied Logic*, **56**: 73–117.

Cobham, A. (1965) The intrinsic computational difficulty of functions, in: *Proceedings of the International Congress Logic, Methodology and Philosophy of Science*, ed. Y. Bar-Hillel, pp. 24–30. North-Holland.

Cook, S A. (1971) The complexity of theorem proving procedures, in: *Proceedings of the 3rd Annual ACM Symposium on Theory of Computing*, pp. 151–8. ACM Press.

Cook, S A. (1975) Feasibly constructive proofs and the propositional calculus, in: *Proceedings of the 7th Annual ACM Symposium on Theory of Computing*, pp. 83–97. ACM Press.

Cook, S A. (1991) Computational complexity of higher type functions, in: *Proceedings ICM Kyoto 1990*, pp. 55–69. Springer-Verlag.

Cook, S. A., and Reckhow, A. R. (1979) The relative efficiency of propositional proof systems, *J. Symbolic Logic*, **44(1)**: 36–50.

Cook, S. A., and Urquhart, A. (1993) Functional interpretation of feasibly constructive arithmetic, *Annals of Pure and Applied Logic*, **63**; 103–200.

Cook, W., Coullard, C. R., and Turán, G. (1987) On the complexity of cutting plane proofs, *Discrete Applied Mathematics*, **18**: 25–38.

Craig, W. (1957) Three uses of the Herbrand–Gentzen theorem in relating model theory and proof theory, *J. Symbolic Logic*, **22(3)**: 269–85.

Davis, M., and Putnam, H. (1960) A computing procedure for quantification theory, *J. ACM*, **7(3)**: 210–15.

Dimitracopoulos, C. (1980) Matijavič's theorem and fragments of arithmetic, Ph.D. Thesis, University of Manchester.

Dimitracopoulos, C., and Paris, J. (1986) The pigeonhole principle and fragments of arithmetic, *Zeitschrift f. Mathematikal Logik u. Grundlagen d. Mathematik*, **32**: 73–80.

Dowd, M. (1979) Propositional representations of arithmetic proofs, *Ph.D. Thesis, University of Toronto*.

Furst, M., Saxe, J. B., and Sipser, M. (1984) Parity, circuits and the polynomial-time hierarchy, *Math. Systems Theory*, **17**: 13–27.

Gaifman, H., and Dimitracopoulos, C. (1982) Fragments of Peano's arithmetic and the *MRDP* theorem, *Logic and algorithmic, Monogr. Enseign. Mathemat.*, **30**: 187–206.

Garey, M.R., and Johnson, D. S. (1979) Computers and intractability. New York, W.H. Freeman.

Gentzen, G. (1969) *The collected papers of Gerhard Gentzen*. Amsterdam, North-Holland.

Gödel, K. (1986) Collected Works I. eds. S. Feferman et. al. New York and Oxford, Oxford University Press.

Gödel, K. (1990) Collected works II. eds. S. Feferman et. al. New York and Oxford, Oxford University Press.

Goerdt, A. (1990) Cutting plane versus Frege proof systems, in: *Computer Science Logic '90*, LNCS 533, pp. 174–94. Springer-Verlag.

Gratzer G. (1979) Universal algebra, 2nd ed. Springer-Verlag.

Grzegorczyk, A. (1953) Some classes of recursive functions, in: *Rozprawy Matematiczne IV*, Warszawa.

Hájek, P., and Pudlák, P. (1993) Metamathematics of first-order arithmetic, *Perspectives in Mathematical Logic*. Springer-Verlag.

Haken,A. (1985) The intractability of resolution, *Theoretical Computer Science*, **39**: 297–308.

Hartmanis, J., Lewis, P. M., and Stearns, R. E. (1965) Hierarchies of memory limited computations, *IEEE Conference Research on Switching Circuit th. and Logic design*, Ann Arbor, Michigan, pp. 179–90.

Hartmanis, J., and Stearns, R. E. (1965) On the computational complexity of algorithms, *Transactions of the American Mathematical Society*, **117**: 285–306.

Hastad, J. (1989) Almost optimal lower bounds for small depth circuits. in: *Randomness and Computation*, ed. S. Micali, *Ser. Adv. Comp. Res.*, **5**: 143–70. JAI Press.

Hastad, J. (1993) The shrinkage factor is 2, in: *Proceedings of the IEEE 34th Annual Symposium on Foundations of Computer Science*, pp. 114–23.

Hausdorf, F. (1978) *Set Theory*, 3rd ed. Chelsea.

Hopcroft, J., Paul, W., and Valiant, L. (1975) On time versus space, *J. Assoc. Computing Machinery*, **24**: 332–7.

Immerman, N. (1988) Nondeterministic space is closed under complementation, *SIAM J. Computing*, **17(5)**: 935–8.

Johnson, D. S., Papadimitriou, C. H., and Yannakakis, M. (1988) How easy is local search? *J. Comput. System Sci.*, **37**: 79–100.

Kadin, J. (1988) The polynomial time hierarchy collapses if the boolean hierarchy collapses, in: *Proceedings of the IEEE 29th Annual Symposium on Foundations of Computer Science*, pp. 278–92.

Karchmer, M. (1993) On proving lower bounds for circuit size, in: *Proceedings of Structure in Complexity*, 8th Annual Conference, pp. 112–19. IEEE Computer Science Press.

Karchmer, M., and Wigderson, A. (1988) Monotone circuits for connectivity require super-logarithmic depth, in: *Proceedings of the 20th Annual ACM Symposium on Theory of Computing*, pp. 539–50. ACM Press.

Karp, R., and Lipton, R. J. (1982) Turing machines that take advice, *Enseign. Mathem.*, **28**: 191–209.

Kaye R. (1991) *Models of Peano arithmetic*. Oxford, Oxford University Press.

Krajíček, J. (1989a) On the number of steps in proofs, *Annals of Pure and Applied Logic*, **41**: 153–78.

Krajíček, J. (1989b) Speed-up for propositional Frege systems via generalizations of proofs, *Commentationes Mathematicae Universitas Carolinae*, **30(1)**: 137–40.

Krajíček, J. (1990) Exponentiation and second-order bounded arithmetic, *Annals of Pure and Applied Logic*, **48**: 261–76.

Krajíček, J. (1992) No counter-example interpretation and interactive computation, in: *Logic From Computer Science*, Proceedings of a Workshop held November 13–17, 1989 in Berkeley, ed. Y. N. Moschovakis, *Mathematical Sciences Research Institute Publication*, **21**: 287–93. Springer-Verlag.

Krajíček, J. (1993) Fragments of bounded arithmetic and bounded query classes, *Transactions of the A.M.S.*, **338(2)**: 587–98.

Krajíček, J. (1994) Lower bounds to the size of constant-depth propositional proofs, *J. Symbolic Logic*, **59(1)**: 73–86.

Krajíček, J. (1995a) On Frege and Extended Frege proof systems, in: *Feasible Mathematics II*, eds. P. Clote and J. Remmel, pp. 284–319. Birkhäuser.

Krajíček, J. (1995b) Interpolation theorems, lower bounds for proof systems and independence results for bounded arithmetic, submitted.

Krajíček, J., and Pudlák, P. (1988) The number of proof lines and the size of proofs in first order logic, *Archive for Mathematical Logic*, **27**: 69–84.

Krajíček, J., and Pudlák, P. (1989a) Propositional proof systems, the consistency of first order theories and the complexity of computations, *J. Symbolic Logic*, **54(3)**: 1063–79.

Krajíček, J., and Pudlák, P. (1989b) On the structure of initial segments of models of arithmetic, *Archive for Mathematical Logic*, **28(2)**: 91–8.

Krajíček, J., and Pudlák, P. (1990a) Quantified propositional calculi and fragments of bounded arithmetic, *Zeitschrift f. Mathematikal Logik u. Grundlagen d. Mathematik*, **36**: 29–46.

Krajíček, J., and Pudlák, P. (1990b) Propositional provability in models of weak arithmetic, in: *Computer Science Logic (Kaiserslautern, Oct. '89)*, eds. E. Boerger, H. Kleine-Bunning, and M.M. Richter, LNCS 440, pp. 193–210. Springer-Verlag.

Krajíček, J., and Pudlák, P. (1995) Some consequences of cryptographical conjectures for S_2^1 and EF, submitted.

Krajíček, J., Pudlák, P., and Sgall, J. (1990) Interactive computations of optimal solutions, in: *Mathematical Foundations of Computer Science (B. Bystrica, August '90)*, ed. B. Rovan, LNCS 452, pp. 48–60. Springer-Verlag.

Krajíček, J., Pudlák, P., and Takeuti, G. (1991) Bounded arithmetic and the polynomial hierarchy, *Annals of Pure and Applied Logic*, **52**: 143–53.

Krajíček, J., Pudlák, P., and Woods, A. (1991) Exponential lower bound to the size of bounded depth Frege proofs of the pigeonhole principle, *Random Structures and Algorithms*, to appear.

Krajíček, J., and Takeuti, G. (1990) On bounded $\sum_{1}^{1}$-polynomial induction, in: *Feasible Mathematics*, eds. S.R. Buss and P.J. Scott, pp. 259–80. Birkhäuser.

Krajíček, J., and Takeuti, G. (1992) On induction-free provability, *Annals of Mathematics and Artificial Intelligence*, **6**: 107–26.

Krentel, M. W. (1986) The complexity of optimization problems, in: *Proceedings of the 18th Annual ACM Symposium on Theory of Computing*, pp. 69–76. ACM Press.

Lesan, H. (1978) *Models of arithmetic*, Ph.D. Thesis, University of Manchester.

Lovász, L. (1990) Communication complexity: a survey, in: *Paths, Flows and VLSI Layout*, eds. Korte, Lovász, Promer, and Schrijver, pp. 325–66. Springer-Verlag.

Lovász, L., Naor, M., Newman, I., and Wigderson, A. (1991) Search problems in the decision tree model, in: *Proceedings of the 32nd IEEE Symposium on Foundations of Computer Science*, pp. 576–85. IEEE Computer Society Press.

Macintyre, A. (1987) The strength of weak systems, in: *Schriftenreihe der Wittgenstein-Gesellschaft*, **13**, Logic, Philosophy of Science and Epistemology, pp. 43–59, Vienna.

Mehlhorn, K., and Schmidt, F. M. (1982) Las Vegas is better than determinism in VLSI and distributed computing, in: *14th Annual ACM Symposium on Theory of Computing*, pp. 330–7. ACM Press.

Meyer, A., and Stockmeyer, L. (1973) The equivalence problem for regular expressions with squaring requires exponential time, in: *Proceedings of the IEEE 13th Symposium on Switching and Automata Theory*, pp. 125–9.

Mints, G. E. (1976) What can be done in PRA? *Zapiski Nauc. Sem. LOMI*, **60**, Steklov Institute, pp. 93–102.

Muchnik, A. A. (1970) On two approaches to the classification of recursive functions, in: *Problems in Mathematical Logic, Complexity of Algorithms and Classes of Computable Functions*, eds. V.A. Kozmidiadi and A.A.Muchnik, pp. 123–8. Mir, Moscow.

Muller, D. E. (1956) Complexity of electronic switching circuits, *IRE Trans. Electronic Computers*, **5**: 15–19.

Nepomnjascij, V.A. (1970) Rudimentary predicates and Turing calculations, *Doklady AN SSSR*, **195**.

Otero, M. (1991) *Models of open indunction*, Ph.D. Thesis, Oxford University.

Papadimitriou, C. H. (1990) On graph-theoretic lemmata and complexity classes (extended abstract), in: *Proceedings of the 31st IEEE Symposium on Foundations of Computer Science (Volume II)*, pp. 794–801. IEEE Computer Society Press.

Papadimitriou, C. H. (1994) On the complexity of the parity argument and other inefficient proofs of existence, *J. Comput. System Sci.*, **48(3)**: 498–532.

Papadimitriou, C. H., and Yannakakis, M. (1988) Optimization, approximation and complexity classes, in: *20th Annual ACM Symposium on Theory of Computing*, pp. 229–34. ACM Press.

Parikh, R. (1971) Existence and feasibility in arithmetic, *J. Symbolic Logic*, **36**: 494–508.

Parikh, R. (1973) Some results on the length of proofs, *Transactions of the A.M.S.*, **177**: 29–36.

Paris, J. B. (1984) O struktuře modelu omezené E_1 indukce (in Czech), *Časopis pěstování matematiky*, **109**: 372–9.

Paris, J., and Dimitracopoulos, C. (1982) Truth definitions for Δ_0 formulae, in: *Logic and Algorithmic, l'Enseignement Mathématique*, **30**: 318–29. Genève.

Paris, J., and Dimitracopoulos, C. (1983) A note on undefinability of cuts, *J. Symbolic Logic*, **48**: 564–9.

Paris, J.B., Handley, W.G., and Wilkie, A.J. (1984) Characterizing some low arithmetic classes, *Colloquia Mathematica Soc. J. Bolyai*, **44**: 353–64.

Paris, J.B., and Kirby, L. (1978) Σ_n-collection schemes in arithmetic, in: *Logic Colloquium '77*, pp. 199–209. Amsterdam, North-Holland.

Paris, J., and Wilkie, A. J. (1981a) Δ_0 sets and induction, in: *Proceedings of the Jadwisin Logic Conference, Poland*, pp. 237–48. Leeds University Press.

Paris, J., and Wilkie, A. J. (1981b) Models of arithmetic and rudimentary sets, *Bull. Soc. Mathem. Belg., Ser. B*, **33**: 157–69.

Paris, J., and Wilkie, A. J. (1984) Some results on bounded induction, in: *Proceedings of the 2nd Easter Conference on Model Theory (Wittenberg '84)*, pp. 223–8. Humboldt Universität, Berlin.

Paris, J., and Wilkie, A. J. (1985) Counting problems in bounded arithmetic, in: *Methods in Mathematical Logic*, LNM 1130, pp. 317–40. Springer-Verlag.

Paris, J., and Wilkie, A. J. (1987a) Counting Δ_0 sets, *Fundamenta Mathematica*, **127**: 67–76.

Paris, J., and Wilkie, A. J. (1987b) On the scheme of induction for bounded arithmetic formulas, *Annals of Pure and Applied Logic*, **35**: 261–302.

Paris, J. B., Wilkie, A. J., and Woods, A. R. (1988) Provability of the pigeonhole principle and the existence of infinitely many primes, *J. Symbolic Logic*, **53**: 1235–44.

Parsons, C. (1970) On a number theoretic choice schema and its relation to induction, in: *Intuitionism and Proof Theory*, eds. A. Kino et. al., pp. 459–73. Amsterdam, North-Holland.

Pitassi, T., Beame, P., and Impagliazzo, R. (1993) Exponential lower bounds for the pigeonhole principle, *Computational complexity*, **3**: 97–140.

Pitassi, T., and Urquhart, A. (1992) The complexity of the Hajós calculus, in: *Proceedings of the IEEE 33rd Annual Symposium on Foundations of Computer Science*, pp. 187–96. IEEE Computer Society Press.

Pudlák, P. (1983) A definition of exponentiation by a bounded arithmetic formula, *Commentationes Mathematicae Universitas Carolinae*, **24(4)**: 667–71.

Pudlák, P. (1985) Cuts, consistency statements and interpretations, *J. Symbolic Logic*, **50**: 423–41.

Pudlák, P. (1986) On the length of finitistic consistency statements in first order theories, in: *Logic Colloquim 84*, pp. 165–96. North-Holland.

Pudlák, P. (1987) Improved bounds to the length of proofs of finitistic consistency statements, *Contemporary Mathematics*, **65**: 309–31.

Pudlák, P. (1990) A note on bounded arithmetic, *Fundamenta Mathematica*, **136**: 85–9.

Pudlák, P. (1992a) Ramsey's theorem in bounded arithmetic, in: *Proceedings of Computer Science Logic*, eds. Boerger E. et. al., *Comp. Sci.*, **553**: 308–12.

Pudlák, P. (1992b) Some relations between subsystems of arithmetic and the complexity of computations, in: *Logic from Computer Science*, Proceedings of a Workshop held November 13 –17, 1989 in Berkeley, ed. Y.N. Moschovakis, *Mathematical Sciences Research Institute Publication*, **21**: 499–519. Springer-Verlag.

Rabin, M. O. (1961) Non-standard models and independence of the induction axioms, in: *Essays on the Foundation of Mathematics*, pp. 287–99. Amsterdam, North-Holland.

Rabin, M. O. (1962) Diophantine equations and non-standard models of arithmetic, in: *Proceedings of the 1st International Congress on Logic, Methodology and Philosophy of Science*, pp. 151–8, Stanford University Press.

Raz, R., and Wigderson, A. (1990) Monotone circuits for matching require linear depth, in: *Proceedings of the 22nd Annual ACM Symposium on Theory of Computing*, pp. 287–92. ACM Press.

Razborov, A. A. (1985) Lower bounds on the monotone complexity of some Boolean functions, *Soviet Mathem. Doklady*, **31**: 354–7.

Razborov, A. A. (1987) Lower bounds on the size of bounded depth networks over a complete basis with logical addition, *Matem. Zametki*, **41(4)**: 598–607.

Razborov, A. A. (1989) On the method of approximations, in: *Proceedings of the 21th Annual ACM Symposium on Theory of Computing*, pp. 168–76. ACM Press.

Razborov, A. A. (1993) An equivalence between second order bounded domain bounded arithmetic and first order bounded arithmetic, in: *Arithmetic, Proof Theory and Computational Complexity*, eds. P. Clote and J. Krajiček, pp. 247–77. Oxford University Press.

Razborov, A. A. (1994) Unprovability of lower bounds on the circuit size in certain fragments of bounded arithmetic, submitted.

Razborov, A. A. (1995) Bounded arithmetic and lower bounds in boolean complexity, in: *Feasible Mathematics II*, eds. P. Clote and J. Remmel, pp. 344–86. Birkhäuser.

Reckhow, R. A. (1976) On the lengths of proofs in the propositional calculus, Ph.D. Thesis, University of Toronto, Technical Report No. 87.

Ressayre, J.-P. (1986) A conservation result for system of bounded arithmetic, unpublished preprint.

Riis, S. (1993a) Making infinite structures finite in models of second order bounded arithmetic, in: *Arithmetic,Proof Theory and Computational Complexity*, eds. P. Clote and J. Krajicek, pp. 289 –319. Oxford University Press.

Riis, S. (1993b) Independence in bounded arithmetic, Ph.D. Thesis, Oxford University.

Riis, S. (1994) *Count(q)* does not imply *Count(p)*, submitted.

Robinson, J., A. (1965) A machine-oriented logic based on the resolution principle, *J. ACM*, **12(1)**: 23–41.

Ryll-Nardzewski, C. (1952) The role of the axiom of induction in elementary arithmetic, *Fundamenta Mathematica*, **39**: 239–63.

Savitch, W. J. (1970) Relationships between nondeterministic and deterministic tape complexities, *J. Computer and System Sciences*, **4**: 177–92.

Shannon, C. E. (1949) The synthesis of two-terminal switching circuits, *Bell Systems Techn. J.*, **28(1)**: 59–98.

Sheperdson, J. C. (1964) A non-standard model of a free variable fragment of number theory, *Bull. Acad. Pol. Sci.*, **12**: 79–86.

Shoenfield, J. (1967) *Mathematical Logic*. Reading, Mass., Addison-Wesley.

Sipser, M. (1983a) Borel sets and circuit complexity, in: *Proceedings of the 15th Annual ACM Symposium on Theory of Computing*, pp. 61–9. ACM Press.

Sipser, M. (1983b) A complexity theoretic approach to randomness, in: *Proceedings of the 15th Annual ACM Symposium on Theory of Computing*, pp. 330–5. ACM Press.

Sipser, M. (1992) The history and status of the *P* versus *NP* question, in: *Proceedings of the 24th Annual ACM Symposium on Theory of Computing*, pp. 603–18. ACM Press.

Smale, S. (1992) Theory of computation, in: *Mathematical Research Today and Tomorrow*, eds. C. Casacuberta and M. Castellat, pp. 59–69. Berlin and Heidelberg, Springer-Verlag.

Smolensky, R (1987) Algebraic methods in the theory of lower bounds for Boolean circuit complexity, in: *Proceedings of the 19th Annual ACM Symposium on Theory of Computing*, pp. 77–82. ACM Press.

Smoryński, C. (1977) The incompleteness theorems, in: *Handbook of Mathematical Logic*, ed. J. Van Leeuwen, *Studies in Logic and Found. of Mathematics* **90**, pp. 821–65. Amsterdam, North-Holland.

Smoryński, C. (1984) Lectures on non-standard models of arithmetic, in: *Logic Colloquium '82*, Stud. in Logic and Found. of Math. **112**, pp. 1–70. Amsterdam, North-Holland.

Smullyan, R. (1961) *Theory of Formal Systems*, Annals of Mathematical Studies, **47**. Princeton, Princeton University Press.

Smullyan, R. (1968) *First-Order Logic*, Springer-Verlag.

Spira, P. M. (1971) On time-hardware complexity of tradeoffs for Boolean functions, in: *Proceedings of the 4th Hawaii Symposium System Sciences*, pp. 525–7. North Hollywood, Calif., Western Periodicals.

Statman, R. (1977) Complexity of derivations from quantifier-free Horn formulae, mechanical introduction of explicit definitions, and refinement of completeness theorems, in: *Logic Colloquium '76*, pp. 505–517. Amsterdam, North-Holland.

Statman, R. (1978) Bounds for proof-search and speed-up in the predicate calculus. *Annals of Mathematical Logic*, **15**: 225–87.

Stockmeyer, L. (1977) The polynomial-time hieararchy, *Theoretical Computer Science*, **3**: 1–22.

Subbotovskaja, B. A. (1961) Realizations of linear functions by formulas using $+, \cdot, -$ (in Russian), *Doklady AN SSSR*, **136(3)**: 553–5.

Szelepcsényi, R. (1987) The method of forcing for non-deterministic automata, *Bull. Europ. Assoc. Theor. Comput.Sci.*, **32**: 96–100.

Tait, W. W. (1968) Normal derivability in classical logic, in: *The Syntax and Semantics of Infinitary Languages*, ed. J. Barwise, pp. 204–36. Springer-Verlag.

Takeuti, G. (1975) *Proof Theory*, North-Holland.

Takeuti, G. (1988) Bounded arithmetic and truth definition, *Annals of Pure and Applied Logic*, **39**: 75–104.

Takeuti, G. (1993) RSUV isomorphism, in: *Arithmetic, Proof Theory and Computational Complexity*, eds. P. Clote and J. Krajíček, pp. 364–86. Oxford, Oxford University Press.

Takeuti, G, and Zaring, W. M. (1973) *Axiomatic Set Theory*, Springer-Verlag.

Tarski, A., Mostowski, A., and Robinson, R. M. (1953) *Undecidable Theories*. Amsterdam, North-Holland.

Tennenbaum S. (1959) Non-archimedean models of arithmetic, *Notices of the A.M.S.*, **6**: 270.

Toda, S. (1989) On the computational power of PP and $\oplus P$, in: *Proceedings of the 30th IEEE Symposium on the Foundations of Computer Science*, pp. 514–19. IEEE Computer Science Press.

Tseitin, G. C. (1968) On the complexity of derivations in propositional calculus, in: *Studies in mathematics and mathematical logic, Part II*, ed. A.O. Slisenko, pp. 115–25.

Tseitin, G. C., and Choubarian, A. A. (1975) On some bounds to the lengths of logical proofs in classical propositional calculus in Russian, *Trudy Vyčisl Centra AN Arm SSR i Erevanskovo Univ.*, **8**: 57–64.

Urquhart, A. (1987) Hard examples for resolution, *J. ACM*, **34**: 209–19.

Urquhart, A. (1992) The relative complexity of resolution and cut-free Gentzen systems, *Annals of Mathematics and Artificial Intelligence*, **6**: 1157–68.

Valiant, L. (1979) The complexity of computing the permanent, *Theoretical Computer Science*, **8**: 189–201.

Van den Dries, L. (1980) Some model theory and number theory for models of weak systems of arithmetic, in: *Model Theory of Algebra and Arithmetic*, eds. L. Pacholski et al., LN Mathematics, pp. 346–67. Springer-Verlag.

Van den Dries, L. (1990) Which curves over **Z** have points with coordinates in a discretely ordered ring? *Transactions of the American Mathematical Society*, **264**: 33–56.

Wagner, K. W. (1990) Bounded query classes, *SIAM J. Computing*, **19(5)**: 833–46.

Wegener, I. (1987) *The Complexity of Boolean Functions*, New York and Stuttgart, Willey/ Teubner.

Wilkie, A. (1978) Some results and problems on weak systems of arithmetic, in: *Logic Colloquium '77*, Studies in the Logical Foundations of Mathematics 96, pp. 285–96. North-Holland.

Wilkie, A. (1980) Applications of complexity theory to Σ_0-definability problems in arithmetic, in: *Model Theory of Algebra and Arithmetic*, LNM **834**, pp. 363–9. Springer-Verlag.

Wilkie, A. (1985) Modèles non standard de l'arithmétique, et complexité algorithmique, in: *Modèles non standard en arithmétique et théorie des ensembles*, *Publications Mathématiques de l'Université Paris VII*, **22**.

Wilkie, A. (1986) On sentences interpretable in systems of arithmetic, in: *Logic Colloquium '84*, Studies in the Logical Foundations of Mathematics 120, pp. 329–42. North-Holland.

Wilkie, A. (1987) On schemes axiomatizing arithmetic, in: *Proceedings of the ICM Berkeley 1986*, pp. 331–7.

Woods, A. (1981) Some problems in logic and number theory and their connections, Ph.D. Thesis, University of Manchester.

Woods, A. (1986) Bounded arithmetic formulas and Turing machines of constant alternation, in: *Logic Colloquium '84*, Studies in the Logical Foundations of Mathematics 120, pp. 355–77. North-Holland.

Wrathall, C. (1978) Rudimentary predicates and relative computation, *SIAM J. Computing*, **7(2)**: 149–209.

Yao, Y. (1985) Separating the polynomial-time hierarchy by oracles, in: *Proceedings of the 26th Annual IEEE Symposium on Foundations of Computer Science*, pp. 1–10. IEEE Computer Science Press.

Žak, S. (1983) A Turing machine hierarchy, *Theoretical Computer Science*, **26**: 327–33.

Zambella, D. (1994) Notes on polynomially bounded arithmetic, preprint.

SUBJECT INDEX

advice 7, 9, 187, 190
ancestor 33
antecedent 32
approximation method 10
axiom scheme 42

basis 9
  de Morgan 9
Bel'tyukov machine 278
binary tree 224, 241
bit graph 219
Boolean algebra 254, 257
  partial 289, 294
Boolean circuit 9, 199, 226, 313
Boolean combination 73, 137, 193
Boolean connective 9, 266
Boolean formula 9
Boolean function 8, 313
Boolean valuation 289, 292, 295, 296, 323, 324
bounded arithmetic 4, 63
bounded query 97
bounding polynomial 145
branching program 15, 22, 27, 38

carry save addition 284
cedent 32
clause 25, 287
clique 13, 189, 233, 306
  function 13
coding 66
communication complexity 16
complete system 244, 252, 313
comprehension
  axiom 88
  bounded 85
concatenation 18
congruence 289

conservativity 78, 79, 96, 111, 136, 204, 207, 218, 280
consistency statement 202, 208, 321
  finitistic 302
constant depth 41, 141, 232, 258, 267, 277, 283, 285, 305
constructive sets 17
counterexample 97, 114, 115, 120, ·128
counting 308
  approximate 310
  function 7, 90, 280, 309
  principle 287
    modular 16, 223, 258, 265, 305
cryptographic 325
cut 65, 104
  in a model 4, 64, 67, 68, 72, 83, 274, 317
  rule 33
cutting planes 61, 279, 281, 283, 286
cut-free 34, 60
cylinder 226

decision tree 15, 243, 288
deduction lemma 46
dense 274
depth 141, 236
  logical 31, 46
  of a circuit 9, 16
  of a formula 31
  of a proof 31
Diophantine equation 65
dyadic numeral 78

eigenvariable 204
equational
  logic 268
  theory 2
end-extension 317, 319

explicit definition 37
exponentiation 20, 65, 309, 324
extended
  Frege
    system 53
    sequence 142
    resolution 57
extension 127, 129
  atom 53
  axiom 53
  rule 53

field 268
finite axiomatizability 185, 193, 202, 311, 317, 318
forcing 175, 254, 272, 274
formula
  bounded arithmetic 4, 18, 21
  free 104
  minor 33
  open 65
  principal 33
  second order 84
  sharply bounded 21
  side 33
Frege
  proof 42
  proof system 42
  rule 42, 256
free-variable normal form 204
function
  Herbrand 113
  multivalued 99, 124
  Sipser 13, 195, 225, 230
  Skolem 102, 171, 319
functional 134
  feasible 2
  interpretation 2
  multivalued 135

generic 274, 275
Gödel number 205
graph 287
  expander 287, 288

Hajós calculus 307
halting configuration 94
hard tautologies 299, 304, 307
height 15, 34
herbrandization 210
hierarchy
  linear-time 6
  polynomial-time 6, 22, 62, 191, 309, 317

Hilbert-style system 61
homomorphism 289

implicationally complete 42
instantaneous description 93
interactive 94
intuitionism 2
interpolant 35
interpretability 62, 66, 209
iteration principle 231

k-complete system 244, 266, 269, 295
k-evaluation 252, 254, 258, 275, 295, 296

language 139, 266
  of arithmetic 3, 17
  context-free 20
  second order 83
limit 294, 295, 296
  direct limit 293
limited extension 26
linear combination 270
linear ordering 223
literal 25
Löb conditions 206

machine independent 2, 75
majority function 13
minimal size 44
model 126, 128, 193, 220, 272, 316, 319
  nonstandard 4, 162, 172, 272, 316, 323
  recursive 4, 65
  standard 4
modus ponens 42
monotone function 13

natural deduction 61
norm 243, 313
NP-completeness 7, 22
number
  of formulas 43
  of steps 42

open problem 10, 24, 31, 125, 185, 243, 262, 267, 272, 286, 300, 311, 312, 317, 321
optimal proof system 299, 300, 302
optimization problem 97, 186

parameter 4
parity function 13
partition 244
part-off quantification 18
path 15, 39

pigeonhole principle 258, 261, 272, 280, 285,
    305, 311
    bounded 308
    weak 28, 118, 213, 216, 218, 223, 225, 232,
        234
polynomial simulation 25, 48, 52, 57, 58, 60,
    158, 164, 168, 186, 281, 283, 285, 303
polynomial-time
    machine 94
    probabilistic 118
principle
    least number 4, 272, 275
    MAX 71
    MIN 70
problem
    NP =?coNP 169, 304, 321, 323, 326
    P =?NP 7, 10, 321
projection 244
propositional proof system 23, 268

quantified propositional
    calculus 58
    formula 58, 145
    proof system 163

recursion on notation 19, 76
refinement 244, 313
reflection principle 158, 167, 171
regular provability 203
resolution 25, 281
    refutation 25
    regular 27, 60
    rule 25
restriction 14, 226, 246
    partial 288
    random 14, 196, 226
RSUV-isomorphism 83, 90, 133, 323
rudimentary
    function 19
    set 6, 18, 19
        extended 22
        positive 18
        strictly 18
        strongly 19
rule
    inference 32, 41
    introduction 148
    IND 103
    LIND 103
    PIND 103

scheme
    choice 74, 88, 90, 136
        dependent 88
    collection 136, 317, 319, 324
        bounded 73
        sharply bounded 73, 90
        strong sharply bounded 74
    induction 4, 69
        polynomial 69
        length 70
    replacement 74
    separation 88
search problem 27, 38, 121, 125, 180, 182, 288
second order oracle 133
sequent 32
    initial 32
sequent calculus 31
sequential theory 66
Σ-depth 239
size 9
soundness 160, 171, 286
space 5
sparse set 304
speed-up 24, 236, 243, 299
    exponential 25
    polynomial 25
    superpolynomial 25
special
    proof 39
    rule 39
st-connectivity 22
subformula property 104
substitution Frege system 53
substitution rule 53
succedent 32
switching lemma 14, 199, 226, 239, 247, 314,
    323

test tree 180, 182
    witnessing 182, 221
theorem
    Beth definability 36
    Bondy 297
    Buss 107
    Cobham 76
    Cook 7, 8
    Craig interpolation 35
    Fermat little 312
    Gentzen (Hauptsatz) 5, 104
    Gödel
        completeness 3
        incompleteness 3, 62, 202, 206
    Graham and Pollak 297
    Herbrand 5, 102, 120, 127, 130

Hilbert Nullstellensatz 269
Lagrange four-square 312
midsequent 60
Nepomnjascij 20
Parikh 65, 323
Ramsey 233, 236, 306
Spira 16, 39, 49, 159, 241
Sylvester 312, 325
Tennenbaum 4, 65
theory
    first order 3
    second order 3
time 5
tournament 233, 236
translation 139, 144, 158
treelike 34, 45, 46, 154, 237
truth
    assignment 159, 313
    partial truth definition 3, 163, 204, 209, 304
    undefinability of 3

Tseitin graph formula 297
Turing machine 5, 93, 133, 304
    oracle 5, 99

ultrafilter 11
ultraproduct 10

variable
    bounded 4
    free 4
vector space 224, 310

witnessing
    formula 106, 156
    function method 105
    oracle 99, 115

Z-part 43

NAME INDEX

Adleman, L. 18
Ajtai, M. 1, 14, 258, 260, 277, 278, 324
Aleliunas, R. 22
Alon, N. 13
Andreev, A. E. 17
Arai, T. 183, 279, 297

Balcazár, J. L. 3, 118, 155
Barrington, D. A. 22
Beame, P. 260, 277, 278
Bellantoni, S. 277
Bel'tyukov, A. 278
Bennett, J. H. 19, 20, 22, 66, 75, 79, 80
Berarducci, A. 312
Beth, E. W. 35, 36
Blake, A. 25
Bondy, J. A. 297
Bonet, M. 61, 297
Boppana, R. 3, 13, 22
Buss, S. R. 1, 21, 60, 61, 62, 63, 78, 80, 83, 84,
    92, 99, 101, 106, 107, 122, 131, 138,
    183, 193, 208, 230, 231, 285, 297, 304,
    325

Cai, J. 277
Chiari, M. 131, 231
Choubarian, A. A. 61
Chrapchenko, V. M. 17
Chvátal, V. 298
Clote, P. 2, 183, 184, 297
Cobham, A. 22, 75, 76
Cook, S. A. 1, 8, 23, 24, 60, 61, 63, 76, 78, 92,
    148, 158, 167, 169, 183, 325
Cook, W. 61, 281, 297
Coullard, C. R. 61, 281, 297
Craig, W. 35

Davis, M. 25, 60
Díaz, J. 3, 118
Dimitracopoulos, C. 62, 66, 80, 83, 118, 209
Dowd, M. 61, 155, 183

Furst, M. 14, 195, 324

Gabarró, J. 3, 118
Gandy, R. O. 278
Garey, M. R. 3, 22, 307
Gentzen, G. 31, 61, 104
Gödel, K. 3, 202, 203
Goerdt, A. 297
Gratzer, G. 289, 293

Hájek, P. 3, 4, 64, 67, 74, 80, 203, 325
Haken, A. 23, 28
Handley, W. G. 278
Hartmanis, J. 5
Hastad, J. 14, 17, 195, 197, 199, 226, 239, 277,
    297, 324, 325
Hausdorf, F. 73
Hay, L. 101
Herbrand, J. 113, 115, 131, 210
Hopcroft, J. 6

Immerman, N. 5
Impagliazzo, R. 277
Intrigila, B. 312

Johnson, D. S. 3, 22, 121, 125, 307

Kadin, J. 208
Karchmer, M. 10, 17, 22, 297
Karp, R. 10
Kaye, R. 4, 62
Kirby, L. 325

Krajíček, J. 2, 25, 37, 45, 60, 61, 92, 98, 99,
        101, 115, 120, 122, 131, 138, 167, 183,
        184, 191, 194, 204, 208, 216, 230, 231,
        239, 277, 289, 292, 297, 298, 304, 307,
        325
Krentel, M. W. 97, 98, 101

Lessan, H. 209
Lewis, P. M. 5
Lipton, R. J. 10
Löb, M. H. 206
Lovász, L. 60

Manders, K. 18
Mints, G. E. 102
de Morgan 9, 161
Mostowski, A. 4, 70
Muchnik, A. A. 76
Muller, D. E. 10

Nepomnjascij, V. A. 20, 22

Otero, M. 65

Papadimitriou, C. H. 98, 121, 125, 126
Parikh, R. 61, 62, 63, 65, 92
Paris, J. B. 1, 4, 22, 62, 65, 66, 67, 68, 80, 83,
        92, 118, 183, 184, 203, 208, 209, 215,
        231, 266, 277, 278, 297, 318, 319, 325
Parsons, C. 102
Paul, W. 6, 7
Peano, G. 3, 251
Pitassi, T. 277, 297, 307
Pudlák, P. 1, 3, 4, 25, 37, 60, 61, 64, 66, 67,
        74, 80, 92, 98, 101, 120, 131, 191, 203,
        208, 209, 231, 277, 302, 304, 307, 325
Putnam, H. 25, 60

Rabin, M. O. 202
Raz, R. 22, 297
Razborov, A. A. 10, 11, 13, 16, 89, 90, 92, 325
Reckhow, R. A. 23, 24, 48, 60, 61
Ressayre, J.-P. 74, 92
Riis, S. 220, 222, 231, 274, 277, 278
Robinson, J. A. 25
Robinson, R. M. 3, 4, 63, 68, 70, 203
Ryll-Nardzewski, C. 202

Savitch, W. J. 5
Saxe, J. B. 14, 195, 324

Shannon, C. E. 10
Sgall, J. 98, 101
Sheperdson, J. C. 65
Shoenfield, J. 3
Sipser, M. 3, 14, 22, 195, 310, 324
Smolensky, R. 16
Smoryńsky, C. 4, 206
Smullyan, R. 6, 17, 18, 61
Solovay, R. 67
Spira, P. M. 16, 39, 49, 160
Statman, R. 60, 61
Stearns, R. E. 5
Stockmeyer, L. 6
Subbotovskaja, B. A. 288
Szelepcsényi, R. 5
Szemerédi, E. 298

Takeuti, G. 5, 31, 60, 89, 90, 92, 101, 102, 104,
        120, 131, 138, 167, 183, 191, 204, 208,
        209, 274, 275
Tarski, A. 3, 4, 70, 163, 171, 205, 304
Tennenbaum S. 4, 65
Toda, S. 7, 309
Tseitin, G. C. 60, 61, 287, 297
Turán, G. 28, 60, 61, 281, 297

Urquhart, A. 60, 287, 297, 298, 307, 325

Valiant, L. 6, 7
Van den Dries, L. 65

Wagner, K. W. 101
Wegener, I. 3, 22
Wigderson, A. 17, 22
Wilkie, A. 1, 21, 22, 62, 65, 67, 68, 80, 92, 131,
        183, 184, 203, 208, 209, 215, 266, 277,
        278, 297, 319, 325
Woods, A. 21, 62, 215, 277, 297, 311, 312, 325
Wrathall, C. 6, 19, 20

Yannakakis, M. 98, 121, 125
Yao, Y. 14, 195, 277, 324

Žak, S. 6
Zambella, D. 131, 208
Zaring, W. M. 274, 275

SYMBOL INDEX

2.1 L_{PA}, Q, IND, LNP, PA, N, Th(N), IE_1, $\subset$, $\subseteq$, $\neg$, $\vee$, $\wedge$, $\equiv$, $\rightarrow$, $O(g(n))$, $\Omega(g(n))$, $\Theta(g(n))$, $o(g(n))$

2.2 Time(f), Space(f), NTime(f), NSpace(f), LinTime, P, NP, L, PSpace, LinSpace, E, EXP, LinH, $\Sigma_i^{\lin}$, Σ_i^p, $\Pi_i^{\lin}$, Π_i^p, coX, $\#R(x)$, $\#P$, P/poly, NP/poly, L/poly, $\Box_i^p$

3.1 $C_\Omega(f)$, Depth$_\Omega(f)$, $L_\Omega(f)$, CLIQUE, $C^+(f)$, $C^m(f)$, $\oplus$, MAJ, $S_d(x_{i_1,\ldots,i_d})$, MOD_p, CC(f)

3.2 L, $\lfloor\frac{x}{2}\rfloor$, $|x|$, $x\#y$, E_i, U_i, Δ_o, $\Delta_o(M)$, Λ, $\frown$, $\subseteq_p$, RUD, SRUD, RUD$^+$, strRUD, TimeSpace(f, g), $V_k(N)$, Σ_i^b, Π_i^b, Σ_∞^b, Δ_i^b

4.1 TAUT, $P \leq Q$, $P \leq_p Q$

4.2 Ext(θ), $\neg\text{PHP}_n^m$, R

4.3 LK

4.4 F, $k_P(\tau)$, $\ell_P(\tau)$

4.5 SF, EF, ER

4.6 Σ_i^q, Π_i^q, Σ_∞^q, G, G_i, G_i^*, TAUT$_i$

5.1 $I\Delta_0$, PA$^-$, $\subseteq_e$, IOpen, Ω_1, $\omega(x)$, Exp, $\omega_k(x)$, $I\Delta_0 + \Omega_1$, $I\Delta_o + \text{Exp}$

5.2 L^+, BASIC, T_2^i, T_2, S_2^i, S_2, PIND, LIND, $\#_k$, MIN, LENGTH–MIN, MAX, LENGTH–MAX, Δ_i^b–IND, $\mathcal{B}(\Gamma)$, $B\Sigma_i^b$, $BB\Sigma_i^b$, BASIC$^+$, $S_2^i(L^+)$, strictΣ_i^b

5.3 PV, $S_2^1(\mathrm{PV})$, PV_i

5.4 Numones(x), $\langle a, b \rangle$, bit(a, i), $i \in a$, Seq(w)

5.5 Log(M), α^t, L_2, $\Sigma_i^{1,b}$, $\Pi_i^{1,b}$, $\Delta_i^{1,b}$, $I\Sigma_0^{1,b}$, V_j^i, V_j,

   U_j^i, U_j, SEP, AC, DC, R_2^i, R_2, R_3^i, Enum(f, u, α), 1–Exp

6.1 $L(\alpha)$, $\Sigma_i^b(\alpha)$, $\Pi_i^b(\alpha)$, $\Delta_i^b(\alpha)$, $S_2^i(\alpha)$,

   $T_2^i(\alpha)$, Comp$_M$, UNIV$_i$

6.2 $\mathrm{P}^{\mathrm{NP}}[O(\log n)]$, $\mathrm{P}^{\Sigma_i^p}[O(\log n)]$, L^{NP}, $\leq_{tt}$ (NP), $\Sigma_0^b(\Sigma_i^b)$

6.3 WitComp$_M$, $\mathrm{FP}^{\Sigma_i^p}[\mathrm{wit}, q(n)]$

7.1 LKB, BASIC^{LK}

7.2 Witness$_A^{i, \bar{a}}$

7.3 ψ_H, PHP(Σ_∞^b), WPHP(f), WPHP(PV_1)

7.5 PLS, F_P, N_P, C_P, PPA, PPAD, PPP, PPA(PV_1)

8.1 $\#_3$, R_2^i, $L(\#_3)$, TIME^Q, SPACE^Q, EXP^Q, PSPACE^Q

8.2 $\mathrm{EXP}^{\Sigma_i^{1,b}}[\mathrm{wit, poly}]$

9.1 $L_{\mathrm{PA}}(R)$, $I\Delta_0(R)$, $\langle \theta \rangle_{(\bar{n})}$

9.2 $n(i)$, $\|A\|_{q(m)}^m$, $\|\Gamma\|$

9.3 Fla(α), Prf$_P(\pi, \alpha)$, Assign(η, α), Eval(η, α, γ),

   $\eta \models \alpha$, TAUT(α), Fla$_d(\alpha)$, $\tilde{\alpha}$,

   $w \models_i \alpha$, Taut$_i(A)$, TAUT$_i$, $P \geq_i Q$, i-RFN(P), Con(P)

10.2 Ω_i

10.4 R_q^+, R_q^-, $\Sigma_{i,m}^{S,t}$

10.5 $I\Sigma_i^0$, Con(T), BdCon(T), RCon(T),

   Tr$_i(x, y)$, STr$_i(x, y)$

11.2 PHP$_n^m$

11.3 MOD$_k(R, S)_a$

11.4 WPHP(a, R)

12.1 RAM$_n$

12.2 $\Sigma_d^{S,t}$, $\Pi_d^{S,t}$, Δ_1^t, $S_{i,j}^{d,n}$,

   $R_{i,j}^+(q)$, $R_{i,j}^-(q)$, Σ–depth, $\neg$ MPHP(A, B, r)

12.3 $\mathcal{M}^{\mathrm{PHP}}$, $\mathcal{M}^{\mathrm{MOD}_a}$, $\|H\|$, $S \times T$, $S(H)$,
$H \lhd S$, α^ρ, H^ρ, $\mathrm{supp}(\rho)$

12.4 $[\alpha]$, $\mathcal{M}^{\mathrm{trivi}}$, ϕ^ρ, H_ϕ, S_ϕ

12.5 Count_n^a, PHP_n, $g\mathrm{Count}^a$

12.6 $Q_a x < t$, $Q_a \Delta_0$, $IQ_a \Delta_0$,
$\mathrm{MOD}_{a,i}$, $F(\mathrm{MOD}_a)$, $\mathrm{LK}(\mathrm{MOD}_a)$, $\vdash_*$, $X \perp Y$

12.7 $L\exists_1$, $\mathcal{P}$, $\|-$

13.1 L_i, L_ω, CP, $\mathrm{PHP}_n^{\mathrm{CP}}$, $I\Delta_0(\alpha)^{\mathrm{count}}$

13.2 $C(G)$, $D(G)$

13.3 $\vdash_\Gamma$, BA, Γ_π^*, Γ^+

14.1 $\mathrm{Taut}(x)$, $c_P(\tau)$, Prf_T, Con_T, $\vdash_m$

14.2 $\geq_P$

15.1 $\oplus P$, $\Box_\infty^p$, $\Sigma_\infty^b \mathrm{PHP}$, $\Delta_0 \mathrm{PHP}$, $I\Delta_0(\pi)$

15.3 $P(M)$, $\Sigma_i^p(M)$, $\Pi_i^p(M)$, $\mathrm{NP}(M)$, $\mathrm{coNP}(M)$,
$\mathrm{PH}(M)$, $\mathrm{P}^{\mathrm{stan}}(M)$, $(\Sigma_i^p)^{\mathrm{stan}}(M)$, $(\Pi_i^p)^{\mathrm{stan}}(M)$, $\mathrm{PH}^{\mathrm{stan}}(M)$,
$(\mathrm{P/\,poly})(M)$, $(\mathrm{NP\,/\,poly})(M)$, $(\mathrm{coNP\,/\,poly})(M)$, $\mathrm{Diag}(M)$, $\Delta_0^{\mathrm{stan}}(M)$,
$\mathrm{Bound}(f)$